Mechanisms in Cell Physiology

Mechanisms in Cell Physiology

By

Michele Mazzanti

Cambridge Scholars Publishing

Mechanisms in Cell Physiology

By Michele Mazzanti

This book first published 2022

Cambridge Scholars Publishing

Lady Stephenson Library, Newcastle upon Tyne, NE6 2PA, UK

British Library Cataloguing in Publication Data
A catalogue record for this book is available from the British Library

ISBN (10): 1-5275-8248-5
ISBN (13): 978-1-5275-8248-4

This book is dedicated to Edoardo, Louis, Giovanni and Arnaldo

TABLE OF CONTENTS

PREFACE

With the advent of new technologies, decisively supported by the computerization of scientific instruments, biological research in general and the study of physiology in particular have made enormous progress in the last 20 years. Technologies alone, however, cannot explain the successes that scientific research continues to achieve. Achieving certain goals requires comprehensive science policy with specific objectives, common efforts and exceptional economic support. Three examples of dedicated international programs bear witness to this. First, the Human Genome Project, coordinated by the National Institutes of Health (NIH), launched in 1990, concluded in 2003, and with an initial allocation of more than $250 million, led to the decoding of the entire human genetic make-up. A second example is the decade dedicated to the brain: 1990-2000 saw enormous resources concentrated on research on the function of the central nervous system. Last is The Human Brain Project (HBP) launched by the European Commission's Future and Emerging Technologies (FET) scheme in October 2013. The HBP has the following main objective to create and operate a European Scientific Research Infrastructure for brain research, cognitive neuroscience, and other brain-inspired science.

There have been several factors contributing to the particular attention paid to scientific research – the opportunity for huge gains, the idea of defeating diseases, the desire to study the biology of humans to satisfy the timeless search for immortality. However, the prevailing reason for the success of biological investigation, we believe, and we hope, is rooted in the human tendency to discover the unknown, in the thirst for knowledge, in the idea of exploring inner nature of the human being in human beings beyond wretched drifts and dangerously unrealizable ideas. It is the combination and convergence of different events that determine the global success of a particular developmental trend.

In the years following the end of the Second World War, a policy of supremacy was pushed by the two superpowers, the United States and the

Soviet Union. The challenge concerned the control of territory, waters and skies. The "frontier" was represented by space conquests, with an enormous economic effort on the part of the two main players. The space enterprises raced each other with exceptional peaks of accomplishment, such as the first man in space in 1961 and the first man's walk on the moon in 1969. However, a sequence of events beginning in the 1980s, including the collapse of the Soviet empire, but also the oil and economic crisis, and the realization that the conquest of space did not give the expected results in economic terms, led to a downsizing of this strategic project. Even the idea of mass commercialization of interplanetary travel and exploitation of extraterrestrial resources has gradually died out under the relentless blows of the prohibitive costs and technologies that this type of conquest requires, which are still probably barriers today.

Therefore, almost naturally, instead of the new frontier being the extremely large, to satisfy the thirst for knowledge, human curiosity and the innate egocentricity of humans, it turned out to be the infinitely small. The investigations into the origin of living beings, the attempt to unveil the secrets of biological "machines" and, above all, the idea of generating and controlling a process central to life have an attraction and an intellectual fascination that few researchers have been able to or wanted to resist.

We believe that for a student who approaches the world of biology and scientific research, it is important to know, in addition to experimental theory and practice, the history of the evolution of biological thought and the consequent technological evolutions that have led to such great success. There are many publications dealing with this topic and we think that some should be obligatory reading for the training of future scientists. In addition, in order to fully understand the physiology of organs, cells and sub-cellular compartments, it is important to have a clear understanding of the fundamental principles of biochemistry, biophysics and molecular biology. Based on these principles, it is possible to understand well the totality of physiological processes that we know today and the hypotheses developed to explain those that are still obscure.

Biological research is a relatively young field of investigation compared to physics, chemistry and medicine for several reasons. The most important factor of all, we believe, is the fact that even in the most basic experiments, there is a need for adequate technology. Beyond some brilliant intuition that we see in the monk Gregor Mendel, in the most sensational example, research on the origin and function of living organisms has gone hand in hand with technology development.

There is also an equally important reason of an almost philosophical nature: biological research is a mixture of chemistry, physics and mathematics. For this reason, the birth of schools of biology had to wait for the concept of a researcher in the field of biology to mature. The process was slow until 1970, when it underwent an acceleration whose push continues today. Biology courses have changed at all levels. In primary and secondary school, the biology program has changed topic from a descriptive observation of living being towards up to date experimental biology capable of interfering with the natural processes (genetic engineer for example). University biology curricula have been enriched with new, more and more thorough courses. Biological knowledge has become increasingly socially, culturally and ethically relevant.

The study of Physiology had, like all disciplines, its golden age. The years between 1940 and 1970 saw a flourishing of schools that took into account the mechanisms underpinning the functions of biological events. Names like Hodgkin, Huxley, Fen, Margaria, and many others, should be known to today's students for the immense contribution of ideas and experiences that they have left us. They not only had the ability to create schools of thought: we believe that their genius was to predict the physical mechanism hidden among the layers of membranes, clusters of proteins, strands of filaments and fluids of all kinds that we call living matter. They knew how it was likely to work, before embarking on any kind of study of the biological element. They studied it from the outside, acquired data on the relationships of the phenomenon with the surrounding world, hypothesized a probable mechanism and only at the end proceeded by opening, dissecting and dividing the object of their research to find support for their hypothesis. Everything was the result of an in-depth logical discussion on the most probable mechanism underlying a certain physiological phenomenon. **Logic was their most powerful weapon of investigation**.

The discussion with their collaborators and other scholars, the responses to their hypotheses and the possibility to modify them allowed them to acquire new elements and reach theoretical conclusions. This prelude to the laboratory work not only gave excellent results, but led to the growth of generations of scholars, first, and researchers, then, which have allowed the great development of research in the field of biology to the present day. We are purposefully talking about ‘scholars’ and not ‘students’. We tend to associate the term ‘student’ more with the young high school pupil who, with few exceptions, acquires information, metabolizes it and retains by heart what is most useful. The scholar is an actor, who does not only learn. The scholar discusses, proposes, enters into contradiction with the tutor and is able to very often modify the final thought and theoretical conclusions that lead to the choice of one experimental path over another.

A university professor loved to repeat often: "To do good research you have to have good teachers, but above all good students". One cannot exist without the other. Several years ago, the old Mr. Honda, the founder of the Japanese multinational company of the same name, declared publicly: "I do not understand my young engineers. I do not understand their ideas and their way of thinking. I am particularly happy about this fact: if I understood their ideas, it would mean that there has been no progress". This is a lesson that everyone should hold very dear, because we believe it is the basis for good research and the most intelligent way to interact with others.

Today our laboratories are full of sophisticated tools and machines that are used far below their potential. More and more often, we find ourselves chasing technology to adapt our ideas to it. We believe that this technological prevalence, which has had a very important function, should be reduced. Today what is missing are the new ideas: the reason lies in the fact that there are no longer the "schools", those aggregations that were used to propose new ideas. There are scientific problems, such as the mechanism of the basis of human thought, which require and will require innovative mental efforts that must be produced by competent and prepared minds, but with a strong sense of scientific integrity. Of course it is a great challenge to study the mechanism of thought and behavior using the same tool, the brain, which generates both thought and behavior. Some people think that we will never succeed, that understanding ourselves is unachievable.

This book has had a difficult "birth". Difficult because there are many good books on Physiology. Difficult because the idea matured over several years of teaching, in an attempt to write something different. Difficult because this book focuses a great deal on the logic of thought. We have observed for years horrified groups of students, facing the study of Physiology, absorb hundreds of sometimes very complex concepts, pouring them out more or less politely in front of exam boards and then purging them from their brains as soon as they passed the exam. This book is an attempt to reduce this disorder. We believe that for a basic Physiology course, the size of the program or the number of systems that are listed is not so important. We believe that with a solid foundation in cytology, biochemistry and biophysics a student has everything needed to understand the concept of dynamic processes in biology. Those who work with Physiology must have clear concepts of structural biology and energy. **WHERE**, **HOW** and **WHY** are essential. However, what discriminates physiology from other biological disciplines and its unifying element is certainly **WHEN**. The concept of time – which very few students grasp – becomes essential for the study of the dynamics of biological processes, which usually requires a physical-mathematical description of phenomena. If the elements that contribute to the biological phenomenon are clear, the mathematical description will be not only easy but also very helpful. Physics and mathematics that can be used by students of Biological Sciences could represent another great goal: not as pure science, but as language.

One cannot conclude a book introduction today without mentioning the relationship with the internet. The internet offers an infinite support to knowledge and therefore an opportunity that must be seized. However, it offers nothing for understanding. Unfortunately, in the last few years, a degeneration of knowledge has taken hold, in which users mistake an informational "warehouse", where anyone can put anything, for a "truth provider". We are witnessing a worrying “leveling” of knowledge, where it seems that all have become experts in any given field. Watching the sea from a jet ski speeding across the surface is not the same as going down in a bathyscaphe in the Mariana Trench. Everyone knows how to use a jet ski. Going down into the Mariana Trench not only takes courage, it takes knowledge, skills and experience.

The internet, with a huge amount of totally uncontrolled data, has a lot of valuable information, but precisely because its access is free to all, it contains an "avalanche of stupidity", sometimes harmless but very often harmful or even dangerous. More authoritative sources than us say that the quantity of false information on the web is far greater than that controlled by experienced staff. The huge success of the "do it yourself" aspect offered by the web lies in the extreme ease of use, in the immediacy of the response, in the lack of contradiction and, no less important, in the possibility of always finding the answer that we like best.

This hint of discussion on the validity of education via the web, on the topicality of printed books and all the way to the role of professors in face-to-face lessons is addressed in an exemplary way by the book "The Death of Expertise: The Campaign Against Established Knowledge and Why it Matters (2017, Oxford University Press). We believe that today this is essential reading for teachers and students. What is attempted is the interruption of the decline in higher education taking place today more than ever. The task is arduous but we hope that our effort is understood for what it is and not for what it lacks.

I would like to express all my gratitude to my "Maestri" of Life and Science: Edoardo Mazzanti, Giovanni Cavagna, Louis DeFelice and Arnaldo Ferroni. My English language knowledge is not enough good to find the proper words to express my gratefulness. Raffaella Tonini who always supported me with love and hate. A special thanks goes to the undergraduate students of the General Physiology Course of the Degree in Biological Science, of the University of Roma and Milan, who were used as guinea pigs without their knowledge and on whom the logic with which this book was conceived was tested. I hope that they received sufficient compensation during my lectures. I am in debt with my colleague and friend Enzo Mancinelli who did most of the work. Without his help this book was not possible. A special thanks to what were more than simply colleagues: Henry Malter, Berth Smith, Steve Ralph, Raffaele DeFrancesco, Antonio Torroni, Aurelio Galli, Federica Bertaso, Laura Bianchi, Gaia Novarino, Marco Foiani, Andrea Locarno, Anna Moroni. Thanks also to my young collaborators Ivan Verduci, Francesca Cianci, Gaetano Cannavale and Federico Brandalise. In one way or in another they all help me physically and psychologically during these trouble years.

M. Mazzanti

CHAPTER 1

Introduction

1.1 The concept of living matter in Physiology
1.2 The concept of separation
1.3 The concept of energy
1.4 The concept of work

The study and understanding of Physiology is based on three fundamental concepts: the possibility of creating different but not isolated environments, the ability of these environments to accumulate energy and the ability to use this energy with maximum efficiency through appropriate structures that transform it into work. The development of even the simplest systems with these characteristics took an enormous amount of time, so much that the oldest example likely able to perform these functions was found in a 1.9-billion-year-old Precambrian flint. This seems to be the oldest organism morphologically comparable with some modern iron bacteria (Schopf et al., 1965). The process that led to this ancient system and later to today's cellular forms was possible because of three important factors. The first is the enormous amount of time that has passed from the origin of the Earth to the appearance of the first forms of life. The second is the presence in the primordial aqueous environment, formed by the cooling of the Earth's crust and the condensation of water vapor, of chemical components in large quantities, such as chlorides, carbonates and

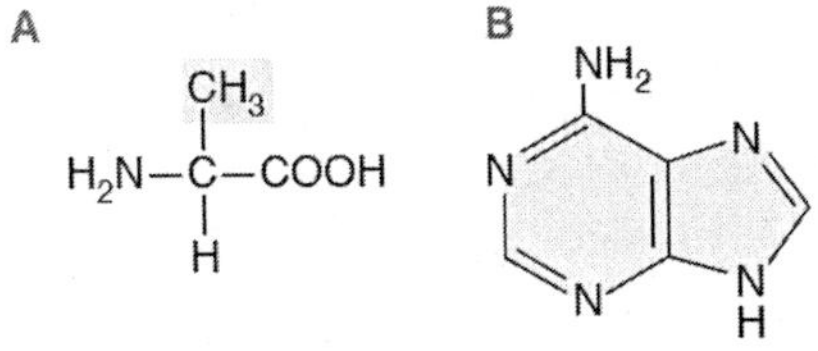

Figure 1.1 Structure formula (A) of alanine, a simple amino acid, and (B) of the nucleotide adenine.

phosphates of sodium, potassium and calcium, formed by the reaction between the pulverized rocks and methane generated by volcanic eruptions. The third is the presence of enormous amounts of energy in the form of electric discharges, visible and ultraviolet light, ionizing radiation and heat sources, in some cases very high, with consequent generation of water vapor. All of these factors contributed to the random synthesis of thousands and thousands of different molecules and more or less complex aggregates, until dynamic molecular systems capable of breaking down and reforming depending on energy conditions and material availability were formed. Thus, complex molecules that are characteristic of living matter, such as amino acids and nucleotides (Figure 1.1), phospholipids, nucleic acids and increasingly complex proteins (Figure 1.2), have been formed in a completely random way.

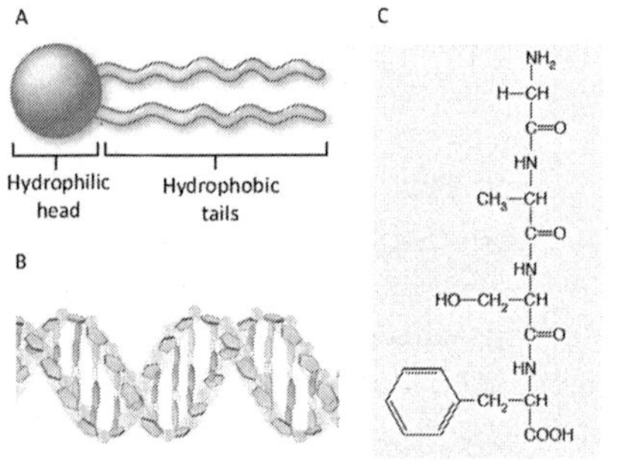

Figure 1.2 Structure formula (A) of alanine, a simple amino acid, and (B) of the nucleotide adenine.

The formation of phospholipids was as important for physiology as the aggregation of nucleic acids was for genetics. The principle on which all physiological processes are based is the presence of aqueous environments with different compositions of solutes. The aggregation of phospholipid molecules in a double layer (Figure 1.3A), due to their characteristic of having a hydrophilic head and a hydrophobic tail, leads to an energetically stable structure that has allowed an evolutionary leap of great importance: the formation of biological membranes. The ability of biological membranes to be in contact with a hydrophilic solute, such as water, through the polar heads, combined with their impermeability due to hydrophobic tail juxtaposition, has given them a fundamental role as universal separators in the biological world. All membranes, whether cellular or intracellular, of cytoplasmic organelles or the nuclear envelope, have an identical basic structure of phospholipids in a double layer: their primary function is to form separate environments with different compositions of solutes.

It is important to understand that the formation of membranes that delimit a restricted three-dimensional space, defined as in the example of Figure 1.3B, has a functional aspect that, in addition to providing different environments, allows interactions between molecules that would be otherwise highly unlikely.

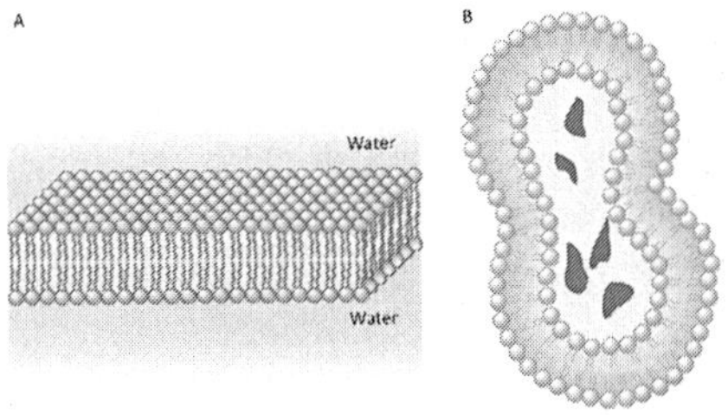

Figure 1.3 A) In water, phospholipid molecules spontaneously arrange themselves in double layer. B) The double layers thus formed can grow three-dimensionally until they close on themselves and form a separate environment from the outside.

All of the reactions that occur in the cellular and subcellular world are the result of random impacts due to thermal agitation. Concentrating molecules in confined spaces decreases the degrees of freedom of the system and greatly increases the probability of interaction. Space management is therefore a means of modulating the reactions between molecules that occur both inside and outside the space delimited by the double phospholipid layer.

While it is true that all biological membranes have the same characteristics, their specificity is ensured by the protein structures that are adjacent to the membrane on both sides or that pass through it (Figure 1.4). Some of these proteins are specialized and are responsible for the passage of material between one environment and another.

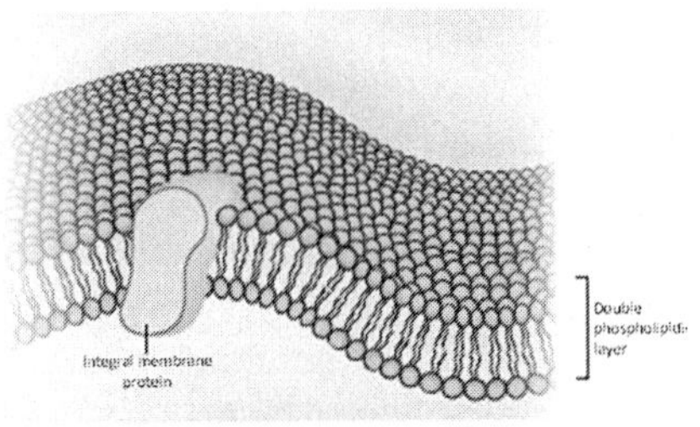

Figure 1.4 Schematic representation of an integral membrane protein. Double phospholipidic layer Integral membrane protein.

Transmembrane proteins contain aqueous pores that determine the selectivity of the membrane itself. There are high-resistance membranes, usually with low protein content, membranes that manage to change their selectivity over time, passively or through energy consumption and, finally, membranes that only perform containment. In general, membranes that enable the movement of solutes from one environment to another must have integral proteins belonging to the category of transporters, specific molecules for this very important function.

The specialization of the phospholipid double layer as a dynamic structure, i.e., able to change its permeability, has created the conditions for membranes to be the main structure responsible for the accumulation and storage of energy. Accumulators are systems that can increase their potential energy and maintain it in a dynamic balance. Potential energy is the fastest form of energy that can be used to perform work. From the point of view of physiological functions, the reactions responsible for maintaining biological systems away from static equilibrium use chemical energy such as adenosine triphosphate (ATP). ATP-controlled reactions are continuously active to ensure a constant accumulation of potential energy. Electrical potential across membranes, ionic accumulations in specific compartments and reserves of complex molecules are the most common ways through which the energy level of biological structures is increased. Systems are constantly maintained at a high level of potential energy in dynamic equilibria based on mechanisms that depend on the hydrolysis of ATP combined with passive events. When the biological system needs to do

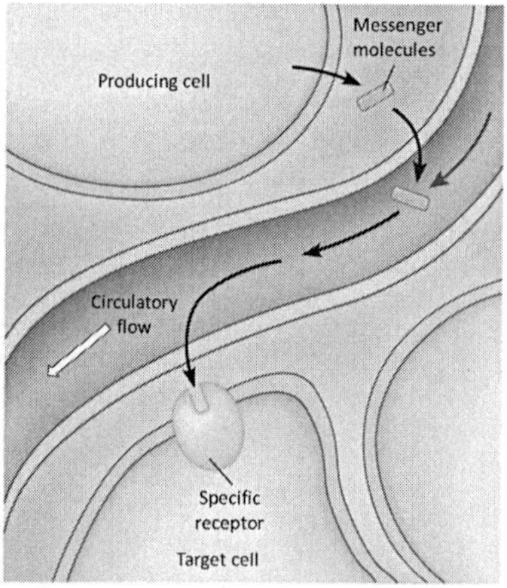

Figure 1.5 Molecular messengers, produced by specialized cells, are carried by circulatory flow (white arrow) near specific receptors of the target cells (black arrows).

some work, the dynamic equilibrium is momentarily disturbed. This occurs when a perturbation of the environment on the biological system removes the constraints that keep the system at a high energy level. This allows the system to "run" spontaneously towards different balances at lower and lower energy levels, rapidly releasing the energy that, from potential, is converted into other forms capable of doing different forms of work. At the end of these processes, active mechanisms, which directly use chemical energy, will restore the dynamic balance of the system with a high potential energy level. However, maximum efficiency is achieved when the dynamic equilibrium can be maintained and re-established in a short period with passive mechanisms and a minimum contribution of chemical energy.

As long as organisms were made up of one or a few cells, cytoplasmic or intracellular communication could be left to passive phenomena such as the diffusion of compounds in the cytoplasm or from cell to cell. The

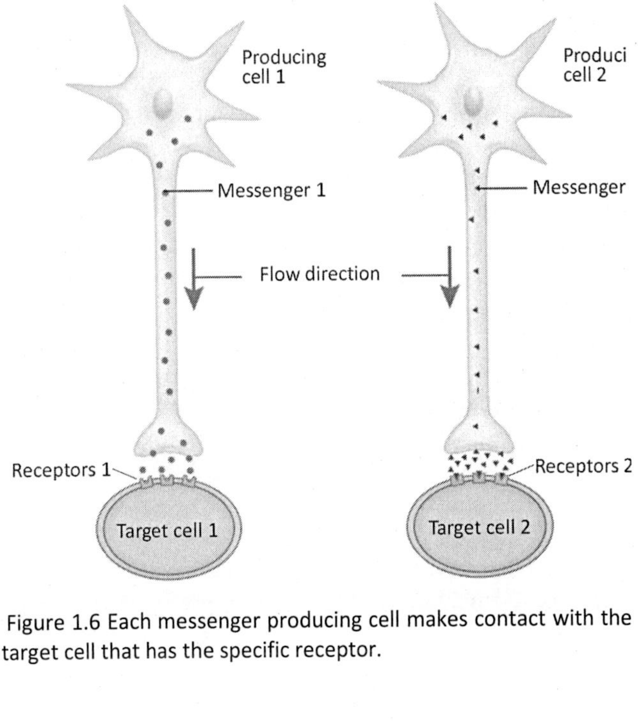

Figure 1.6 Each messenger producing cell makes contact with the target cell that has the specific receptor.

increased complexity of organisms necessarily posed the problem of communication between and control of the various locations within the multicellular system in formation (Figure 1.5). Controlling the different parts as well as other interacting systems, decoding signals from the surrounding environment, and generating instructions to coordinate a better survival strategy represents a further great evolutionary leap. The synthesis and release by cells, in response to an adequate stimulus, of diffusible molecular messengers was probably the primordial extracellular form of a communication system. This system, however, presents a fair degree of complexity because since there is a diffusible molecular messenger, there must also be the ability to intercept the message by means of specific receptors present on target cells. To increase the probability that the message meets its receptor, there are several structures that can be functionalized. First, releasing a diffusible molecule could be a slow process depending on the distance of the target. A solution is represented by the circulatory systems: a preferential method that uses a liquid carrier to move a great variety of substances from the manufacture site to structures with specific receptors. A communication system of this type, consisting of a sufficiently high number of molecules and capable of interacting with all cells that have specific receptors on their membrane, can be, for example, the control system for the growth and harmonious functioning of a multi-cellular organism. A specialization of this communication system is represented in Figure 1.6. A cell produces, at the level of the cell body and in response to an adequate stimulus, a molecular messenger, which is carried using molecular engines near the target cell, which in turn has the specific receptor for the messenger on the membrane.

With further specialization and with clear functional advantages, the messenger can be synthesized by the producing cell directly near the target cell. This cellular organization allows direct communication between two cells and has two major advantages: the accuracy of the message, as it establishes a direct and precise relationship between a producing cell and a target cell, and the control and speed of the response, as there is no need for circulatory flows and large areas of dispersion. Rather, the diffusion takes place in extremely small spaces.

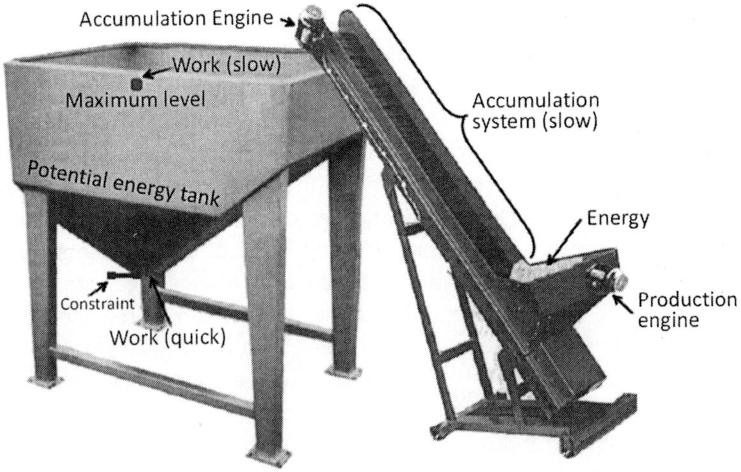

Figure 1.7 The diagram represents the mechanisms of production, storage and use of potential energy. Slow work maintains the body's homeostasis, while fast work responds to a stimulus inside or outside the body and requires the removal of the constraint.

The different communication systems just described have been adopted by biological organisms, either only one or all of them. Depending on the complexity of the organism, these communication systems are adaptable to various situations. Apart from direct contact between cells, the paradigm used by living organisms to communicate is always composed of a stimulator apparatus, which produces and releases messenger molecules, and a receptor apparatus. The organism is able to use the potential energy released by the appropriate stimulus to transform it into kinetic, chemical or mechanical energy and produce work. In this case, the apparatus that converts potential energy into work is, in fact, transformers. There homeostasis are many systems capable of accumulating potential energy and transforming it, directly or indirectly, into work. These systems can be very different from one another. In particular, the most important axis of diversity concerns the factor of time, which impacts the primary use of chemical energy and/or the interval to perform given work.

The mechanisms that the cell has selected to produce and store energy are represented schematically in Figure 1.7. The organism uses a "motor" that produces potential energy through a slow and continuous process, in greater quantity during periods of inactivity, and with chemical energy consumption mainly represented by ATP. The energy produced is stored as potential energy in a "reservoir" with an equally slow process, again by a "motor" that consumes ATP. The overproduction of potential energy is usually sufficient to maintain the basic functions of a living organism. In the presence of an external stimulus, the stored energy can be released. The system uses the stimulus energy to remove the constraints implemented to maintain a high potential energy level. From a physiological point of view, a biological organism is considered "alive" when it reaches a state of dynamic equilibrium with a high potential energy level. When an immediate response is required, such as a voluntary action or a response to an external stimulus, whether useful or harmful, the system is able to respond with a passive mechanism transiently perturbing the dynamic equilibrium. The combination of morphological and functional arrangements allows the reestablishment of equilibrium. In the process, the potential energy is transformed into several other forms (electrical, chemical or kinetic), sustaining the performance of all of the cellular process. The mechanism reaches its maximum efficiency when the potential energy lost in the process is easily and quickly restored when the transient system imbalance is passed.

The sequence of events begins when the voluntary or external stimulus has enough intensity to remove the constraint that keeps the system in dynamic equilibrium with high potential energy. Without constraints, the system would tend towards static equilibrium, dispersing all the energy in the surrounding environment. In biological systems, the achievement and stabilization of this form of equilibrium represents an irreversible condition that decrees the end of the system itself. For a system to be compatible with biological life, the transformation of potential energy must take place in a controlled and reversible way. In order for the system to respond quickly, efficiently, and, above all, continuously, the mechanism must have a very low energy dispersion. The timing of chemical energy generation is incompatible with the response to a stimulus. Thus, long-term recharging mechanisms are useful only to accumulate potential energy. Therefore, to respond to stimuli for a sufficiently long time, the system must increase its efficiency.

There are several artifices put into practice by biological systems to increase efficiency. The energy used to start any physiological function is an infinitesimal fraction of the stored potential energy. Secondly, the potential energy transformed to perform a function must be partly used to restore the initial conditions. Biological systems, in this case, have achieved impressive processes for energy recovery by exploiting a combination of structure and functional relationships that are still in many cases inimitable today because of their efficiency and complexity.

The work that is produced by energy conversion serves very important relationships of the organism with both the internal and external environments. In the internal environment, most of the work is used for communication between the various departments of the organism and is therefore essentially a coordination task; this task becomes increasingly important the more complex the organism.

The transformation of potential energy allows different uses for the relationship of the organism with the external environment. The energy is used to receive and respond to various stimuli that characterize an active biological organism. In order to perform the function of receiving external stimuli, specialized systems are needed that can pick up signals of different natures. Sensors are used for this task, highly specialized cellular structures that perform the function of interfacing between the organism and the surrounding environment. The responses that are then generated depend on the stimulus and are highly varied, from actions related to survival, such as the supply of nutrients, to the reaction to harmful stimuli.

In the following chapters, we will define and analyze the mechanisms by which the elements just described are able to perform and control the main functions of living organisms. Knowing the basic principles of physiological processes can help to understand each mechanism, from the simplest to the most complex. More importantly, it can help to understand still-obscure or only-partially-explored physiological processes. This path is fundamental for the training of biological research operators, who aim to study physiological processes whose function is not yet clear.

1.1 The concept of living matter in Physiology

Explaining what is meant by a living organism is extremely complex and understandably delicate, as it risks slipping into fields of knowledge that are outside the topics to be dealt with in the present context. We will limit ourselves, therefore, to the elaboration of concepts that are supported by scientific data and considerations that specifically concern the scientific field of physiological processes.

A definition that can be valid for all biological disciplines is that life occurs when an organism, however simple it may be, has the ability to reproduce itself. For disciplines such as Biochemistry, Genetics and Physiology, which deal with molecular processes, there are fundamental stages that underpin such a complex event as reproduction. These fundamental stages must be acquired by an organism to establish its "living being". From a physiological point of view, a living organism is one that is able to accumulate energy and transform this energy into work to feed, grow, defend and, ultimately, reproduce itself. From the evolutionary point of view, a biological organism reaches the highest chance of survival once it has reached the highest degree of adaptation to the surrounding environment.

Several conditions must be achieved for an organism to be physiologically viable. Some of these conditions are fundamental:

1. the ability to create separate environments, different from each other in content and composition of solutes, and at the same time to ensure communication strategies between different environments;
2. the physical-chemical characteristics of the interfaces between the different environments must allow the accumulation and maintenance of energy in the form of potential energy;
3. the ability to use electrical, chemical or mechanical potential energy to carry out work.

1.2 The concept of separation

What can be considered a “physiological living organism” must be able to create different environments, to maintain them as they are and with the fundamental characteristic of maintaining communication between them. Of what does this diversity consist? How can we hypothesize a process that leads to the formation of closed structures, such as the one presented in Figure 1.3 B. How is it possible that such structures manage to maintain different conditions between the two environments and at the same time communicate with the outside world without this leading, over time, to the formation of homogeneous environments?

The availability of time in the order of tens or hundreds of thousands of years, large quantities and a variety of materials in the elementary state and energy of various natures in enormous quantities have made the probability of events useful for the formation of complex structures relatively high. Secondly, it is also necessary to remember that there probably have been hundreds of thousands of attempts with different forms of proto-organisms, but that most have not exceeded the level of prototype. The simplest structures known today able to realize a separation of environments are the bacteria, followed by nucleated cells.

From the physiological point of view, the division between prokaryotes and eukaryotes can still be considered valid. In fact, while prokaryotes show only a division between internal and external to the organism, eukaryotes, in agreement with the etymology of the term, "perfect nucleus", have at least two internal environments. These internal environments are the cytoplasm and nucleoplasm, different with respect to the composition of solutes and the distribution of subcellular components. Not only that, but there are also two environments, the interior and the exterior, which differ in the distribution of ions. The sodium ion is abundant outside the cell and is present inside at a concentration ten times lower; on the other hand, the potassium ion is abundant inside the cell and dilute outside of it.

Therefore, in the course of evolution, the roadmap to build separations starts from the creation of two or three environments with different ionic compositions. But what were the conditions that favored this process? The analysis of rocks a few billion years old suggests that in primordial waters, potassium was weakly concentrated, from 10^{-3} to 10^{-9} M/L, while sodium and chlorine ions were at high concentrations, overall about 500×10^{-3} M/L. Similarly, if we analyze seawater, an environment where the first forms of life are believed to have evolved, sodium is about 500×10^{-3} M/L. Potassium is from 1 to 5×10^{-3} M/L. Between the two main ions present in the water, potassium thus represents the most suitable to be accumulated in a separate structure. Accordingly, the randomness of events has preferred potassium as the ion to be concentrated within defined structures to create differences in the concentration compared with the outside environment. Constitution of an environment where a particular ion species has a different concentration compared with the surroundings represent the first step to generate a chemical potential.

The creation of two different environments is more energy efficient if one of them is small compared to the other, as in the case of cells. In addition, if the interface between the two environments allows a mutual exchange of information, it must be equipped with selected structures able to discriminate the solutes that can pass through it. These structures are not simple holes integrated into the separation barrier. In addition to working like a filter, their conformation is in equilibrium between at least two states. If they constitute an aqueous pore spanning the separator, the equilibrium, depending on thermal agitation, can switch between a closed and an open state. The general conditions of the environment, the energy level of the system, several chemical modulators and finally, yet importantly, time are all variables that can modify the probability of being in one state or the other. The filter and the open and closed conditions represent one of the simplest examples of constraints.

The size and the shape of the smaller environment is fundamental from a physiological point of view and the relationship between surface area and volume therefore becomes important. In fact, the increase in size of a spherical organism leads to an increase in membrane surface area according to the square of the radius and an increase in cell volume according to the cube of the radius. The need to exchange information with the surrounding environment follows this principle. Therefore, a progressive increase in size leads to an overload of the exchange surface. From this simple consideration, it can be understood that there is an optimal size limitation for a living organism in order to obtain functional and valid control of internal as well as external communications.

The creation of two or more distinct environments, the presence of physical separators, the optimal size of the organism, the characteristics of the interface and the elements present on the separator to ensure proper functioning of the system are a significant example of the close correlation between structure and function, which in the study of Physiology is of primary importance.

1.3 The concept of energy

After having theoretically solved the problem of creating environments with different concentrations of solutes, such as sodium and potassium ions, it remains to be clarified how this could have contributed to the generation of a system that can stay "alive". In practice, the question that must be asked is this: how can a difference in concentrations and a different distribution of solutes on each side of a separator be used to produce chemical, electrical or mechanical work?

To understand how this can happen, it is important to become familiar with the concept of permeability. Because the biological membranes are impermeable to charged molecules and polar substances, solutes move through the membranes in aqueous pores. They are constantly moving by thermal agitation. If in their motion they encounter a membrane, the number of collisions of the solutes and membrane will be proportional to the concentration of the solutes and to the temperature, which increases the thermal agitation of the particles. If the membrane has aqueous pores whose diameter does not affect the passage of a particular solute, there is a certain probability that some particles of that solute cross the membrane, proportional to the concentration of the solute and to the temperature. Therefore, if the sodium ion concentration is equal to the two sides of a membrane, there is a certain probability that a sodium ion passes from one side to the other. However, there is the same probability that another sodium ion follows the reverse path, so there is no net passage of substance through the membrane. In order to obtain a net ion flux, sodium ions must be evenly distributed across the membrane. As long as this difference is maintained, there is a high probability that sodium ions cross the membrane in one preferential direction. In this case, we are in the presence of a chemical potential for the sodium ion. The same reasoning applies to potassium and any other ion that has the ability to permeate through the membrane.

The differences in concentration create, for each permeable ion, a chemical potential. Chemical separation constitutes the elementary reserve of potential energy of the cell. This potential energy can be used to do work. The exploitation of energy from the chemical potential of a solute is possible only in the presence of aqueous pores and requires a time period compatible with the free diffusion of solutes, which is in the order of thousandths of seconds. The utilization of chemical potential to perform work is very important in several cellular functions. However, it would be unsuitable for functions involving, for example, the response to an external stimulus or signal integration between cell networks, like neuronal signals or heart contraction. The use of chemical potential is suitable in very restricted spaces (e.g., neurotransmitter diffusion in chemical synapses), second messenger activity (e.g., calcium ions in muscle contraction) or the accumulation of messenger RNAs. To realize particular cell functions for which the execution time must be in the millisecond range, there must be a different type of energy involved. The fastest form of energy that could be utilized in the biological system is electric. The physical characteristics of the separators (lipid bilayers) have been the starting point for the evolution of an alternative way to accumulate energy. The chemical potential established by different solute concentrations produces an accumulation of electric potential when the separation across the membrane involves charged particles, in particular ions.

In solid conductors like metals, the current flows after the establishment of a potential difference. Electron current speed approximates that of light (about 300×10^6 m/s). In saline solutions, the current is carried by ions at a slightly lower speed than the current in solid conductors and in any case, several orders of magnitude faster than the speed of solute diffusion in water.

The establishment of an electric potential implies that the membrane has very precise and unusual physical-chemical characteristics. The establishment of a difference in electrical potential across a membrane implies electrical charge separation. It is important to remember that to respect the principle of electrical neutrality of solutions, communicating environments must remain electrically neutral. How could this be reconciled in the case of a lipid membrane with open pores permeable to charged particles and with different ionic concentrations on either side? This is possible due to the characteristics of the double phospholipid layer (Figure 1.3 B). The thickness of biological membranes (about 5×10^{-9} m) allows the immobilization of charges of opposite sign on the two sides, so that they neutralize each other even if physically separated. The accumulation of charges of opposite signs across the membrane generates an electrical potential. The membrane shows, in this case, one of its most important functional characteristics, which is to behave like a capacitor that can realize a separation of charges while maintaining electrical balance and at the same time accumulating potential energy. If you calculate the linear voltage density spanning the double phospholipid layer, you realize that the value reaches ~10 mV/nm in a neuron, which can also be written as ~10 megavolts/meter, to provide a tangible idea of the enormous amount of energy that a cell can store across its membrane. In addition, in this case, it can be observed that the relationship between structure and function is fundamental for the efficiency of physiological processes.

When then, following an appropriate stimulus, aqueous pores become permeable, the ionic current generated by the ions not in equilibrium will instantly flow through the membrane, modify the potential difference and become the fastest biological phenomenon when compared to any other cellular process. The greater the difference in potential, and therefore the electrical potential across the membrane, the more pronounced the phenomenon. It is important to point out that the speed at which the electrical potential spanning the membrane changes does not at all affect the performance of any biological function involving chemical reactions and permeability of substances through the membrane.

1.4 The concept of work

Every organism, whether unicellular or complex, must do work in order to achieve its purpose, which is to survive and reproduce. It can be said that work at the cellular level is generally divided into two types: work to maintain homeostasis and work useful for relationships with the external environment. While the first is essential, the second increases the probability of survival of the organism and allows it to have greater evolutionary success in developing into a more complex organism.

The maintenance of homeostasis is achieved by the combination of chemical energy consumption resulting from exergonic reactions, the exploitation of accumulated energy and passive mechanisms. The aim is to maintain a difference between the organism's own environment and the external environment, a difference that will be more or less accentuated depending on the function that a given system has inside the organism;

moreover, a greater difference means a greater capacity to accumulate potential energy.

In the biological world, the dynamic balances that are established are a combination between mechanisms that directly consume chemical energy and passive events due to the structure of the organism. The structure/function balance in this case must guarantee that creating different environments is possible and keep them in a dynamic balance with a minimum of work and therefore with a minimum expenditure of energy. The capacity of the system to accumulate and maintain potential energy will be all the greater the more adapted structures become to this task.

From the point of view of organism homeostasis, which does not involve large and fast oscillations in the amount of energy that is used, a greater or lesser accumulation of potential energy is mostly irrelevant. For this reason, if an impediment of any kind impairs the basic metabolism, the dynamic balances are maintained for some time, giving a chance of survival even in moderately unfavorable conditions. However, with the persistence of the blockage, the mechanism that guarantees homeostasis degrades and the system inexorably moves towards homogeneity between the different environments, which, as already mentioned, from a physiological point of view is synonymous with death.

The mechanisms that oversee homeostasis are in themselves sufficient for the survival functions of unicellular organisms or those composed of just a few elements. When the organism increases in size and begins to have necessary interactions with the surrounding environment, the mere maintenance of homeostasis is no longer sufficient to ensure its development and function. The different parts of a complex organism must be coordinated in their growth and controlled in their actions. An organism that is confronted with the surrounding environment must be able to relate to it, must be able to perceive stimuli and generate actions towards that environment. Therefore, in addition to a mechanism of maintenance, it is necessary to establish a mechanism of communication and control that stimulates, coordinates, synchronizes and diversifies the operations that transform energy into work.

It is also necessary that the organism is able to respond to the stimuli and stresses that come from the surrounding environment. In this case, the work that will be done by the organism will be divided between the transfer of information and the transformation of the message into action. The execution times of both functions must necessarily be short, since it is no longer a matter of maintaining a system in a stationary state, rather the priority becoming that of reaction times and execution of specific work. In all these cases, the energy used is essentially due to chemical potential, electrical potential or in most cases, both. For a correct sequence of information transfer and execution to take place, the two functions must have different timings. In particular, the transfer of information must not be a limiting factor to perform a job, as several functional situations confirm. For example, the case exists where the same signal to carry out a particular task must reach different locations. In this case, the action in one location must be synchronized with that in all the others. For this to happen, the time in which the message will arrive at its destination must be infinitesimal compared to the time of execution. In this way, the time of execution of the work will be synchronous in all parts of the body.

A second extremely important point in the biological world is control. In order to control performance, systems must continuously inform the source of the primary message about the progress of the work. The primary stimulus has to be reinforced or repressed if the communication failed or the work is done in excess, respectively. In addition, once the work is completed, there should also be a stop signal for the stimulus.

These mechanisms are information exchange circuits called feedback or feedback systems (Figure 1.8). Feedback can be positive or negative depending on whether it amplifies a signal or reduces it. It is easy to see that an information system in which there is a close relationship between stimulus and stimulus control, with a feedback loop that is able to influence the source of the stimulus, will be very precise in its task. Even the feedback system therefore requires an information exchange system that is much faster than the specific work to be done: the less time between stimulus and control feedback, the more efficient the system will be.

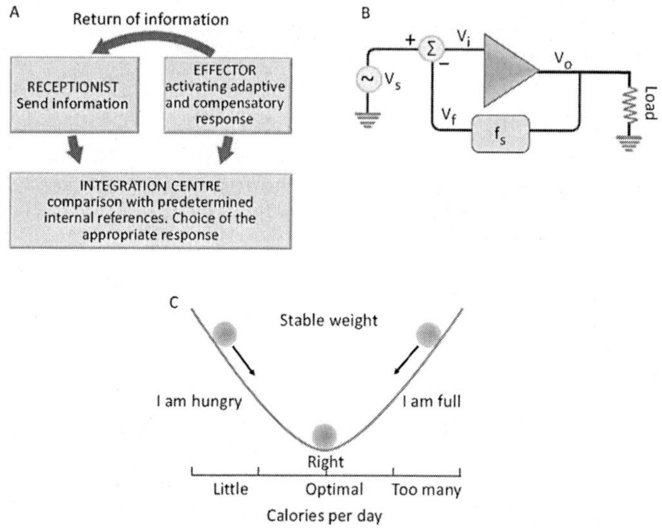

Figure 1.8 A) Schematic sequence of a feedback system. B) Simple electric amplifier circuit with feedback control (fs). C) Feedback mechanism applied to body weight control.

For tasks of information transfer, both chemical and electrical potential could be used. The discriminating factor remains the time with which the two forms of energy are transformed into useful work. Consequently, the method adopted to stimulate and to control will be adapted and related to the possibility of being synchronized and controlled by the speed of the information network. The information systems that take advantage of the chemical potential is based on diffusion processes. Even if in biological organisms there are structures able to direct the chemical messages and speed them up, the message will still be carried physically by the molecule that moves in fluid. The system that is based on diffusion must take into account the mass of the particle, the viscosity of the liquid in which it is immersed and the consequent friction that slows down the particle.

A different method is using electrical potential, in which the information transfer only assumes that the particles carrying the information must be charged. The solutions are second species conductors, in which the current flow occurs through ions. The speed of energy transfer is slightly lower than the electron transfer in the current flow that takes place in solid conductors. As shown in Figure 1.9A in "Newton's Cradles", the ball moves physically from 1 to 2 with high energy consumption and significantly high execution time. Instead, in the case of ion current, the process takes place by charge transfer as a variation of the momentum (Figure 1.9B), therefore very quickly and with low or very low energy consumption.

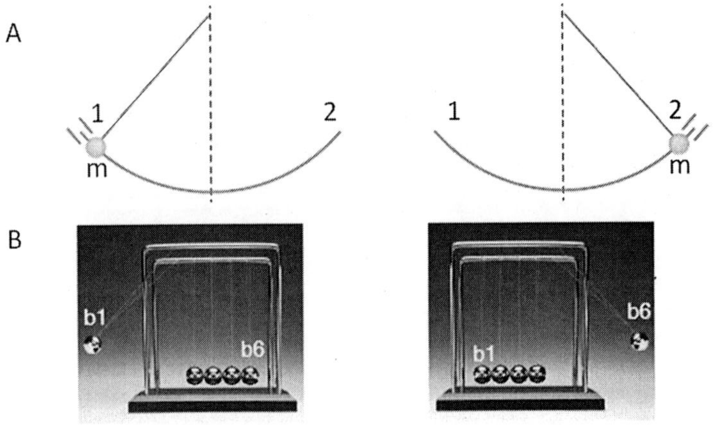

Figure 1.9 Mechanical diagram of the difference in runtime between chemical and electrical potential. A) The message m to go from 1 to 2 must go all the way of the pendulum. B) The movement of b6 happens by energy transfer through the series of elements that compose the system. The potential energy stored by m in 1 is equal to that of b1, but the time taken by m to go from 1 to 2 is greater than that required for the movement of b6 by the potential energy transferred by b1.

From the point of view of the ability to utilize the potential energy to perform work and therefore be able to build more complex physiological systems, the milestones can be summarized as:

1. acquisition of homeostatic mechanisms capable of storing chemical potential energy;
2. possibility of communication of chemical messages between the cells of a complex organism;

3. structuring communication pathways that direct diffusion processes;
4. use of electrical potential and ionic currents for information exchange processes;
5. direct relationship between cell and cell as the ultimate specialization of information transfer;
6. close relationship between structure and function for an efficient transformation of the message in the execution of a task.

Bibliography

Schopf JW, Barghoorn ES, Maser MD, Gordon RO. Electron microscopy of fossil bacteria two billion years old. Science (1965). 149 (3690): 1365-1367.

CHAPTER 2

Separators

By separators, we mean those structures that, in the biological world, serve to divide two environments that can be distinguished. The creation of separate environments has undoubtedly been one of the fundamental stages in the evolution of organisms capable of autonomous life. However, structural and functional problems had to be solved in order to achieve this objective. From a structural point of view, the biological separator must have a stable composition, exploiting not only the molecular cohesion of the particles that comprise it, but also the interaction it has with the

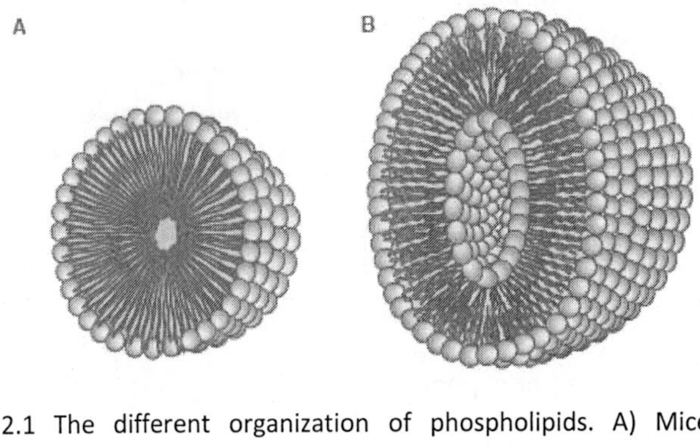

Figure 2.1 The different organization of phospholipids. A) Micelles. B) Liposomes.

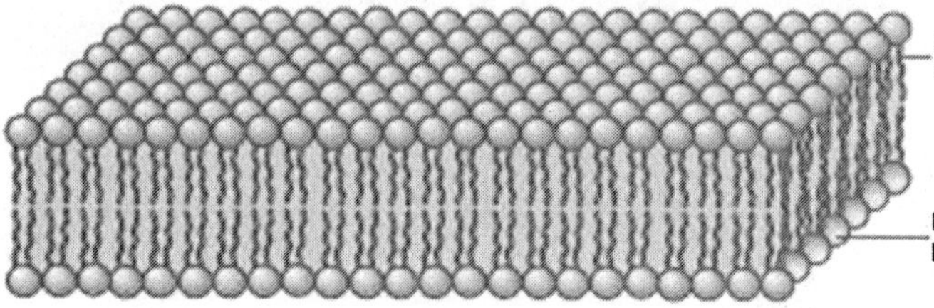

Figure 2.2 The phospholipid bilayer consists of two hydrophilic layers facing the solution and a double layer of hydrophobic lipid tails.

surrounding aqueous environment. Functionally, the separator has two major tasks: to constitute a permanent barrier so that it can form two sufficiently distinct environments and at the same time be dynamically permeable to solutes to ensure sufficient communication between the spaces it encloses and the external environment. These objectives are met by three characteristics specific to cell membranes.

The first characteristic is the thickness, which has a fixed value of about 10 nm ($10 \cdot 10^{-9}$ m) in any biological membrane, whether it is between the inside and outside of the cell or between intracellular environments. Biological membranes are evolutionarily the best way in which lipid molecules can be organized to form a barrier that ensures impermeability to all polar solutes without a rigid structure and, at the same time, allow the insertion of transmembrane proteins specialized in the passage and transport of hydrophilic substances.

The second structural characteristic is the molecular composition of the membrane, which is composed of **amphipathic molecules**, i.e., formed by a component with an electric charge, which is functionally **hydrophilic** and a neutral component, which is functionally **hydrophobic**. The selection of these mixed composition structures required biochemical elaborations of considerable complexity, which in turn took a very long time and an enormous number of attempts.

Thermodynamically stable structures made up of amphipathic molecules, such as **phospholipids**, can be assembled in different ways. These range from the simplest, a single layer enclosing a hydrophobic environment (Figure 2.1A), to the most complex that can delimit hydrophilic environments (Figure 2.1B), in both cases with the formation of mainly spherical structures, to real membranes capable of forming primitive "cells" that can lead to more complex forms of biological membranes as we know them today.

This stable structure is obtained by juxtaposing the hydrophobic part of two single layers of phospholipid with the exclusion of water (Figure 2.2). The interaction of the water solvent with the symmetrical phospholipid double layer, or bilayer, plays a fundamental and effective role in the stabilization process of the bilayer, as the charges of the hydrophilic part of the phospholipid molecules interact with the water molecules that are dipoles, with the oxygen atoms that constitute the negative part and the hydrogen atoms that constitute the positive part.

The third structural characteristic of the phospholipid double layer membrane is its ability to reach considerably extended lengths, thus being able to separate large volumes of solutions. Due to its peculiar mechanical stability and its remarkable flexibility, the membrane could have dimensions tending to infinity. It must be remembered, however, that the membrane must not only be a containment structure, but an active structure through which communications and all exchanges between the external and the internal cytoplasmic environment take place. The extent of these exchanges is determined by the amount of reactions and biochemical work performed by the cell, an amount of work that obviously increases as the cell volume increases.

There is a limit to the increase in cell volume, which is determined by the ratio of the membrane surface area and cell volume. If you approximate a cell to a sphere, the volume increases with the cube of the radius while the surface increases with the square of the radius. It is clear that a continuous increase in cell size would lead to the critical situation of a cell volume that is disproportionate to the ability to exchange through the surface. The ratio between surface and volume has, in biology, a functional value, because the higher this ratio, the more active a cell is. The two extreme examples are the neuron and the adipocyte (Figure 2.3). While the former generally presents itself with numerous branches and has a high metabolism with a high ratio between surface area and volume, the latter only acts as a container, is practically spherical and only stores fat, so it does not have to make many exchanges with the outside world.

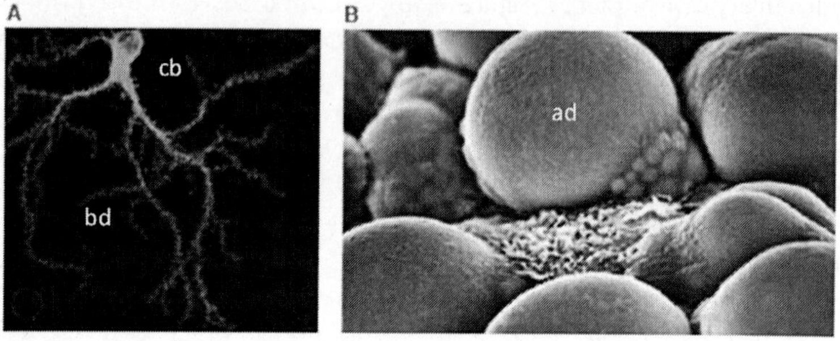

Figure 2.3 A) Fluorescence microscope image of a neuron consisting of a cell body (cb) and numerous branched dendrites (bd). B) Adipocytes (ad) observed under a scanning electron microscope.

The former must have as little volume as possible; the latter has no volume limitations.

In short, the membranes surrounding all the cells of each organism and those of all the intracellular organelles are formed by the bilayer, a structure 7-10 nm thick, thermodynamically stable, impermeable to ions, which depending on its lipid composition can be more or less rigid, but with characteristics of elasticity and tensile strength that allow it to respond even to relatively strong stresses of both an osmotic and mechanical nature. The double phospholipid layer acts as a support for numerous enzymes and proteins of various kinds, such as, for example, those that form pores and ion channels which, in turn, form low-electrical-resistance openings through which it is possible to exchange water and selected ion species between environments, or those that form the specific conveyors of different compounds.

To the structural characteristics of phospholipid membranes, it is necessary to add a special function that makes them unique: the electrical properties that play a particularly important role, performing an irreplaceable task for the functionality of all cells. One energy source that would be able to guarantee the functionality of the cells could be to create separations of electrical charges in solutions. However, this is not possible, as large amounts of energy would accumulate, which would inevitably lead to the destruction of the cell. Moreover, in a certain environment, the principle of electron neutrality must always be respected, according to which every charge of a certain type, positive or negative, must correspond to one of the opposite sign.

In the course of evolution, one of the most exploited sources of energy from an organism's cells has been selected for work. This is the potential energy that is obtained through the progressive separation and accumulation of electrical charges by the membranes that separate the two aqueous environments (Figure 2.4 A). This phenomenon can occur if, with a higher concentration of cations and anions in one environment than in the other, the membrane has transiently **selective gaps**, for example for cations (Figure 2.4 B). In these conditions, there is a flow between one environment and the other according to the concentration gradient, with the creation of an imbalance of charges due to accumulation, for example, of positive charges near one of the two sides of the membrane (Figure 2.4 C). In order to respect the principle of electron neutrality, an anion must be held close to the other side of the membrane (Figure 2.4 D). For this to happen, the separator must have specific physical characteristics such that the ions of opposite sign are able to attract each other. It is therefore clear that one of the functional characteristics of the phospholipid membrane concerns its thickness. The measurement of 10 nm is not a value chosen at random but rather the thickness of the insulation comprised by the grouping of phospholipid tails that functionally allows the interaction between

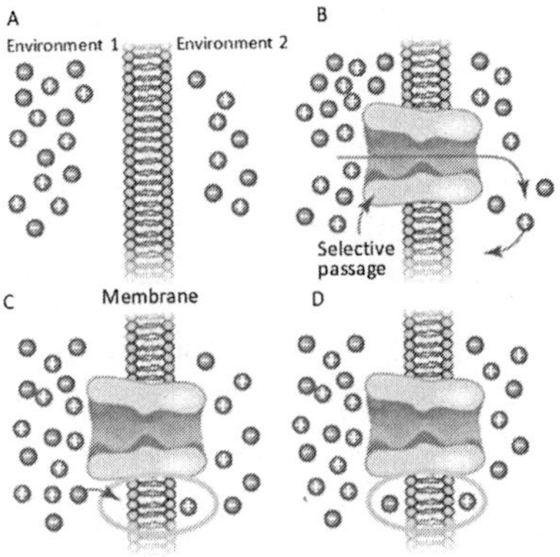

Figure 2.4 Biological membranes are capable of holding separate charges of opposite sign. A) The membrane is completely impermeable to ions. B) A selective gap for cations runs through the membrane. C) There is diffusion of cations, retained on the outer face of the membrane, and generation of an imbalance of charges. D) The imbalance is eliminated by an anion retained on the inner face of the membrane.

oppositely-charged ions straddling the membrane. It can be hypothesized that the possibility of interaction between cations and anions on either side of the membrane was one of the factors that contributed to the selection of the thickness of the phospholipid double layer.

2.1 The electrical membrane model

From an electrical point of view, the membrane has the characteristics of a capacitor and a resistor. The capacitor (C in Figure 2.5 A) is an electronic component made up of two parallel metallic conducting armatures separated from each other by a dielectric which may be air, ceramic, paper or, more often, non-conducting synthetic materials. The resistor (R of Figure 2.5 A) is a conductor made of a special material that

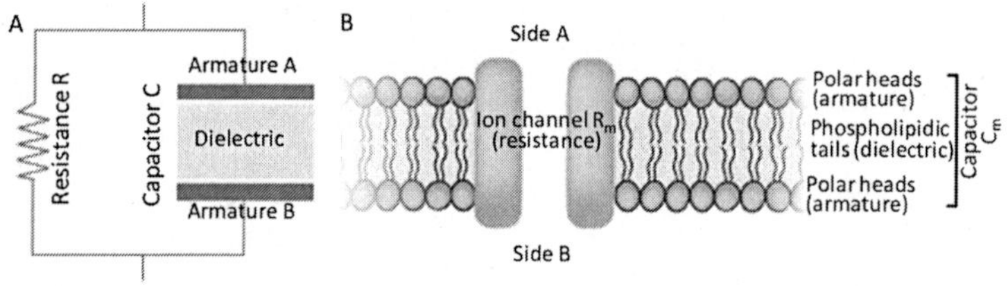

Figure 2.5 Schematic representation comparing a circuit consisting of a resistor and a capacitor placed in parallel (A) and a section of phospholipid membrane with an ion channel (B). The ion channel R_m behaves like the resistor R and the phospholipid bilayer C_m behaves like the capacitor C: the polar heads are the armatures and the lipid tails are the dielectric of the capacitor.

controls the passage of electrical charges. Capacitor and resistor are connected in parallel: charges, and therefore current, can pass either through the capacitor or through the resistor depending on the state of the two components. The laws of physics describing the behavior of resistors and capacitors in electrical circuits can therefore be used to characterize the capacitance and resistance of biological membranes (Data sheet 2.1).

$$\tau = R_m \cdot C_m, \qquad (2.1)$$

From a structural point of view, the polar heads of the membrane phospholipid, consisting of phosphate and glycerol, correspond to the capacitor armatures, while the dielectric is represented by the hydrophobic tails (Figure 2.5 B). This structure gives the functional property of accumulating charges and maintaining a potential difference between the two sides A and B of the membrane. The accumulated potential energy is readily dissipated when a short-circuit is caused between the two sides, e.g., by opening the membrane's selective gaps, that is, the ion channels (Data sheet 2.1). Charging the membrane capacitor is one of the fundamental steps during the excitation process. It will be seen below that the physical characteristics of this essential passive component of biological membranes discriminate between excitable and non-excitable cells and play an important role in the exchange of information between cells.

First of all, we need to examine the particular way in which the membrane capacitor works. At time $t = 0$, when the cell is at rest and the electrical parameters are stable, the resistance of the membrane capacitor C_m is zero, while the membrane resistance R_m has a value that is definitely not zero. A difference in potential applied between side A and side B of the membrane generates a current that can flow through both the membrane resistance R_m and the membrane capacitor C_m as they are arranged in parallel. However, the current will only affect the capacitor, in accordance with the physical principle whereby the current flows where it finds the least resistance, not flowing through the membrane due to the presence of the phospholipid tails. The charges can only deposit on the polar heads of one side of the membrane, inducing on the other side as many charges of opposite sign (Data sheet 2.1). The more that charges are deposited on the two surfaces and the capacitor becomes charged, the more the potential difference across the membrane changes and the capacitor's resistance increases. At the same time, the current across the membrane resistance R_m increases, reaching its maximum when the capacitor does not accumulate further charges and the potential difference across the membrane is kept constant.

The capacity for a cell to accumulate charges depends essentially on two factors: the surface area and the resistance of the membrane. The greater the capacitance of the capacitor C_m, and hence the greater the

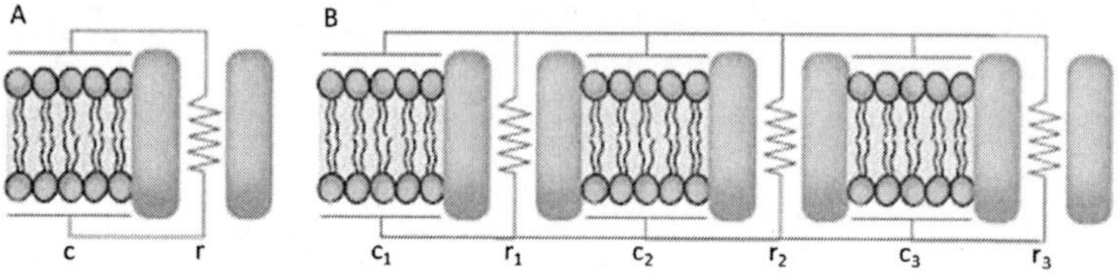

Figure 2.6 A) If the membrane surface is S, you have C_m=c and R_m=r. B) If the surface is equal to 3S, you have C_m=c·3 and R_m=r/3.

surface area of the membrane, the greater the charge accumulation time of the capacitor (Data sheet 2.1, equation S2.3). Similarly, the greater the membrane resistance R_m, the greater the current, which, as stated, flows in the component with lower resistance (the capacitor), which will thus have more influence on the variation time and on the value of the membrane potential. It can therefore be written:

where τ is the time constant and characterizes the charging time of the membrane capacitor and therefore the time required for the potential to change significantly.

It is clear at this point that the larger the cell, the greater the membrane capacity C_m and the lower the membrane resistance R_m. In fact, if the capacitances c_1, c_2 and c_3 and the resistances r_1, r_2 and r_3 of the individual components that form the membrane (Figure 2.6B) are respectively equal, C_m is equal to $c \cdot 3$ and R_m is equal to r/3 moreover, the greater the surface of the membrane, the more time is needed to charge and discharge the capacitor, and therefore to vary the membrane potential. This is because during the charge or discharge of the capacitor, much more current flows through the resistance, which is lower than that of a smaller cell (Figure 2.6 A). Moreover, with the same R_m, smaller cells are quicker to change potential following a stimulus. In fact, as has been shown above, at time zero, the resistance of the membrane capacitor is zero and therefore the time that the variation of the potential takes depends essentially on the charge of the capacitor.

The resistive components of the membrane, such as ion channels, aqueous pores and carriers (Figure 2.8), have a basic resistance by definition. In addition to this, the protein components electrically involved are subject to modulations, such as to modulate their permeability in a cell type manner of each type of cell.

From a structural point of view, biological separators can be divided into three main categories: cell membranes, intracellular membranes and epithelia. While the first two are specializations of the same molecular structure, the epithelia are macroscopic formations with ordering tasks at a higher level of complexity.

2.2 The plasma membrane

The existence of the plasma membrane (or cell membrane or plasmalemma) was indirectly demonstrated in the middle of the 19th century by Karl Wilhelm von Nägeli (Kilchberg 1817 - Munich 1891), who added dye to a container containing cells in a solution of water and salts. He observed that the cells did not or only faintly took up the color, while the whole solution was uniformly colored (Figure 2.7). He also observed that a cell placed in a diluted solution swelled due to water ingress, while when placed in a concentrated solution, it wrinkled due to water loss. Von Nägeli, therefore, had observed a series of phenomena, the coloring of the cytoplasm, the swelling and the wrinkling, that led him to suppose the existence of a barrier between the inside and the outside of the cell, but he was unable to see it. Only with the use of the electron microscope did J. David Robertson (1955) discover that the cells were surrounded by a three-layer membrane about 10 nm thick.

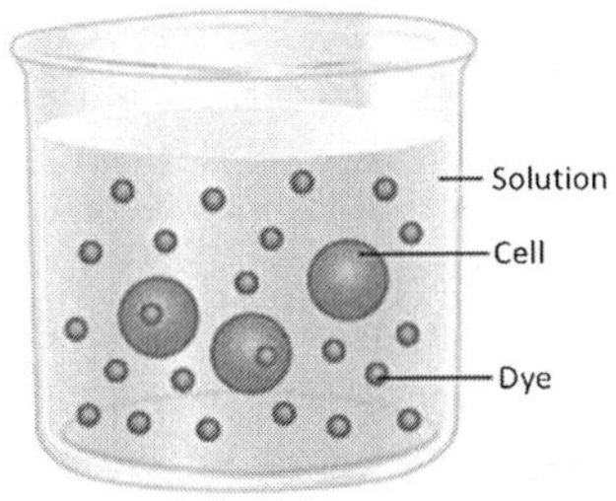

Figure 2.7 The dye is spread evenly in the solution but not in the cells.

von Nägeli's experiments indirectly demonstrated the existence of the plasma membrane, but also two of its fundamental functions. The first is the **barrier function**, which keeps the cell's internal environment separate from the outside and prevents the entry of substances unsuitable for its operation, such as dye, and the uncontrolled escape of useful substances. This property is due to the composition of the plasma membrane which makes it impermeable to water-soluble molecules.

The experiment with concentrated or diluted solutions, on the other hand, demonstrates the **selective function** of the membrane. In fact, with both diluted and concentrated solutions, only water moves through the plasma membrane and not the solutes present inside and outside the cell. This means that the plasma membrane is able to select which molecules can pass through it, generally thanks to specific proteins present on the membrane itself.

The proteins present on the external face of the plasma membrane can also be receptors for hormones and neurotransmitters, making the membrane itself perform the **function of communication** between the outside and inside of the cell. Numerous protein molecules with enzymatic functions can function correctly because they are in an orderly sequence on the internal face of the plasma membrane, such as the protein sequence

receptor → G-protein → adenylate cyclase

or the sequence

receptor → G-protein → phospholipase C→ phosphatidyl inositol diphosphate → diacylglycerol

thus performing an important **biochemical function**.

The widely-accepted plasma membrane model is the **double phospholipid fluid mosaic layer** (Singer and Nicolson, 1972). If we imagine looking at a cell from the outside, we observe a heterogeneous mosaic of single molecules, or small groups of molecules, of a protein nature, distributed more or less uniformly within the phospholipid double layer. It has been calculated that an average of 10^6 phospholipid molecules per μm^2 of membrane and as many protein

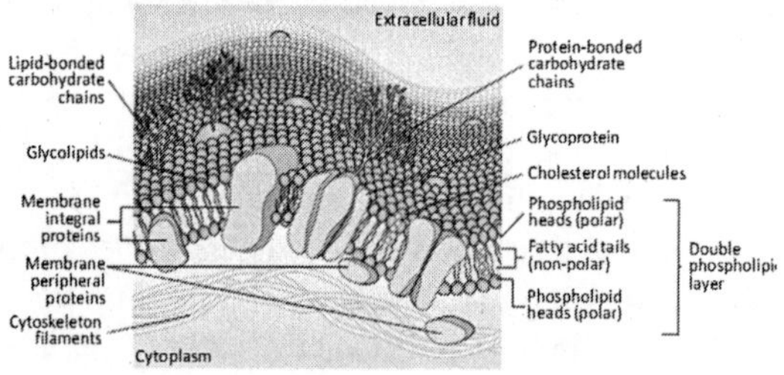

Figure 2.8 The plasma membrane consists mainly of phospholipids and proteins.

molecules are present.

The fluidity of the membrane is due to the fact that the forces holding the tails of the molecules in the two layers and those of each layer together are the relatively weak van der Waals forces. These forces allow phospholipid molecules to move rarely, on average once every thirty days, from one layer to the other and more frequently, it is believed, about 10^7 times per second, within a layer. It is because of this fluidity that the membrane can repair small tears that may form.

The plasma membrane consists almost entirely of two types of molecules: phospholipids and proteins (Figure 2.8).

2.2.1 Phospholipids

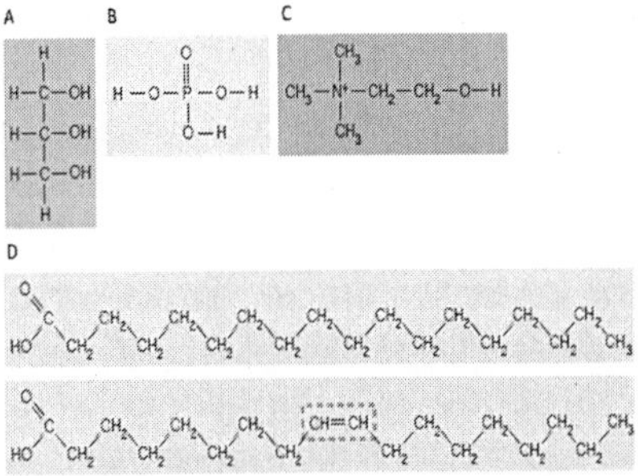

Figure 2.9 Phospholipid constituents. A) Glycerol. B) Phosphoric acid. C) Choline. D) Stearic (above) and oleic (below) fatty acids.

A phospholipid consists of: a **glycerol** molecule, a carbohydrate with three carbon atoms (Figure 2.9A); a phosphoric acid molecule (Figure 2.9B); a molecule such as **choline** which also contains a nitrogen atom (Figure 2.9C); and two **fatty acid** molecules, each with 16-20 carbon atoms, one of which often has at least a double bond (Figure 2.9D). The glycerol molecule forms the skeleton of the phospholipid. The two adjacent alcohol groups esterify two fatty acids and the third alcohol group esterifies phosphoric acid, which in turn has esterified the choline molecule containing the nitrogen atom (Figure 2.10). An oxygen bound to phosphorus takes on a negative charge fraction and a nitrogen

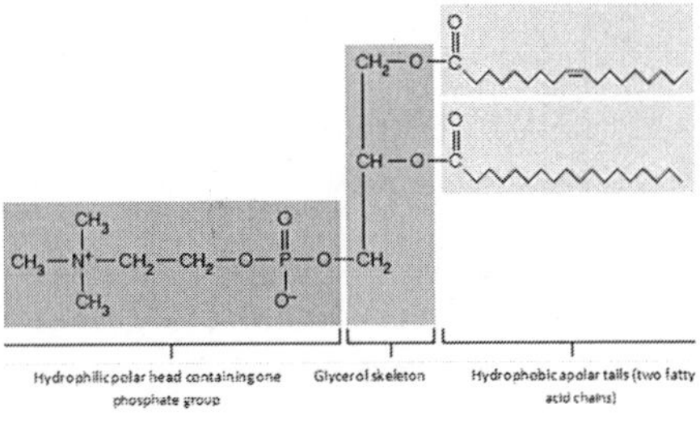

Figure 2.10 A phospholipid molecule.

adjacent to phosphorus a positive charge fraction, thus forming a **dipole**. The phospholipid molecule is therefore **amphipathic**: the part that forms the dipole is hydrophilic and constitutes the **head**, and the two fatty acid chains that are hydrophobic constitute the **tail** (Figure 1.2). The individual C—C bonds of each fatty acid molecule can rotate freely and randomly by thermal agitation, but when the complete phospholipid molecule is formed, the fatty acids tend to be arranged in parallel in each layer of the membrane. The double bonds C==C, in contrast, do not rotate and therefore give the molecule a certain rigidity.

The composition of phospholipids in the plasma membrane can vary according to species and cell type. For example, the unsaturated fatty acid in Figure 2.9 D is oleic acid, while the saturated fatty acid is stearic acid, but in terms of saturated fatty acids, there may also be palmitic acid with 16 carbon atoms and arachidic acid with 20 carbon atoms; in terms of unsaturated fatty acids, there is linoleic acid with 18 carbon atoms and two double bonds. In addition, colamine or serine may be present instead of choline.

The percentage of phospholipids compared to proteins is also variable in the different membrane types. For example, phospholipids are 80% of the myelin sheaths lining the axons of nerve cells, 50% in liver cells, 40% in erythrocytes, 35% in the epithelial cells of intestinal microvilli and up to 20-30% in the membranes of cytoplasmic organelles. The selection of different percentages of phospholipids in cells of different tissues has characterized the structure of cell membranes over the course of evolution. The membrane structure, in turn, is closely related to the function of each cell.

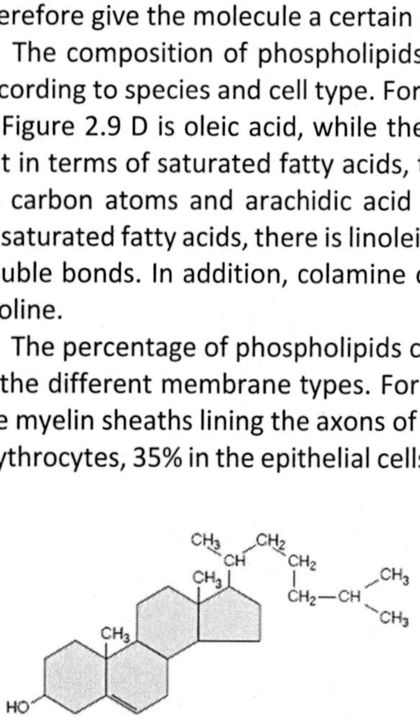

Figure 2.11 Structure formula of a cholesterol molecule.

Most mammalian cell membranes, but not the internal membranes or those of bacteria, actually contain another lipid-like molecule: **cholesterol** (Figure 2.11). Cholesterol, which is slightly amphipathic because it has a large hydrophobic steroid portion and hydrophilic hydroxyl, is inserted with the large hydrophobic portion between two adjacent fatty acid chains, with which it forms weak bonds, limiting its movements (Figure 2.8).

In most membranes, about 18% of the lipid fraction is cholesterol, but there are some membranes that are almost free of it and are characterized by high fluidity and low mechanical resistance. Others have the same number of cholesterol and phospholipid molecules, and in this case, each phospholipid is bound to the adjacent one by cholesterol and the membrane almost completely loses fluidity but acquires considerable mechanical resistance.

2.2.2 Proteins

While there are few, as we have seen, lipid species that make up the plasma membrane, proteins are represented by at least 20 different types, with a molecular weight ranging from 25 to 260 kDa.

Membrane proteins can be grouped into two categories: **integral proteins** and **peripheral proteins** (Figure 2.8); the former can have both primary α-helix and quaternary globular structure, while the latter are mainly α-helix proteins.

Of the 20 amino acids that can form proteins, 6 are strongly hydrophobic (alanine, phenylalanine, glycine, isoleucine, leucine, valine), 7 are weakly hydrophobic (asparagine, cysteine, glutamine, methionine, proline, tyrosine, tryptophan) and 7 are hydrophilic (aspartic acid, arginine, glutamate, histidine, lysine), depending on the type of radical that forms the amino acid (Figure 2.12). The amino acid composition makes the protein **hydrophilic**, **hydrophobic** or **amphoteric**. If hydrophilic, the protein cannot be immersed in the phospholipid double layer and will therefore be a peripheral protein. Generally, 47% of proteins of this type consist of hydrophilic amino acids arranged at the periphery of the molecule. The hydrophobic amino acids are arranged towards the

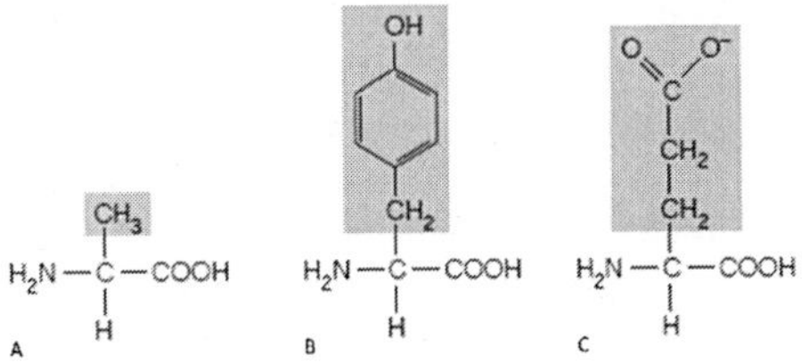

Figure 2.12 A) Alanine, a hydrophobic amino acid. B) Tyrosine, a weakly hydrophobic amino acid. C) Glutamate, a hydrophilic amino acid.

inside of the molecule, which does not come into contact with water; they are bound to the hydrophilic heads of phospholipids with electrostatic-type bonds and very often are also bound to an integral protein. A typical example of such a protein is the red blood cell spectrin, a protein consisting of an α chain of 255 kDa and a β chain of 220 kDa bound together. It is located on the cytoplasmic face of the membrane, makes up about 33% of the membrane proteins and serves to hold the other proteins in place.

A hydrophobic protein, on the other hand, will be completely immersed in the phospholipid double layer and will be an integral protein. An amphoteric protein will also be an integral protein and will have one part immersed in the phospholipid double layer and the hydrophilic parts outside it, towards the cytoplasmic side or towards the external part of the membrane, where it will be able to easily interact with molecules in solution or ions present on the sides of the membrane.

2.2.3 Carbohydrates

Carbohydrates, present exclusively on the external side of the plasma membrane, are branched or filamentous oligosaccharides consisting of monosaccharides such as glucose, galactose, mannose and others. They bind with covalent binding to membrane proteins, forming **glycoproteins**, or to lipids forming **glycolipids** (Figure 2.8). Glycolipids are abundant in nerve tissue but their role in the functioning of the synapse is still unknown (paragraph 6.3). Those present on the membranes of erythrocytes are involved, together with glycoproteins, in the histocompatibility of blood groups and in cell recognition linked to tissue growth, as well as in the immune response in general. In some cases, glycolipids are involved in the anchoring systems of certain extracellular matrix proteins to the membrane.

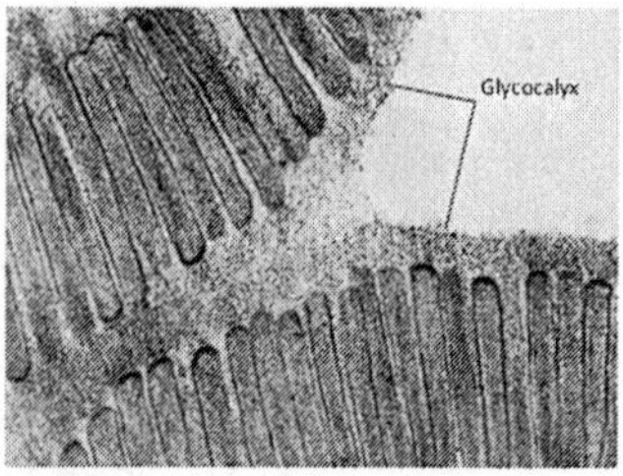

Figure 2.13 Example of glycocalyx arrangement observed under an electron microscope.

Glycoproteins constitute, in some tissues, the **glycocalyx**, with protective and cell anchoring functions (Figure 2.13), as for example in intestinal microvilli, where the glycocalyx is made up of short, branched carbohydrates intertwined with other long, filamentous ones.

2.3 Intracellular membranes

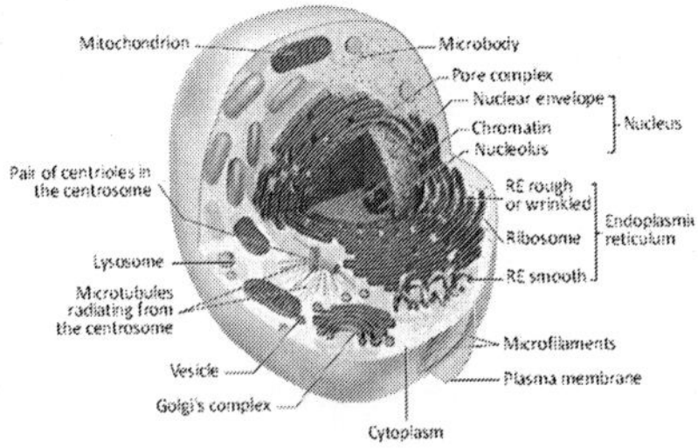

Figure 2.14 The interior of a generic cell and the intracellular membrane complex.

The intracellular membranes are a varied series of separators with very different structures, compositions and functions. The importance of the membranes inside a cell is once again to create protected and restricted environments where the various cellular components are more *likely to interact with each other* or where large quantities of solutes in the form of chemical potential energy can be captured (Figure 2.14). Two classic examples are the membranous systems that make up the nuclear envelope and the mitochondria.

The nuclear envelope consists of a double membrane surrounding the genetic material and in eukaryotic cells, it has essentially a containing and protective function (Figure 2.15). There is free communication between nucleoplasm and cytoplasm for ions and small solutes. The genesis of the nuclear envelope, which is a specialization of the endoplasmic reticulum, is

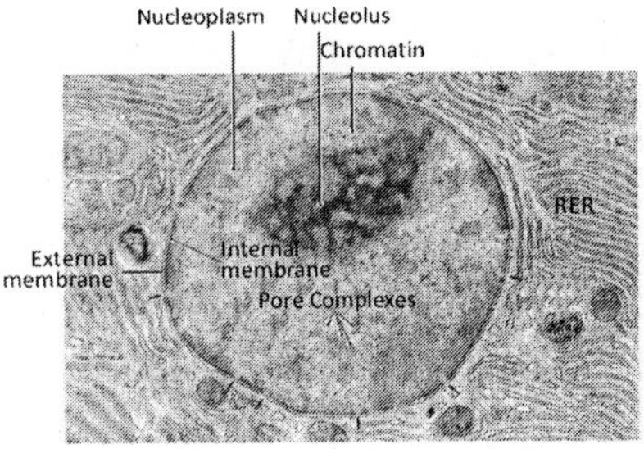

Figure 2.15 Electron micrograph of a portion of a cell. The cell nucleus, the external and internal membranes forming the nuclear envelope, some pore complexes on the nuclear envelope, the section of some pore complexes (black arrows), the wrinkled endoplasmic reticulum (RER) and some cytoplasmic organelles, probably mitochondria, are visible.

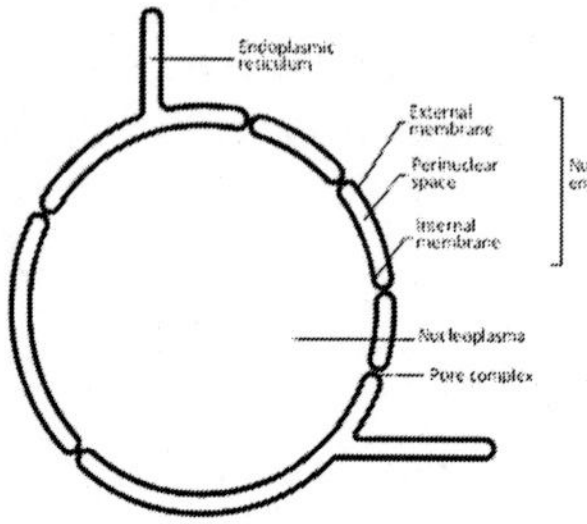

Figure 2.16 Structural composition of the nuclear envelope.

unclear; in fact, the outer membrane has ribosomes on its surface and evident continuity with the reticulum cisternae (Figure 2.16). On the contrary, the inner membrane is in close contact with a dense protein network composed of type A and type B laminin, which acts as an anchor for the DNA when it is in its interphasic form.

Various functions are ascribed to the nuclear envelope. There is experimental evidence that perinuclear cisternae, i.e., the nuclear envelope and the associated wrinkled endoplasmic latticework, are able to accumulate calcium ions and control their concentration in the nucleoplasm. This task is very important because fluctuations in the concentration of calcium ions can be very harmful to the genetic material: a prolonged increase in the concentration of this ion causes the activation of enzymes such as endonucleases which, by fragmenting DNA, are involved in the apoptotic process.

A very peculiar structure of the nuclear envelope and object of countless studies is the protein complex of the nuclear pore, or the **pore complex** (Figure 2.17), which is formed by different proteins with a total molecular weight up to 1200 kDa. Among these proteins,

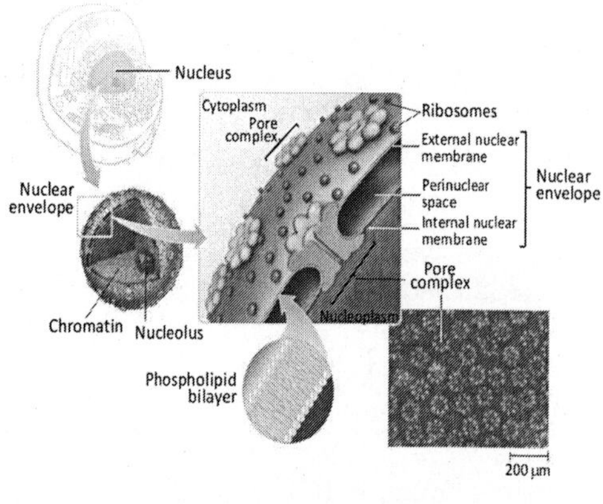

Figure 2.17 Diagram of the pore complex in a nucleus of a generic eukaryotic cell. The diagram in the center shows the structure of the nuclear envelope. In the inserts are electron microscopic images of the phospholipid bilayer and the surface of the nuclear envelope, showing the density and spatial distribution of the pore complexes.

the presence of ATPase and acto-myosin complexes has been demonstrated. The pore complex is anchored to the introflexions that the nuclear envelope presents for the purpose of allowing insertion of the pore. Therefore, the discontinuities of the nuclear envelope at the pores do not correspond to an interruption of either the external or internal membrane, but only to the folds that ensure the continuity of the double phospholipid layer (Figure 2.16). Under physiological conditions, nucleocytoplasmic communication is guaranteed by a pore opening of about 9 nm, which allows free diffusion of ions and small solutes up to a molecular weight of 40 kDa. The pore complex appears to be mainly involved in the transport of RNA from the nucleus to the cytoplasm, although the molecular mechanism has not yet been clarified. The use of a peptide, WGA (Wheat Germ Agglutinin), that is able to interact with the pore complex interrupts the passage of RNA.

The other membranous complex presenting a double phospholipid layer is the mitochondria (Figure 2.18). Many scholars believe that mitochondria were originally autonomous bacterial organisms that later became symbionts with cells during evolution. Their main function is to produce ATP through the Krebs cycle using glucose. The enzymes of the respiratory chain are supported by the internal mitochondrial membrane, which folds into the characteristic "ridges" and increases the useful surface area. The outer membrane has several permeation pathways through the single phospholipid bilayer. There are also proteins that pass through both membranes at the points where they pack together, areas called contact sites.

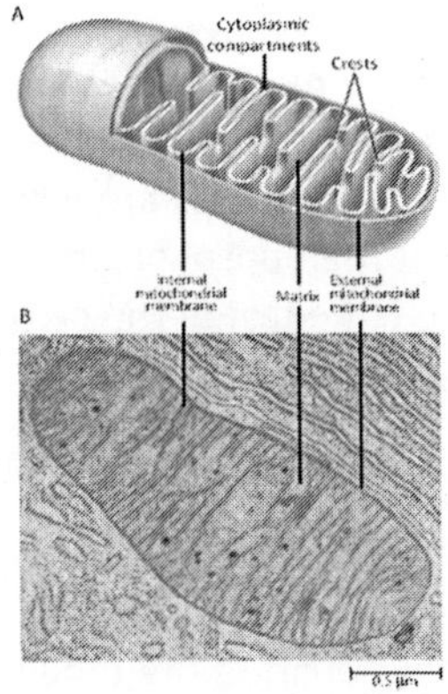

Figure 2.18 Structural scheme (A) and electron microscope ultrastructure (B) of a generic mitochondrion.

These proteins are real ionic channels called VDAC, or Voltage-Dependent Anion Channels, and ensure the modulation of

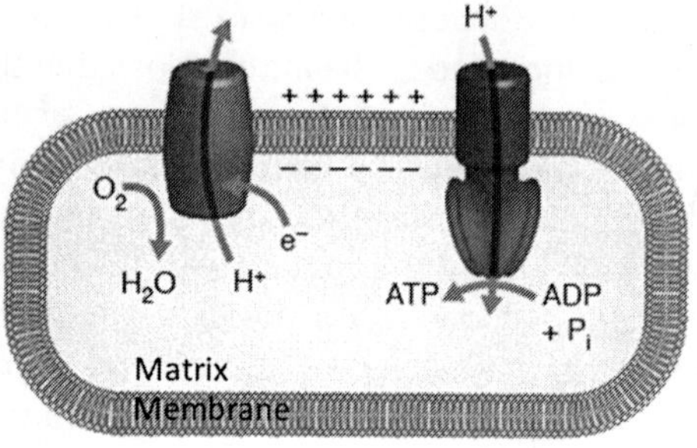

Figure 2.19 Scheme of the exploitation of the mitochondrial potential difference for ATP production.

solute exchanges between the cytoplasm and mitochondrial matrix. The importance of these permeation pathways is manifested in the enormous potential energy that the mitochondria are able to accumulate, in the form of an electrical potential difference between the two membranes that reaches values close to -200 mV (Figure 2.19) and which guarantees sufficient energy for the production of ATP.

There are also several other membranous formations inside the cell, such as the endoplasmic reticulum, which can have many specializations in cells with different functions. In particular, the ability of the endoplasmic reticulum to accumulate calcium ions to be used as an intracellular messenger will be considered (Chapter 4). The Golgi apparatus, lysosomes and endosomes are just some of the many other intracellular structures that use phospholipid membranes to isolate environments within the cell from the rest of the cytoplasm, in order to diversify the functions that are required in an adult cell to ensure its proper function.

2.4 The epithelia

Epithelia, as we know, are tissues formed by one or more layers of cells. In some organs and apparatus, such as the digestive and urinary apparatus, they separate the organism from the external environment, forming the organ **lumen**.

The epithelia perform, on one hand, the function of **absorbing** useful substances present in the lumen, with highly specific mechanisms capable of selecting with great precision the molecules to be transported and bringing them close to the capillary circulatory system. On the other hand, they also function as part of the **secretion** system, to eliminate the substances produced by cell metabolism. The cells of the epithelium

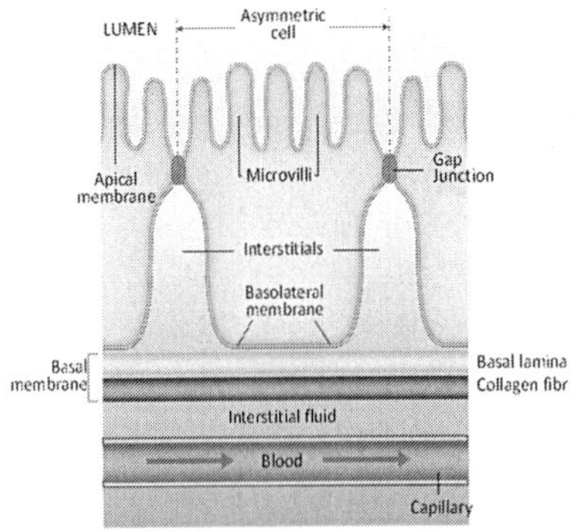

Figure 2.20 A monolayer epithelium of asymmetric cells joined by numerous gap junctions.

(Figure 2.20) are **polarized**, as they have an **apical membrane** on the mucosal side of the epithelium, facing the lumen, and a **basolateral membrane** on the **serosal** side of the epithelium facing the blood capillaries; the two types of membranes sometimes have very different properties, even if their fundamental structure is that already described, with a double phospholipid layer.

Between the basolateral membrane and the blood capillaries, there is a **basal membrane** formed by a **basal lamina** of mucopolysaccharides, collagen that is not organized into fibrils and a layer of collagen fibrils joined to the basal lamina. The basal membrane acts as a support for the epithelial cells, but does not interfere with the flow of substances, as it has large pores with a diameter of one to several tens of nanometers.

The epithelium is also characterized by the presence of numerous gap junctions (Data sheet 6.1), fundamental for giving continuity to the barrier towards the lumen, by the presence of microvilli, often localized on the mucosal side and fundamental for increasing the exchange surface, and by the presence between one cell and another of the relatively voluminous interstices, important, as we will see later, for water absorption.

Bibliography

Robertson JD. The ultrastructure of adult vertebrate peripheral myelinated nerve fibers in relation to myelinogenesis. J. Biophys. Biochem. Cytol. (1955) 1(4): 271-278.

Singer SJ and Nicolson GL. The fluid mosaic model of the structure of cell membranes. Science (1972) 175(4023), 720-731.

Data Sheet 2.1

The electric membrane model

In Figure S2.1, the electrical model of the plasma membrane is represented as a circuit formed by: a) a resistor, the ion-conducting channels; b) a capacitor, the phospholipid bilayer with an insulating component (dielectric), constituted by the hydrophobic tails, and two conducting plates, constituted by the hydrophilic heads (Figure 2. 2); c) a current generator, representing an electrical stimulus produced, for example, by a neurotransmitter at a synapse (section 6.3) or imposed by the experimenter; d) a switch, any short circuit between internal and external caused, for example, by the opening of a high-conducting membrane channel (Figure 2.4) or by a stimulus instrument. When the switch is closed (ON of Figure S2.2), the generator produces a current, in the model of only positive charges (Figure S2.1), which rises instantaneously to a maximum value and then returns to zero instantaneously when the switch is opened (OFF of Figure S2.2). From the generator (1 of Figure S2.1), this current runs through the circuit up to side A of the model (2 of Figure S2.1) and must necessarily run through the branch with the capacity C (3 of Figure S2.1), generating the capacitive current I_C, and the branch with the resistors R (9 of Figure S2.1), generating the resistive current I_R. The total current I_{TOT} will therefore be:

$$I_{TOT} = I_R + I_C \tag{S2.1}$$

The capacitive current. At the time t=0 when the switch is closed (ON of Figure S2.2), the capacitor C is completely discharged. It can accumulate on side A a number tending to infinity of charges and therefore has zero resistance. The resistance R, on the other hand, has a value other than zero which depends on its physical characteristics. The current will be, at the time t=0, only capacitive because the capacitor, with zero resistance unlike the resistance R, is able to interact with a very large quantity, tending to infinity, of charges. These charges accumulate on the plate of side A (4 of Figure S2.1) and, not being able to cross the insulator to go through the whole circuit, induce the formation of as many negative charges on the plate of side B (5 of Figure S2.1) by the removal of as many positive charges (6 of Figure S2.1). The positive charges travel through the rest of the circuit

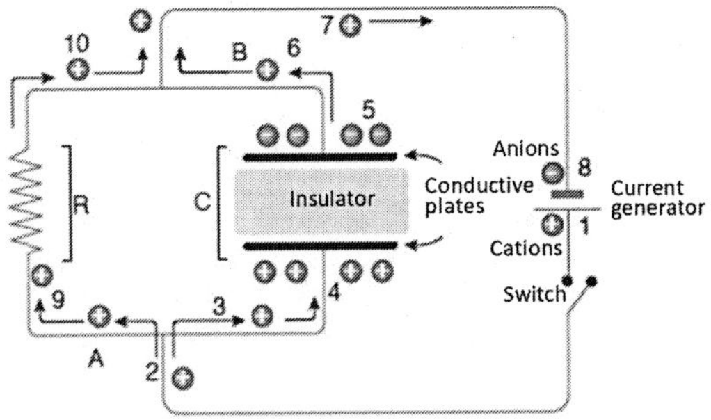

Figure S2.1 The electrical model of a plasma membrane consists of a resistor R (the ion channels) and a capacitor C (the double phospholipidic layer), arranged in parallel, a current generator (the potential difference that can be set across the membrane e.g. by a battery or an external generator) and a switch.

(7 of Figure S2.1) and neutralize the negative charges of the current generator (8 of Figure S2.1) which reduces its charge. The capacitive current I_C, therefore, travels through the circuit from 1 to 8 of Figure S2.1 without the charges, having been able to physically cross the capacitor due

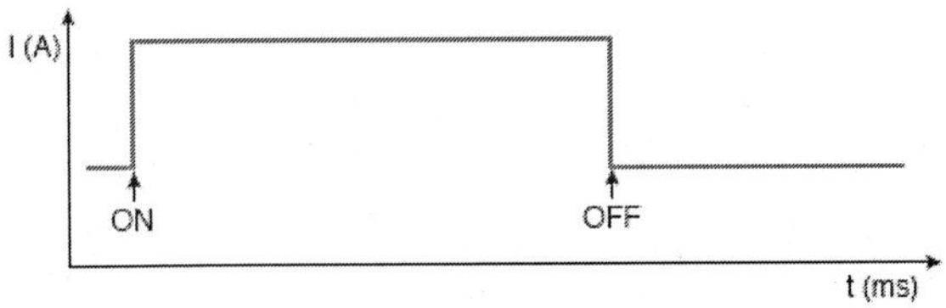

Figure S2.2 Temporal trend of the total I_{TOT} current between side A and side B of the model in Figure S2.1. ON and OFF indicate the moment of closing and opening of the switch, respectively.

to the presence of the insulator. Naturally, if the current were carried by negative charges, it would travel through the circuit from point 8 to point 1 of Figure S2.1.

The capacitive current I_C, i.e., the amount of charge that affects the capacitor, is directly proportional to dV, the variation of potential that generates the current between the two plates of the capacitor, as a function of time dt according to the relation:

$$I_C = C \cdot \frac{dV}{dt} \qquad (S2.2)$$

The proportionality constant C in equation S2.2 is the membrane capacity, expressed in Farad (F). It indicates the amount of charge that the capacitor can accumulate, is directly proportional to the area A of the overlapping part of the two plates (since the greater the area, the greater the number of charges that the capacitor can accumulate) and is inversely proportional to the distance d between the plates, since a smaller distance allows a greater force of attraction or repulsion between the charges of opposite sign:

$$C = \varepsilon_r \varepsilon_0 \cdot \frac{A}{d}; \qquad (S2.3)$$

ε_0 is the dielectric constant, which in a vacuum is 8.85×10^{-12}, while ε_r, is the relative dielectric constant of the insulating material and a dimensionless number greater than 1.

At the time dt=0 of closing of the switch (ON of Figure S2.3), the term (equation S2.2) tends to infinity because in the infinitesimal time dt, there is an instantaneous increase of the potential dV created by the current generator. The amount of charge that can be accumulated at side A of the capacitor, which tends to infinity, depends on the capacitance C of the capacitor (equation S2. 2) and in reality, the capacitive current I_C has a peak up to a finite value. As time passes, the number of charges that can be accumulated decreases and I_C decreases, because the positive charges already accumulated tend to oppose the accumulation of other charges, until I_C becomes zero when the capacitor is fully charged. The variation of potential dV becomes equal to zero and its variation in the infinitesimal time dt becomes equal to zero (I_C curve of Figure S2.3).

When the switch is turned OFF, the current ceases instantaneously and there is an instantaneous decrease of the potential dV in the infinitesimal time dt. This produces a peak of negative capacitive current that decreases with a time trend comparable to that of the ON and which, through the resistor R, tends to neutralize the charges accumulated on the two plates of the capacitor. Therefore, a resistive current I_R is generated, which decreases with time (Figure S2.3) and decreases the difference of potential between side A and side B of the capacitor.

The resistive current. At time dt=0 when the switch is closed, the current affecting the circuit is, as stated, all capacitive and the resistive current I_R is zero (ON of Figure S2.3). As time passes, the number of charges that can accumulate on the capacitor plates decreases and I_C decreases. The charges, which are continuously supplied by the current generator, begin to flow through the branch of the circuit with the resistor R (9 in Figure S2.1) and thus generate the resistive current I_R. Unlike the capacitor, the resistor causes the current to flow continuously from side A to side B in a time-independent manner, but, due to the presence of the capacitor, the resistive current I_R increases in a time-dependent manner. I_{TOT} remains constant from the ON point to the OFF point in Figure S2. 2, so while the I_C current decreases, the I_R resistive current increases up to a maximum value, becoming constant when the I_C current becomes zero (Figure S2.3 and equation S2.1). The charges complete the circuit (10 and 7 of Figure

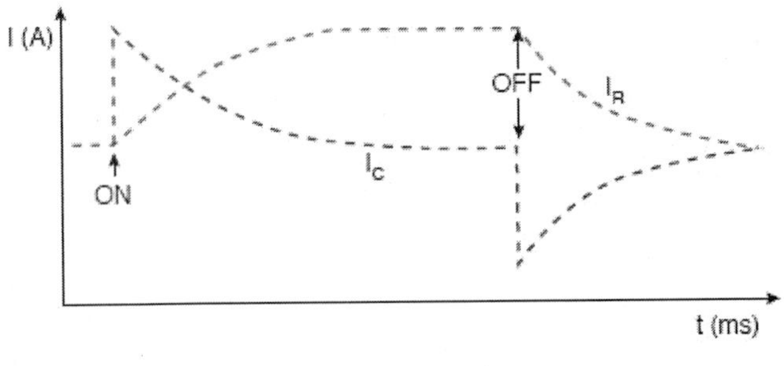

Figure S2.3 Temporal trend of the I_C capacitive and I_R resistive current between A and B side of the model. ON and OFF indicate the moment of closing and opening, respectively, of the switch shown in Figure S2.1.

S2.1) and go to neutralize the negative charges of the current generator (8 of Figure S2.1), which reduces its charge. As said, at the OFF of Figure S2.2, a resistive current I_R is generated which decreases in time (Figure S2.3).

I_R (dimensionally Ampère, A) is directly proportional to the potential V (dimensionally Volt, V) given by the current generator and inversely proportional to the resistance R (Ω, Ohm), or directly proportional to the conductance G (dimensionally Siemens, S), according to Ohm's law:

$$I_R = \frac{1}{R} \cdot V = G \cdot V \,. \tag{S2.4}$$

The presence of a capacitive current I_C and a resistive current I_R, with different time courses, makes the potential vary in time $V_{(t)}$ between side A and side B of the circuit with an exponential trend (Figure S2.4), described by the equation:

$$V_{(t)} = V_{t=\infty} \cdot \left(1 - e^{\frac{-t}{\tau}} \right), \tag{S2.5}$$

where $V_{t=\infty}$ is the potential at time t=∞. Solving equation S2.5 for t=τ, we have:

$$V_{(t=\tau)} = V_{t=\infty} \cdot \left(1 - e^{-1} \right) = V_{t=\infty} \cdot (1 - 0.368) = V_{t=\infty} \cdot 0.632 \,; \tag{S2.6}$$

so τ is the time required for the membrane potential to reach 63.2% of the potential at infinite time (Figure S2.4), according to the equation S2.7.

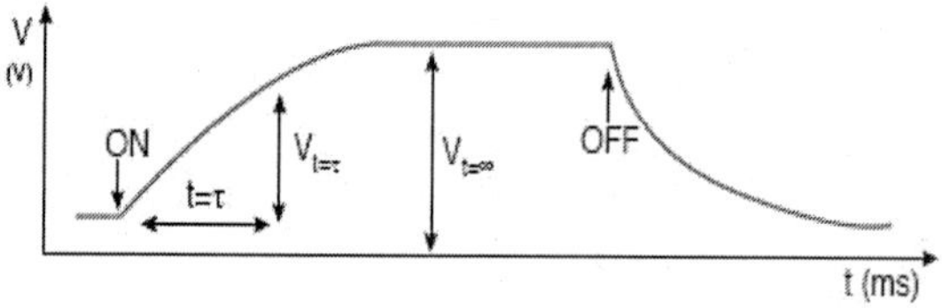

Figure S2.4 Temporal trend of the potential difference between the A and B side of the model. ON and OFF indicate the moment of closing and opening, respectively, of the switch shown in Figure S2.2.

$$\tau = R \cdot C \tag{S2.7}$$

where t is the time constant (seconds), R is the membrane resistance (Ω, Ohm) and C is the capacitance (F, Farad).

CHAPTER 3

Accumulators

All cells in a biological organism are in a state of **rest**, which consists of a **dynamic equilibrium** between the cell and its environment. The dynamic equilibrium, also characterized as homeostatic conditions, is maintained by expending chemical energy. The cell, to reach these conditions, takes advantage of the "symbiosis" between the **separators** (Chapter 2) and the **transporters** (Chapter 4). During these resting conditions, the system is constantly maintained at a high level of potential energy, the magnitude of which is directly related to the particular duty assigned to that particular cell.

There are constraints, controlled by structures and by functional mechanisms, which depend on various parameters and which are able to keep the biological system in a state of dynamic equilibrium, away from static equilibrium. At the precise moment when the system needs to utilize the potential energy stored by transforming it into other forms of energy, and eventually into work, these constraints are removed, allowing the system a momentary **fall** towards static equilibrium. In order to remove the constraints, a perturbation of the system is required, most of the time coming from the surrounding environment, which can be envisioned as an **activation energy** or an **appropriate stimulus**. The stimulus can be electrical, chemical, mechanical or a variation of physical parameters, such as temperature or hydrostatic pressure, or chemical parameters, such as osmolarity and the concentration of polar and neutral solutes. The energy required for the removal of constraints is most often a negligible portion of the energy that is released from the system. The removal of constraints will result in the loss of potential energy at the maximum rate for the needs of the organism, which is often essential for its survival.

The advantage of a system designed in this way is the possibility of both modulating the intensity of the response to any stimulus and having available energy in excess. Moreover, the time that the transfer of energy takes to trigger any physical processes involved in the production of work can be extremely small (for responses to electrical stimuli) or greater (in response to chemical, mechanical or physical stimuli).

Highly efficient active or passive mechanisms have the task of re-establishing resting conditions. Ideally, the dynamic equilibrium at high potential energy levels is reached with a minimum of energy consumption. If these mechanisms are missing or malfunction, the system is unable to restore the initial conditions and heads inexorably towards static equilibrium, which in biological systems means death.

From a physiological point of view, a biological system can be considered **alive** as long as it is able to accumulate energy. Potential energy is used for the primary reaction to a stimulus and necessarily occurs in the zone of functional contact between two environments that are at different energy levels. This is why in biological systems, all reactions to stimuli are mediated by the variation of parameters at the level of phospholipid membranes, which not only have an important structural role, but also fundamental functional tasks. In the following paragraphs, we will discuss how cells are able to increase their potential energy level and keep it high in dynamic equilibrium. During evolution, different ways of storing energy have been experienced and developed, and we will see that they have led to the creation of energy reserves that differ in terms of time of use and thermodynamic efficiency in producing work.

3.1 Energy storage in biological systems

The mechanism most frequently used by cells to respond to stimuli exploits previously accumulated **potential energy**. The potential difference across the membrane is characterized by the possibility of rapid transfer into work with timescales in the order of hundreds of microseconds. The system has a great capacity to accumulate electrical charges across the plasma membrane, in a range from 10 to 100-120 mV, which corresponds to a massive electric field of about $10 \cdot 10^6$ V/m. The aim is to maintain the widest possible charge separation using a mechanism that is a highly efficient combination of selectively permeable barriers and ATP-consuming ion transporters. In this case, the energy stored is in the form of an electric potential difference, the use of which occurs when any given stimulus succeeds in removing the constraints responsible for maintaining the dynamic equilibrium. Their removal causes an immediate lowering of the membrane's electrical resistance, creating a short circuit that leads to the dissipation of a certain amount of potential energy. The physical phenomenon of the passage of electric current is practically immediate, since it is a movement of charges comparable to that in metallic conductors, and therefore at a speed close to that of light ($3 \cdot 10^5$ km/s). It should be noted that the electric current, induced by a difference in electrical potential and produced by the flow of ions through the membrane, takes advantage of the passage of charges from one environment to another without the ion necessarily having to physically cross the membrane (Data Sheet 2.1); an approximate estimate of the time it takes an electric charge to cross the plasma membrane is about $0.3 \cdot 10^{-15}$ seconds.

A mechanical model that can describe how and how fast a charge passes through a plasma membrane is the Newton Collision Ball Device. The energy, in the form of **momentum**, of the incident ball (1 of Figure 3.1)

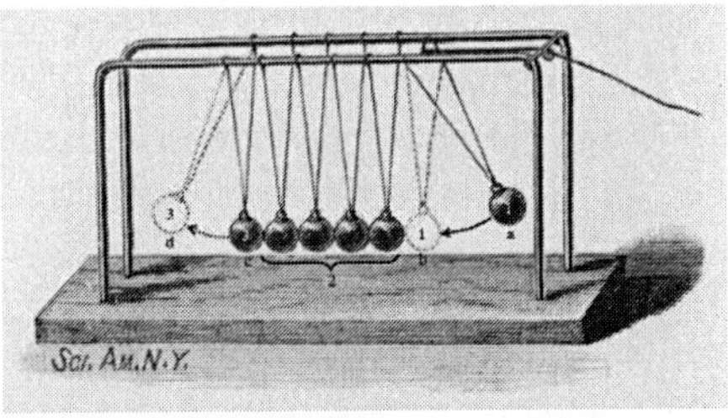

Figure 3.1 In the Newton Collision Ball Device, the momentum of the incident sphere (1), which falls by gravity from position a to position b and hits the central spheres (2), is transferred to the last sphere of the series (3) which passes from position c to position d, without the central spheres moving.

is transferred instantaneously and entirely, except for energy losses in the form of heat, from one ball to another (2 of Figure 3.1) until the last ball (3 of Figure 3.1), which moves upwards against the force of gravity.

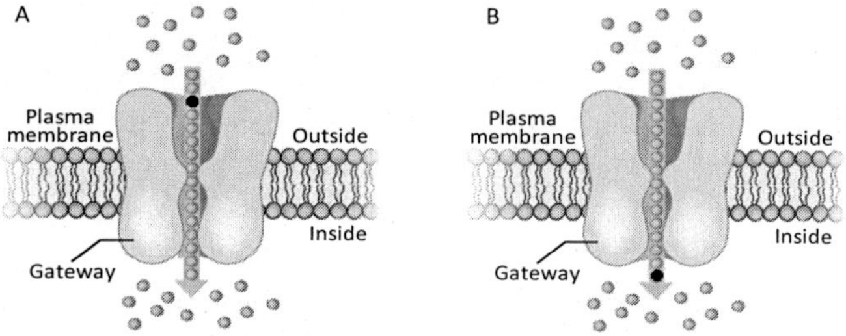

Figure 3.2 The permeation of a plasma membrane by a molecule with or without an electric charge. A) A single particle bounces randomly in the channel's external aperture (black dot). In active transport and/or in electrochemically favorable conditions, the physical passage of one particle to the other side of the membrane (B, black dot) takes a time period that is a few orders of magnitude higher than that for charge permeability.

In ion channels, the charge of a single ion is transferred according to a similar energetic process: each charge, which randomly bounces against the plasma membrane due to thermal agitation, is a candidate to go through the ion channel. If the ion channel opens and the thermodynamic conditions favor its passage, the ion induces the release of an identical charge on the opposite membrane side. The purpose of this phenomenon is to generate a **local and momentary variation in the electrical potential** across the membrane, which is used as a **primary signal** for the cell itself and for different cells.

A second cellular mechanism is used when the accumulation of energy occurs through the sequestration of particular ionic species. The storage of **calcium ions** in intracellular reservoirs is the most common example. By an endoplasmic reticulum membrane transport mechanism that uses ATP, calcium ions are actively stored in intracellular stores (section 3.3). Potential energy is stored inside the stores as a **chemical gradient** with respect to the outside, where the concentration of calcium ions can reach a concentration of the order of 10^{-3} M/L, several orders of magnitude higher than in the cytoplasm, which is of the order of 10^{-9} M/L. The constraint that keeps the system in dynamic equilibrium is again represented by the membrane's resistance to the passage of ions. Once the membrane, following a stimulus, becomes permeable to calcium ions, they are free to diffuse into the intracellular environment. In this case, calcium ions physically cross the membrane through gaps (Figure 3.2). This necessarily entails longer timescales than the passage of electrical charges. In fact, it can be roughly calculated that a generic particle, whether neutral or charged, takes about $50 \cdot 10^{-9}$ seconds (50 ns) to cross the phospholipid bilayer, whereas an electric charge generating a current through the membrane takes about 10^{-6} ns.

Calcium ions are essential as intracellular messengers to initiate many cytoplasmic functions. Given the high reactivity of calcium ions, due to the presence of two highly reactive free positive charges, the release time and cytoplasmic concentration must be finely regulated to prevent the cell from undergoing serious degenerative phenomena. This type of potential energy, which affects a very large and varied group of intracellular molecules, has, precisely because of its heterogeneous characteristics, action times ranging from a few tens of milliseconds to a few seconds. The two forms of energy storage, an electrical gradient and a chemical gradient, are characterized by different mechanisms of accumulation and use, and above all, by very different timelines by which accumulation and use take place.

It should be kept in mind that from a physiological point of view, the overproduction and accumulation, both in the cytoplasm and in the nucleus, of **messenger RNA** as well as various **proteins,** can also be considered accumulations of potential energy. However, this is a form of potential energy that has response times to stimuli with a different order of magnitude compared to electrical potential and the chemical gradient of calcium. This type of accumulation will not be examined further. However, the common feature of these three different ways of storing energy is that they maintain the state of the cells in a dynamic equilibrium with high potential energy levels, always ready for a more-or-less rapid reaction in response to a stimulus generated by changes in the surrounding environment.

3.2 The accumulation of potential energy

The mechanism by which cells are able to accumulate potential energy depends on two fundamental factors: the presence of two environments with different biophysical characteristics and a selectively permeable membrane. The size of the potential difference that can be achieved depends essentially on how different the ion concentrations are in the two environments and how selective and permeable the membrane is. Selective permeability allows the separation of only certain charges in sufficient quantities to produce a sufficient potential difference across the plasma membrane. In principle, even in the absence of a membrane, it is possible to create environments with different biophysical characteristics, and in particular with different ion concentrations, and thus accumulate electric potential. However, the membrane dividing the two environments with very specific bioelectric characteristics makes the system functionally efficient and constant over time.

3.2.1 Creating different environments in ion composition

Potassium chloride and sodium chloride, derived from the melting of rocks that were very rich in them, were almost certainly the prevailing salts in water when it accumulated in a liquid state on the earth's surface, a process that occurred 3.9 billion years ago due to the cooling of the earth. In that liquid environment, the first living organisms are thought to have formed between about 3.9 and 3.5 billion years ago, as these are the ages attributed to some microbial structures at a Greenland site (Nutman A., et al, 2016) and fossil cyanobacteria at a site in Western Australia (Campbell NA and Reece JB, 2004), respectively. In the salt, each positive ion is surrounded by a number of negative ions, each of which is in turn surrounded by a number of positive ions, forming a compact and highly ordered crystal structure (Figure 3.3 A). From a biophysical point of view, water molecules are dipoles with a negative charge fraction due to the oxygen atom and two positive charge fractions due to the two hydrogen atoms. The salt dissolves when the water molecules, moving freely by thermal agitation in a solution with a high concentration of water, have a high probability of inserting themselves in an oriented manner between the negative ions and the positive ions of the salt (Figure 3.3B). The hydrated cations and anions diffuse in all directions due to thermal agitation and are distributed randomly and homogeneously throughout the solution (section 4.1.1). This is done by respecting the principle of electro-neutrality of a solution, according to which every positive charge must correspond to a negative charge, and vice versa. Each ion has a characteristic **hydration number**, which indicates how many water molecules surround it, and this modifies, and in part determines, biophysical characteristics such as the mobility of the ion in solution and the permeability through the aqueous pores of a plasma membrane.

Consider a saline solution containing the same concentration of sodium chloride and potassium chloride: the sodium, potassium and chloride ions are distributed randomly and homogeneously throughout the solution (① in Figure 3.4). If a nucleus with predominantly negative fixed charges is introduced into this system, for example a protein aggregate (② Figure 3.4), the degrees of freedom of the system decrease because the fixed charges of the negative protein aggregate tend to restrain the sodium and potassium cations, which will form a layer around it.

From a random and homogeneous distribution, we are now in a system able to separate charges. The excess of positive charges will now be compensated for by another layer of negative charges, consisting of chloride ions, which in turn will be compensated for by a layer of sodium and potassium ions. The amount of positive and negative charge forming the alternating layers in the solution decreases as one moves from the negative protein nucleus towards the periphery of the system, until the

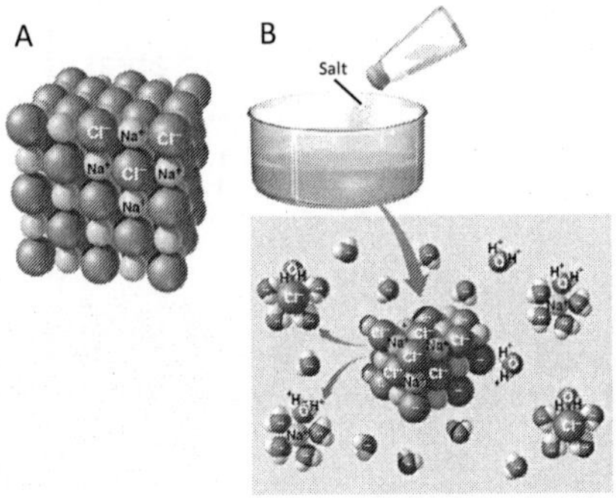

Figure 3.3 If a salt, which has an ordered crystalline structure (A), is placed in an aqueous solution (B), the positively charged cations separate from the negatively charged anions because they are surrounded by the dipoles of the water molecules.

random distribution of anions and cations once again prevails. Ultimately, to build an area at a different potential from its surroundings, it is sufficient to have a charge center formed, in which the positive or negative elements are immobilized (② in Figure 3.4).

These compensation mechanisms make the system stable as the solution reaches equilibrium, even if there are differences in the micro-environments that make up the system. Differences in the number and concentration of electrical charges constitutes **potential energy accumulations**. However, this system is not capable of transforming the accumulated energy into other forms of energy and ultimately into work because immediately after the introduction of the negative nucleus, a new stable equilibrium is instantaneously reached. In order to have the possibility of accumulating potential energy and, at the same time, be able produce work, the system must be in a dynamic equilibrium between two or more interchangeable phases, in which the state differences are well-defined. In a system consisting mainly of water, these different phases will necessarily have to be differentiated by ionic concentrations.

Among the ions contained in seawater, where living organisms probably appeared, there is a high concentration of chloride and sodium and a relatively low concentration of potassium and calcium (Table 3.1). If, in such an environment, it is necessary to create a concentration gradient between the negative protein nucleus and surrounding areas, it is essential that the nucleus has a high affinity for a specific ion, further reducing the system's degrees of freedom. The sodium ion, given its high concentration

Ion	Concentration		% free
	g/kg H_2O	mmol/L	
Cloride	19.37	558	100
Sodium	10.77	479	98
Magnesium	1.30	54	89
Sulphate	2.71	29	39
Calcium	0.41	10	99
Potassium	0.39	10	98
Bicarbonate	0.12	2	80

Table 3.1 Seawater composition (average salinity: 35‰)

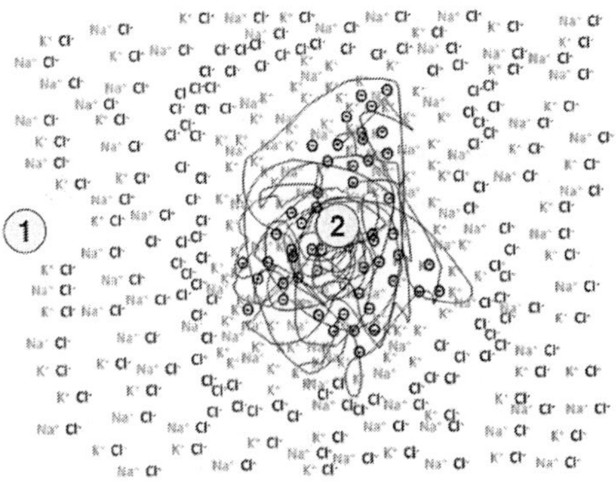

Figure 3.4 In a salt solution, the ions are homogeneously distributed (①), but the introduction of a negative protein nucleus (②) induces the formation of alternating layers of cations and anions.

(① in Figure 3.4), is not suitable for this purpose, as it would have to reach too high a concentration in the environment close to the protein core (② in Figure 3.4). Chloride is excluded, like sodium, because of its high concentration (Table 3.1) and the ionic situation close to the negative protein nucleus (a high concentration of fixed negative charges that functions to neutralize the charges of the cations). A significant increase in calcium ions is not feasible because it would compromise the biophysical characteristics of the system, as calcium compounds precipitate and form crystals, given their low solubility.

Therefore, the only ion suitable for accumulation near the protein core is potassium. If we assume that the negative protein nucleus, due to its biophysical characteristics, has a high probability of retaining potassium ions relative to other cations, the negative fixed charges will be neutralized mainly by this ion, which will form a first layer of cations (② in Figure 3.5). Moving away from the negative protein nucleus, layers of chloride ions and alternating layers of sodium and potassium ions will form. The layering becomes less and less evident due to the progressive decrease of the influence of the negatively charged nucleus. This continues until a random distribution of the ions typical of ① in Figure 3.5 is again established. Also, in this case, the electro-neutrality of the system will be respected, if we consider sufficiently small micro-environments.

With the higher affinity of potassium ions for the negative nucleus, the ensemble of ① and ② are now two different environments with a real ion concentration difference, even if the whole system overall is in equilibrium. Potassium is more concentrated in ②, close to the protein

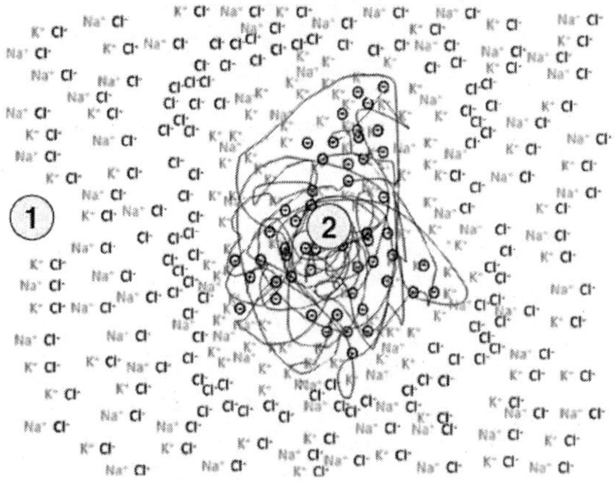

Figure 3.5 The selectivity for the potassium ion of the negative nucleus means that it is mainly this ion that neutralizes the charges (②). Going towards the periphery (①), the layers are randomly formed by the sodium, potassium and chloride ions.

nucleus, whereas sodium is practically absent (② in Figure 3.5). Sodium is more concentrated in the rest of the solution (① in Figure 3.5). This system, however, does not have the possibility of modifying the distribution and the ionic concentration, except by modifying the consistency and the properties of the nucleus. Such a modification would require a lot of energy and relatively, a lot of time, as the synthesis of other proteins would be necessary. Furthermore, although a difference in ion concentrations for each cation has been realized between environments 1 and 2 (Figure 3.5), the system is once again static and incapable of dynamically modifying itself and therefore of transforming the potential energy into work. It should be added that, despite local and small differences in electrical potential, in a system such as this, there is no real separation of charges. Even where a transient accumulation of electrical charges would occur, they would be immediately buffered by an ion with opposite charge to ensure the electro-neutrality of the solution.

3.2.2 A selectively permeable phospholipid membrane

Let us consider an initial situation similar to that described in Figure 3.6, in which a negative nucleus with a high affinity for potassium is immersed

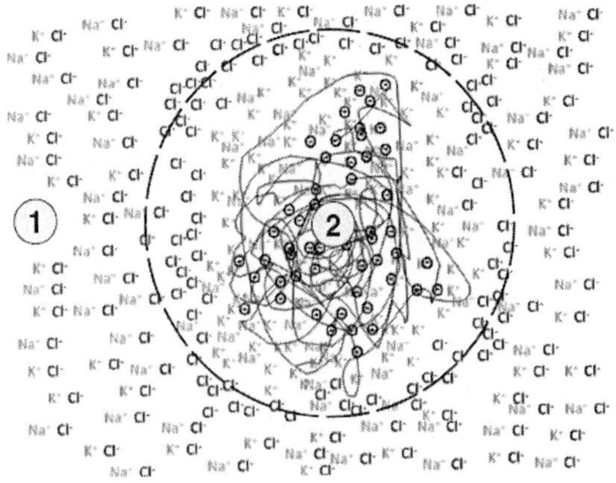

Figure 3.6 The presence of a porous membrane selective only for water and small ions can allow a transient separation of charges, but this is immediately cancelled by the free diffusion of anions and cations.

in a solution of sodium chloride and potassium chloride, with a concentration ratio of 50 to 1, similar to that of seawater (Table 3.1). In this hypothetical situation, the protein cluster is surrounded by a phospholipid bilayer membrane with interruptions that constitute non-specific aqueous pores. Through these apertures, water and small ions such as sodium, potassium and chloride can move, even if with a reduced number of degrees of freedom, between ① and ② (Figure 3.6). Furthermore, consider the volume of ① to be infinite compared to the volume of ②. In this case, it is easy to identify ① as the external side and ② as the inside of a primordial biological system.

At time zero, the negative fixed charges of the central core have a high probability of being compensated for by potassium ions, due to the high affinity for this ion. Sodium chloride and potassium chloride are present on the outside of the membrane, with a concentration ratio of 50 to 1 as mentioned. The amount of potassium present on the inside is higher than in ①, but its concentration is low; it is trapped by the high-affinity protein core. Potassium, which enters through the membrane by concentration and electrical gradient, pulls in the chloride ions that are momentarily unbalanced. The charges of the excess chloride ions are readily neutralized by both new potassium ions and sodium ions entering freely into the internal environment. The system as a whole achieves an immediate balance of charges and concentration, even though there is a concentration gradient for the potassium ion.

A system such as that illustrated in Figure 3.6 is still unsuitable for charge separation and the accumulation of energy. Even if a semi-permeable **separator** is present, the general conditions are identical to those described for Figure 3.5. The small excess of electrical potential due to the imbalance of potassium ions is promptly buffered by other ions. In fact, in an open aqueous environment, i.e., in which water and small molecules diffuse freely, even if locally we can find differences in both ionic concentrations and minimal differences in electric charges, it is not possible to realize a real and consistent accumulation of potential energy. Accumulation can only be achieved if the system does not establish a static but rather a dynamic equilibrium and if there is a real separation of charges, while respecting the principle of electro-neutrality of solutions. Biological systems have resolved this apparent contradiction by interposing a barrier between the two environments of interest that has very special biophysical characteristics: the **plasma membrane**.

First of all, the plasma membrane must be a non-rigid structure, in order to cope with volume variations due to the movements of water; these movements are necessarily created to rebalance the osmolarity of the solution following the movement of ions. The membrane must also be, at least as a first approximation, permeable only to water and selected small ions, to prevent free diffusion. From a strictly functional point of view, it must also have the characteristics of a capacitor, an element that is able to accumulate charges on one side and balance them with opposite charges on the other side (Chapter 2 and Figure 2.5). Once charged, the membrane capacitor can maintain its charge for an infinite period of time if the dielectric is a perfect insulator: only a short circuit between the two sides of the membrane, resetting the dielectric resistance to zero, would allow the flux to distribute the charges again in a random and homogeneous manner.

Biological membranes can separate opposite charges because lipids are highly hydrophobic, and therefore impermeable to ions. Their thickness, as mentioned above, of 7-10 nanometers, is comparable to the distance between the armatures of a capacitor that accumulates charged particles with dimensions similar to those of ions in biological solutions.

If biological membranes were made up solely of a homogeneous phospholipid bilayer, they would behave like a capacitor, but their function would be only that of an inert barrier. In order for the system to be dynamic and react to changes that may occur on either side of the membrane, it is also necessary to have **gaps** through which water and selected ionic species can permeate and become distributed in an appropriate manner. The simultaneous presence of a capacitor in biological membranes, which is able to accumulate charges and offers the possibility of modulating ion permeation, allows not only an accumulation of electrical potential, but also to modulate it according to internal and/or external demands.

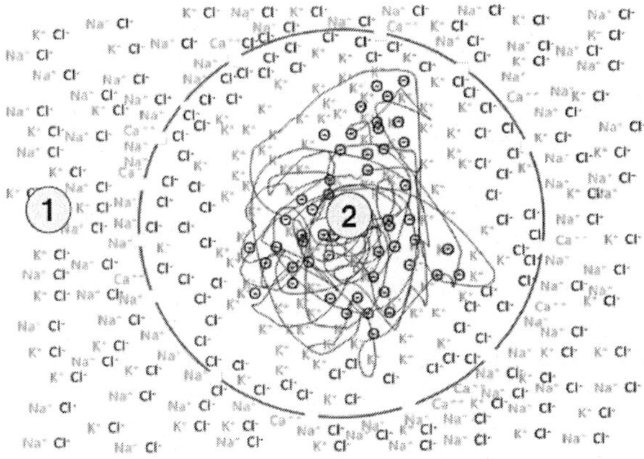

Figure 3.7 The presence of a porous membrane that is selective only for potassium and chloride ions allows the formation of a concentration gradient and an electrical gradient between the external ① and internal ② environments.

Three major points about membrane capacitors should be remembered (Chapter 2.1). The functional contribution of the capacitor occurs only during a change over time of the membrane potential. Furthermore, at time zero, the plasma membrane, behaving as a capacitor, has an electrical resistance equal to zero. Thus, ionic current, as any dynamic flux, always chooses the lowest resistance path. From a functional point of view, this means that the first choice of any uncompensated charge is to get to the membrane, increasing the number of charges present on the membrane capacitor armatures, thus contributing to modifying the membrane potential. It should also be remembered that as the charge of the capacitor increases, the probability that other charges of the same sign will be deposited on the armatures decreases, as they are repelled by those already accumulated. From a functional point of view, during voltage transition (dV/dt), the capacitor increases its electrical resistance over time.

In order to have this dual function, as a charge accumulator and as an electrical potential modulator, it is necessary for the membrane not only to play a structural role, but also to assume an active functional role of **selective permeability only for small ions**. In this case, permeability is for potassium and chloride ions and water, maintaining a low or even zero permeability for all other ions, in particular sodium and calcium. Under these experimental conditions, events take place as described in Figure 3.6. The excess chloride ions, pulled in by the entry of potassium ions, cannot be counterbalanced by the entry of sodium or calcium, since the membrane is impermeable to these ions. The chloride ions, therefore, follow the path of least resistance, i.e., they settle on the inner face of the membrane capacitor (for example, side A of Figure 2.5) which has, at time zero, zero electrical resistance. The excess chloride ions can be neutralized by the positive sodium, calcium, and to a small extent, potassium ions, which accumulate on the outer side of the membrane (side B of Figure 2.5). At the end of the process, we have a system in which the membrane has a predominantly negative charge, even if it is only tens of millivolts near the inner face, and a predominantly positive charge near the outer face. However, considering sufficiently large micro-environments, there is electro-neutrality across the membrane. In addition, both the outside and the inside are in equilibrium chemically and electrically, since the excess of potassium on the inside is compensated for by the accumulation of negative charges on the inside of the membrane and, despite the presence of gaps in the membrane, the net flux across the membrane is zero. The capacitor-like properties of the membrane enable the effective separation of charges, while respecting the principle of electro-neutrality of the solution, and at the same time enable the accumulation of potential energy as a voltage difference across the membrane; this accumulation only affects the micro-environment close to the membrane.

A model such as that shown in Figure 3.7 is capable of producing a momentary separation of charges and does not require the intervention of any form of energy, but only a different combination and distribution in space of the ionic concentrations, charged elements immobilized in one of the two environments and a selectively permeable phospholipid membrane.

This condition is known as **Donnan equilibrium** (tab. 3.1) and, depending on the cell in question, is responsible for a potential difference across the membrane in the order of −10, −15 mV, with the inside environment negative.

3.2.3 Cells use different forms of energy depending on the function

The electrical difference generated in a system bounded by a phospholipid membrane selective for potassium ions, and that contains a negative core with a high affinity for this ion, has a limited ability to accumulate potential energy. In fact, the mechanisms that lead to charge separation are essentially passive,

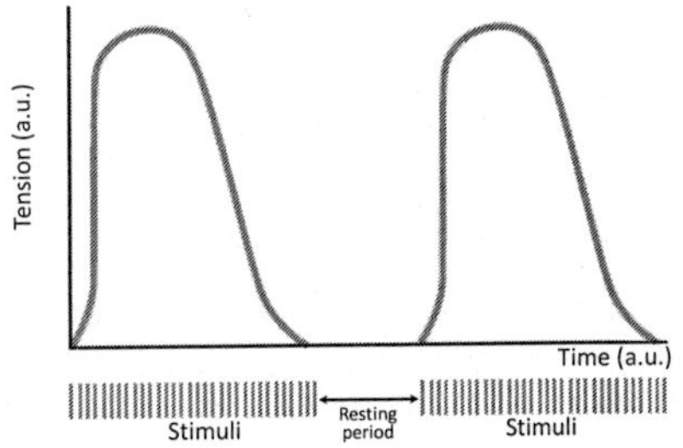

Figure 3.8 The motor unit of a skeletal muscle (tab. 7.1) which contracts intensely needs a rest period, probably to allow the individual cells to rebuild their reserves of potential energy. In the meantime, another motor unit is recruited that can continue the contraction of the muscle.

with low efficiency from a thermodynamic point of view. Moreover, the accumulated potential is exhausted in a short time, since the membrane is not a perfect insulator due to its fluidity. Therefore, slowly but surely, the differences in ion concentration are dissipated, bringing the system back to static equilibrium. However, the little energy accumulated in a passive way may have been sufficient, in the course of evolution, to perform a minimum amount of work. Although small, this amount of energy may have allowed biological systems to build elements capable of generating even greater potential energy accumulation. The possibility of accessing greater amounts of energy has been instrumental to the construction of increasingly complex biological organisms. The different types of cells that make up a biological organism can be grouped, from a functional point of view in terms of use of energy, into two distinct categories. There are cells that carry out **acute, high-intensity work**, exploiting the energy accumulated by the system in a short or very short time, such as the cells of skeletal muscles (Figure 3.8). There are also cells that carry out **chronic work of low intensity**, prolonged in time or even continuous, such as the cells of an absorbent epithelium (Figure 3.9). The two types of cells have evolved by developing the ability to increase the availability of potential energy, but with different timescales and ways of using it.

If a cell has an intense activity, working in an acute mode, it needs to use a large part of the available energy in a short time, dissipating it rapidly. For this reason, the systems that work in acute mode, in addition to using chemical energy to restore potential energy, have developed mechanisms to partially recover the energy used, to reduce consumption, for example, by alternating motor units in the muscle (Figure 3.8).

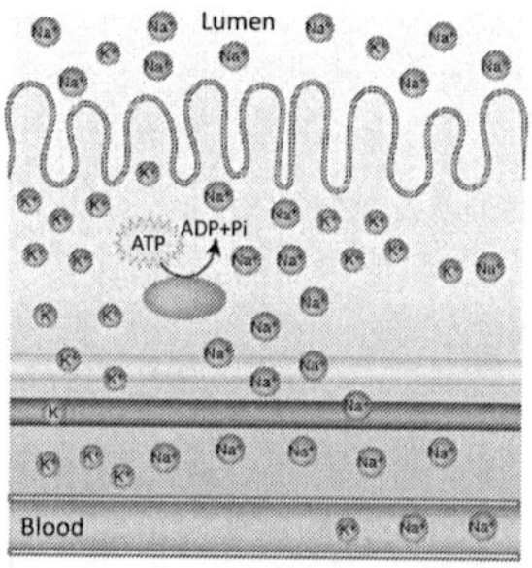

Figure 3.9 An absorptive epithelium, such as the intestine, must continuously transport sodium (red dots) and potassium (blue dots) with an active system (purple oval) that consumes ATP (section 4.3).

For chronic work, the continuous use of stored potential energy is supported by continuous production and consumption of chemical energy, which always keeps the system in a dynamic equilibrium with high potential energy and the ability to perform work. In both cases, the cell ensures that there is always an excess of potential energy compared to basic needs, either to be able to perform work even in the presence of a temporary block of chemical energy production, or to be able to cope with work of higher intensity if required.

In order to perform the work required for complex biological functions, cells must be able to increase their potential energy stores. This has been accomplished by generating molecular structures able to maintain the dynamic equilibrium and at the same time increase the potential energy of the biological system. Over time and the course of evolution, billions of random events have occurred, during which increasingly complex molecular structures have been selected. The close link between structure and function may have allowed the selection of the **primordial cell** shown in Figure 3.7, which could be a reasonable model. It is also plausible that the initial low potential energy passively accumulated by the primordial cell was used to form chemical bonds and to synthesize increasingly complex and specialized molecules. Such molecules could have been able to increase the potential energy accumulated across the membrane, in turn allowing increasingly complex molecules to be built by means of a positive feedback mechanism (Figure 3.10).

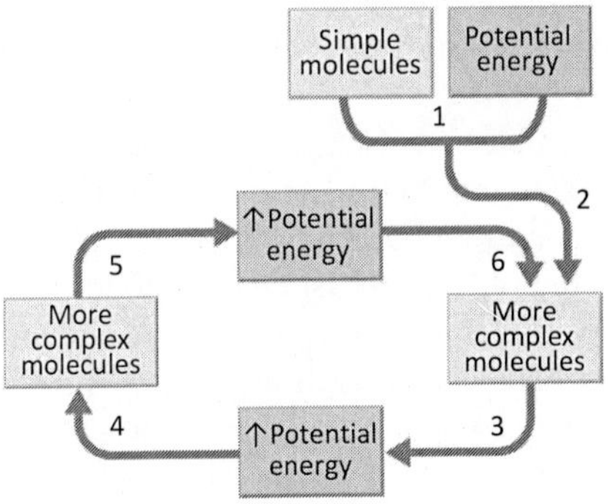

Figure 3.10 Simple molecules (1) can give rise to more complex ones (2) which increase the potential energy (3) with which increasingly complex and specialised molecules are built (4) which in turn increase the potential energy (5) with which increasingly complex and specialised molecules are built (6).

An effective way of increasing the potential energy of the cell, and thus the possibility of performing more and more complex work, is to increase the differences in ion concentration across the membrane. This can be achieved using specialized molecules which, using chemical energy, can move ions across the membrane against their concentration gradient.

The event that enabled the cell to make a quantum leap in energy production was probably the simultaneous selection of complex molecules for the production of chemical energy in the form of ATP, stored inside the cell, and molecules stored in the lipid membrane that can use this energy to perform work. In addition, in order to ensure continuous production of energy in the form of ATP, the selection of intracellular structures capable of performing this task with high efficiency was once again favorable. According to the most widely accepted current hypothesis, these structures are nothing more than protobacteria that have been incorporated into cells and become symbionts, evolving into **mitochondria**.

It has already been discussed that among the ions available in primordial seawater (table 3.1), those most suitable for creating a difference between two environments divided by a phospholipid membrane are sodium and potassium (paragraph 3.2.2). In particular, potassium was the most suitable to be the main element in the internal environment. Cell dimensions are in a range of tens of micrometers (10^{-6} meter) for many reasons. As an example, a restricted space allows a higher probability of interactions between molecules. A limited volume also makes it easier to control ionic concentrations. In any case, to increase the potassium in the internal compartment, there must be energy-dependent ion transport because the movement is against the natural gradient. In this way, the concentration difference across the membrane can be increased and the ionic gradient maintained, compensating for passive second-gradient exchanges across the membrane or those due to imperfect membrane selectivity. It is therefore necessary for the mitochondria to be able to continuously produce sufficient amount of ATP. To do this, they need a constant and consistent supply of chemical energy-rich molecules that can be used to produce ATP, which must necessarily enter the cell from the external environment. From the millions of different molecules that have formed randomly over the course of evolution, sugars have been selected as the most suitable because they are extremely rich in high-energy chemical bonds, relatively simple and small in size, and therefore, despite being hydrophilic, are easy to transport across the hydrophobic phospholipid membrane.

There are now all of the conditions to build a primordial cell: a selective membrane that is mainly permeable to potassium ions, bordering a cytoplasm-containing protein cluster and being rich in mitochondria capable of synthesizing small high-energy molecules such as ATP. In order for the mitochondria to produce energy, sugars must be transported into the cell. Specific membrane transporters for sugars are needed to carry out the transport. The initial low potential energy stored passively at the sides of the membrane and the asymmetric distribution of the sodium, potassium and chloride ions ensure a concentration and charge gradient for sodium ions, which is certainly sufficient to supply the cytoplasm with sugars, albeit in limited quantities.It has long been known that glucose is transported within the cell by a sodium-dependent mechanism, but how the transporter works, from a molecular point of view, is still a matter of hypothesis. What is possible is to construct a plausible model from **an energetic point of view**.

3.2.4 The transport of glucose

First of all, it must be assumed that the transporter is a transmembrane protein. There therefore must be a specific pore allowing sodium ions to pass through the membrane in order to exploit the potential energy, with a portion capable of binding glucose molecules.

If we consider proteins, it is well known that they can take on an infinite number of conformations completely at random as a result of temperature and thermal agitation, with frequencies as high as 1000 transitions per second. It can be assumed that because of these conformational changes and in the absence of external conditioning, the transporter has a high probability of finding itself with the glucose binding sites (schematically three sites in Figure 3.11) facing inwards into the primordial cell, with the glucose binding site saturated and the sodium ion pore predominantly closed (1). Glucose is released because in this position, at a lower level of potential energy, glucose binding sites are at low affinity (2). Glucose release triggers the opening of the sodium channel (3). The inflow of sodium ions from outside gives the energy to cause a conformational

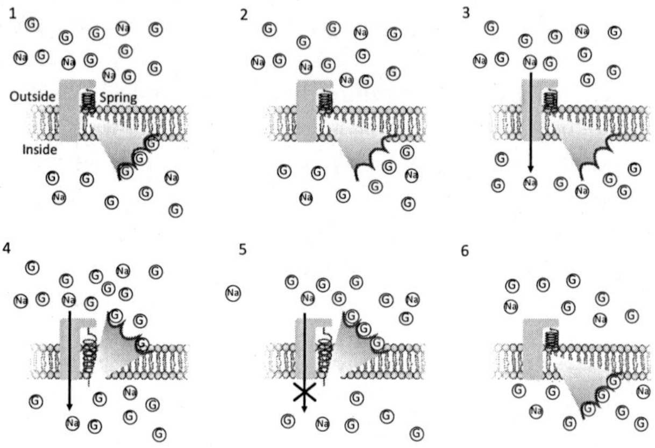

Figure 3.11 Functional model of the membrane element capable of transporting glucose (G) inside the cell. The cycle starts with the transporter at its lower level of potential energy, facing the internal side of the membrane and with glucose molecules saturating the binding sites (1). In this condition, glucose is released due to the drop in the affinity of the binding sites (2). This causes the opening of the Na channel (3) that induces a conformational change in the transporter (4). In this position, at a high potential energy level (stressed spring), there is an increase of the binding site affinity for glucose. Once all the sites are occupied, the channel closes (5) and the system is free to revert to the original, low potential energy position (6).

change in the transporter. The spring in Figure 3.11 represents the intrinsic energy level of the transporter. In (4) the transporter is at its maximum potential energy, with the binding sites facing the external side of the membrane. In this condition, the affinity for glucose increases. After the binding sites are saturated (4), the channel closes (5) and the transporter returns to its original position at a low level of potential energy (6). This molecular model of glucose transport respects the physiological principle that biological systems are always in a state of dynamic equilibrium. The transporter is maintained at high potential energy by the continuous flow of sodium ions. Binding site saturation is the trigger to induce the ion channel to close and to allow the transporter to return to its original conformation at low potential energy.

The sequence of events just described is one of the possible models of a molecular mechanism designed to ensure a constant supply of **fuel**, glucose, which is continuously consumed by the mitochondria to produce energy in the form of ATP. A mechanism such as this is called **co-transport** (paragraph 4.3.4).

3.2.5 Restoring gradients: sodium ions

The continuous inward flow of sodium, which is indispensable to supply the cell with the glucose necessary for ATP production, slowly leads to a

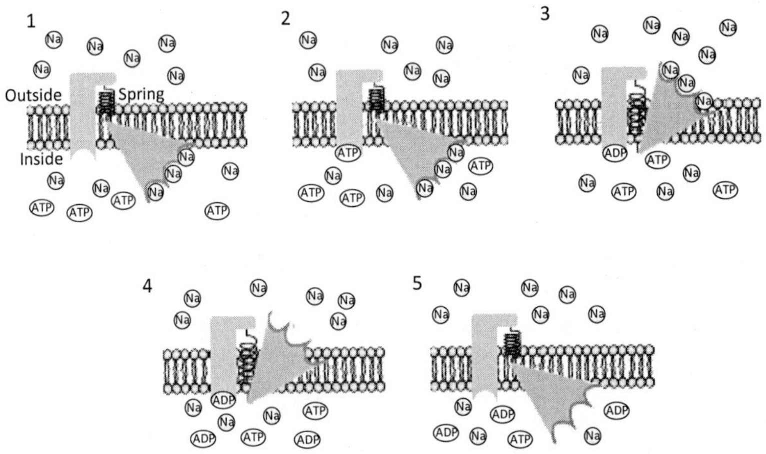

Figure 3.12 Hypothetical working model of the protein capable of transporting sodium ions outside the cell against the gradient. Sodium ions (Na) are transported by means of specific high affinity sites using the energy provided by an ATP molecule. The cycle starts with the transporter in a position at low potential energy (1) (spring discharge) with the high affinity binding site facing the internal side of the membrane. Na ions binds the high affinity binding site, allowing ATP attachment (2). Following saturation, the phosphorylation reaction ATP-ADP gives the energy to cause a conformational change in the transporter (3), taking the transporter to a high potential energy level (stressed spring). In this position, there is a drastic decrease of the binding strength for Na ions, which are released outside (4). The free binding sites and ADP detachment are the triggers for releasing the potential energy accumulated in the spring, with the transporter moving back to the starting configuration (5).

decrease in concentration differences between the outside and the inside of the membrane, with a progressive decrease of the electrical and chemical gradients. The consequence of this process is that the cell is no longer able to produce energy, loses its ability to do work and dies.

In order to overcome this impediment and maintain optimal function over time, the glucose transporter must be paralleled by a membrane protein that is able to maintain low sodium concentration inside the cell, moving sodium outwards against its gradient. This mechanism could be seen as basic maintenance of cellular homeostasis. In this case, the energy needed for the work is fully furnished by the cell in the form of ATP.

The model of the membrane protein, out of an infinite number of possible models, must have a) a high-energy ATP binding site for the molecule, b) high-affinity sites for the specific ion to be transported, and c) allosteric sites other than the specific sites and the ability to be modulated by conformational changes in parts of the protein itself.

Like sodium/glucose co-transport, the model is based on a repetitive cycle. In the resting state, the most likely conformation of the transporter must be that at the lowest level of potential energy (Figure 3.12; spring at rest). The specific high-affinity binding sites are facing into the cell (Figure 3.12, 1). In this condition, the transporter has a high probability of being saturated despite the low concentration of sodium ions within the cell. Saturated sites can lead to the uncovering of an ATP binding site (Figure 3.12, 2). Phosphorylation of the transporter induces a conformational change, exposing the binding sites loaded with sodium ions externally (Figure 3.12, 3), in a position of high potential energy (stressed spring). As a consequence of the conformational changes, there is a drastic decrease in the sodium binding sites' affinity. The sodium ions are released outside the cell, despite its high concentration (Figure 3.12, 4). Finally, ADP detaches from its specific site, allowing the transporter to change conformation back to its initial position, due to the potential energy accumulated in the stressed spring (Figure 3.12, 5).

3.2.6 Restoring gradients: potassium ions

The observations made for the sodium ion can be repeated for the potassium ion. Potassium is subject to a continuous outwards flow according to the gradient, which is essential to maintain the few tens of

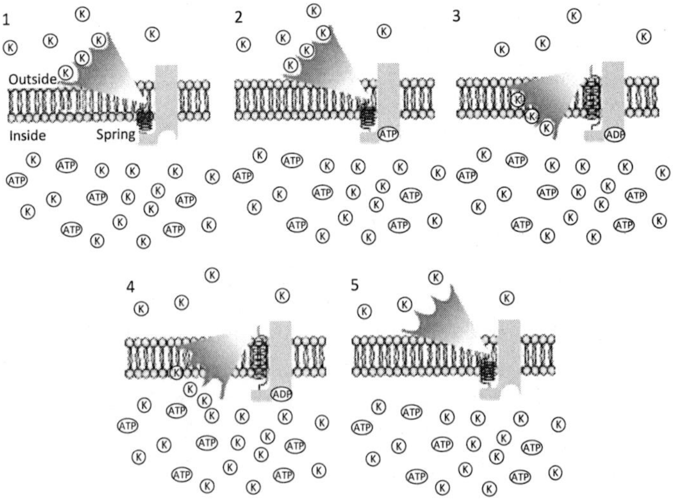

Figure 3.13 Hypothetical working model of a protein capable of transporting potassium ions inside the cell against the gradient. Potassium (K) ions are transported by means of specific high affinity sites using the energy provided by an ATP molecule. The cycle starts with the transporter in a position at low potential energy (1) (spring discharge) with the high affinity binding site facing the external side of the membrane. K ions bind the high affinity binding site, allowing ATP attachment (2). Following saturation, the phosphorylation reaction ATP-ADP gives the energy to cause a conformational change in the transporter (3), taking the transporter to a high potential energy level (stressed spring). In this position, there is a drastic decrease of the binding strength for K ions, which are released inside (4). The free binding sites and ADP detachment are the triggers for releasing the potential energy accumulated in the spring, with the transporter moving back to the starting configuration (5).

millivolts of membrane potential (paragraph 3.5) and probably to maintain overall electrical homeostasis of the cell. In fact, potassium permeability, once in electrochemical equilibrium, moves ions in and out of the membrane to maintain the membrane potential value negative. Since the natural tendency of a biological system is towards stable equilibrium, potassium ions are continuously leaking outwards to maintain a dynamic equilibrium. The leakage of potassium needs to be replaced. The cell will have to select a membrane protein that is able to transport potassium from the outside to the inside against the gradient, with a mechanism similar but opposite to that already hypothesized for sodium. The potassium ion transport cycle starts with the transporter at the lowest level of potential energy and with specific high affinity sites facing outwards from the cell (Figure 3.13, 1). Saturation with potassium ions allows ATP to bind to its specific site (Figure 3.13, panel 2). Phosphorylation of the transporter protein causes a conformational change, taking the transporter to a high potential energy and exposing the binding sites saturated with potassium ions outside (Figure 3.13, panel 3). The potassium ions are released into the internal environment, despite their high concentration, due to the drastic decrease of the specific sites' affinity (Figure 3.13, 4). Finally, ADP detaches from its specific site, allowing the transporter to change conformation back to its initial position (Figure 3.13, 5) due to the potential energy accumulated in the stressed spring.

3.2.7 Restoring gradients: sodium ions and potassium ions

A cell that could select a membrane protein capable of coupling transport against the gradient of the sodium ion and the potassium ion would be evolutionarily advantaged, in that it could use only one ATP molecule for each transport cycle of both ions, increasing the differences in ion concentrations across the membrane and hence, ultimately, the chemical gradient for each ion. Cell physiologists, many years ago, were able to identify and functionally characterize a ubiquitous membrane protein termed the Na/K pump. However, even for this membrane protein, the molecular mechanism is still obscure. Nevertheless, an effective functional model can be built that combines the transport mechanism of the sodium ion (Figure 3.12) with that of the potassium ion (Figure 3.13). The transporter protein at rest must have the specific sites for potassium and sodium ions facing outwards and inwards from the cell membrane, respectively (Figure 3.14, 1). The sites also need to be high affinity, given the low relative concentration of ions in the environment towards which they are facing. Saturation of the binding sites uncovers A TP binding sites (Figure 3.14, 2). Phosphorylation of the transporter protein promotes a conformational change able to charge both the saturated binding sites at a maximum level of potential energy (stressed spring), exposing binding complexes externally for Na and internally for K (Figure 3.14, 3). At the end of the conformational change, a drastic decrease in the affinity of the

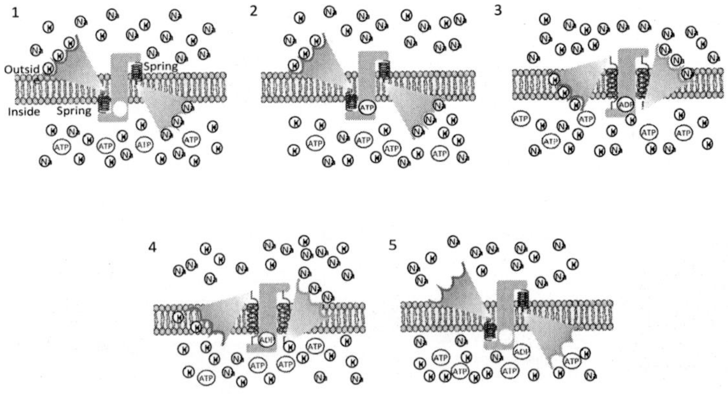

Figure 3.14. Hypothetical working model of the sodium/potassium pump. Sodium (Na) and potassium (K) ions are transported by means of specific high affinity sites using the energy provided by an ATP molecule. The cycle starts with the coupled transporter in a position at low potential energy (1) (spring discharge) with the high affinity binding sites facing the external side for K and the internal membrane face for Na. Na and K ions bind the high affinity site, allowing ATP attachment (2). Following saturation, the phosphorylation reaction ATP-ADP provides the energy to cause a conformational change in the transporter (3), taking the transporter to a high potential energy level (stressed spring). In this position, there is a drastic decrease of the binding strength for Na and K ions, which are released outside and inside respectively (4). The free binding sites and ADP detachment are the triggers to release the potential energy accumulated in the spring, with the transporter returning to the starting configuration (5).

specific sites allows K and Na ions to be released despite the high concentrations in their respective environments (Figure 3. 14, 4). Finally, ADP detaches from its specific site, allowing the transporter to change conformation back to its initial position, due to the potential energy accumulated in the stressed spring (Figure 3.14, panel 5).

This coupled transport is, from an electrical point of view, neutral. During its work, the distribution of charges across the membrane and the electrical potential remains unchanged. As we discussed previously (paragraph 3.1), the chemical gradient built by passive mechanisms for each permeable ion involved is responsible for a membrane potential difference in the order of -15, -20 mV. The system is in a dynamic equilibrium, with potassium the only permeable ion. If sodium also becomes permeable, the potential difference will dissipate and sodium and potassium will soon be equal inside and out. Even if active transport is able to increase the chemical gradient of the two main ions (Figure 3.14),

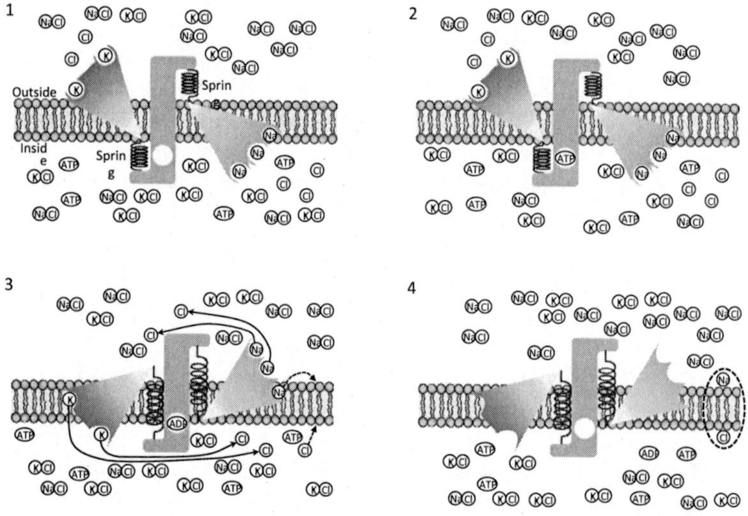

Figure 3.15 Associated transport model of sodium and potassium ions with an electrogenic function. Sodium (Na) and potassium (K) ions saturate their respective binding sites on the transporter (1), releasing the corresponding negative charges (Cl^-). Once the transporter sites are saturated, ATP is allowed to bind to its specific site (2). Hydroxylation promotes phosphorylation of the transporter molecule, which changes conformation (3). The excess of positive (external) and negative (internal) charges is compensated for by the membrane capacitor (dotted oval). Na/K ATPase electrogenic transport is one of the main contributors to increasing the potential difference across the membrane in most cells.

the membrane electrical potential will remain unchanged. To accumulate more potential energy across the membrane, it is necessary to select a membrane protein capable of asymmetrical ion transport.

Suppose that the cell has selected a protein capable of simultaneously transporting 3 sodium ions to the outside of the cell and 2 potassium ions to the inside (Figure 3.15). The transporter protein mechanism is the same as that already described (Figure 3.14). However, now for each cycle of the transporter, there is a negative charge left behind on the inner side of the membrane. Two of the three chloride ions in the internal environment compensate for the two potassium ions released by the transporter protein. Two chloride ions in the external environment compensate for two of the three excess sodium ions (continuous arrows in Figure 3.15B). The excess positive charge outside and the negative charge inside the cell create an unstable condition, as the principle of electro-neutrality of the solution is not respected. The two charges attract each other close to the membrane (dotted arrows of Figure 3.15B) and stabilize by exploiting the capacitive properties of the plasma membrane and compensate for each other, even if they are physically separated (Figure 3.15C, dotted oval). The electro-neutrality of the solution is respected and the consequence is an increase in charge across the membrane.

The work performed by the asymmetrical active transport across the plasma membrane just described, which uses the chemical energy of ATP, is actually present in cells and is termed the **sodium/potassium pump** or more correctly **Na^+/K^+ dependent ATPase** or simply **Na^+/K^+ ATPase** (paragraph 4.3.2). The number of ions transported is unknown and in the

Solute	Intracellular (mM)	Extracellular (mM)
K^+	140.0	4.0
Na^+	15.0	145.0
Mg^{2+}	0.8	1.5
Ca^{2+}	≈10^{-5}	1.8
Cl^-	4.0	115.0
HCO_3^-	10.0	25.0
H_2PO_4	40.0	2.0
Aminoacids	8.0	2.0
Glucose	1.0	5.6
ATP	4.0	0.0

Table 3.2 Mean intracellular and extracellular concentrations of various solutes in a generic mammalian cell

models of Figure 3.15, 3 binding sites for sodium and 2 for potassium have been used for simplicity. These numbers reflect the correct ratio Na/K of 3:2 operated by the transporter during repetitive cycling. The transport performed by the Na^+/K^+ ATPase is **electrogenic** in that it produces, in addition to counter-gradient transport, an electrochemical gradient for sodium and potassium through a continuous accumulation of charges on the membrane capacitor and thus a large amount of potential energy in the form of electric potential difference. It is also one of the main mechanisms responsible for the negative voltage across the membrane.

The concentration of ATP inside the cell is constantly high due to continuous synthesis by the mitochondria. The Na^+/K^+ ATPase can therefore perform its electrogenic active transport action, which could result in an infinite increase in the negative charges inside the cell and a progressive and infinite hyperpolarization. This does not occur because the active electrogenic transport is compensated for mainly by the background potassium permeability of the plasma membrane. In addition, Na^+/K^+ ATPase activity is voltage- and concentration-dependent. This results in a non-linear relation for the pump kinetics (Data Sheet 4.3).

3.2.8 The electrochemical gradient

Consider a cell with an internal potassium ion concentration of 140 mM, immersed in an external solution with a potassium ion concentration of 4 mM (Table 3.2). There is a large difference in concentration, hence a large chemical gradient. This would imply a high probability that a potassium ion can cross the membrane if we consider only the chemical gradient. However, due to the non-permeable negative charge present inside the membrane and the work of the Na/K ATPase, the asymmetrical ionic distribution is compensated for by the charges accumulated across the membrane. The excess negative charge inside and the excess positive charge outside is balanced by the membrane capacitor, which is able to buffer the uncompensated charges and to restore the electro-neutrality of the solution (Figure 3.15). This creates a negative potential difference between inside and outside.

This behavior is precisely described by the **Nernst equation**, which is valid for all ions capable of crossing a plasma membrane in the presence of specific channels: > $$Ex = \frac{RT}{zF} \cdot \ln \frac{[x]_o}{[x]_i} \quad (3.1)$$

where x is a generic ion, Ex is the membrane potential related to ion x, R is the gas constant, which is 8.314472 Joules/(K.mole), T is the absolute temperature in Kelvin, z, dimensionless, is the valence of ion x, F is Faraday's constant, which is 96485.309 Coulombs/mole and represents the number of electric charges per mole of substance and $[x]_o$ and $[x]_i$ are the external and internal concentrations of ion I, respectively. The RT/zF fraction of Equation 3.1, dimensionally expressed in Volts (tab. 3.2), takes into account the electrical gradient of the cell, while the dimensionless concentration ratio $[x]_o/[x]_i$ takes into account the chemical gradient.

Equation 3.1 defines, for each ion freely permeable through a membrane and for a given ion concentration ratio between the inside and outside of the cell, the potential at which ion x has zero flux. For this reason, it is commonly defined as the **equilibrium potential** (for that particular ion) or even **reversal potential**, because positive or negative values of the membrane potential promote opposite movements of the ion x through the lipid bilayer.

In Equation 3.1, the relevant element is the equal sign: it associates the electrical potential E proper to ion x with its biophysical characteristics, defined by the ratio RT/zF and the ion concentrations inside and outside the cell, defined by the ratio $[x]_o/[x]_i$. The electrical equilibrium potential E for each permeable ion is therefore determined by the combined action of the electrical gradient and the chemical gradient, i.e., it is equal to the **electrochemical gradient**.

When, as a result of the flux of ion x, the membrane potential becomes equal to E (Equation 3.1), ion x has zero net flux because it is in electrochemical equilibrium with respect to the electric potential and the chemical potential.

Each ion capable of crossing the membrane through specific channels due to the presence of an electrochemical gradient has its own equilibrium potential, which is constant at constant concentrations and can be calculated using Equation 3.1 (tab 3.2).

Take, for example, the sodium ion. Its equilibrium potential, calculated according to Equation 3.1 with the concentrations given in Table 3.2, is +57.1 mV. If specific channels are present, sodium moves inwards with a membrane potential more negative than +57.1 mV (hyperpolarized). On the contrary, sodium moves outwards if the membrane potential is more positive than +57.1 mV (depolarized). At a membrane potential equal to +57.1 mV, sodium ion net flux is null. The potassium ion, with an equilibrium, or reversal, potential of -89.6 mV (tab 3.2), has an outward flux for more depolarized membrane potentials, inward movement for more hyperpolarized membrane potentials and has zero net flux when the equilibrium potential and the membrane potential coincide. The same behavior can be applied to all the permeable ions. The ratio $[Ca]_o/[Ca]_i$ is very high and E_{Ca} very positive (tab 3.2). The chloride ion is an anion. z in the Nernst equation is negative and the ratio must be calculated with $[Cl]_i/[Cl]_o$. By convention, the direction of the current takes into account positively charged ions. Inward charge movement is a negative current, and outward current is positive. Chloride ion movement out from the cell is like positive charge inward and thus the ion flux results in a depolarization of the membrane.

The role of chloride current, similarly to the potassium ion flux, has and inhibitory effect. If a chloride channel opens according with its equilibrium potential, negative charges enter in the cell with an hyperpolarizing effect. Chloride has a negative equilibrium potential (tab 3.2), while the calcium ion, with a very positive equilibrium potential (tab 3.2), is only able to enter the cell in the physiological range of membrane potentials.

3.2.9 The membrane potential difference

The membrane potential of −10, −15 mV given by the Donnan equilibrium (tab. 3.1) represents the minimum potential energy content of the cell. Certainly, the functions that this low amount of energy is able to activate are limited. In order to increase the potential energy and, as a consequence, the possibility to perform work, the electrochemical potential must be increased by the electrogenic transport of the Na^+/K^+ ATPase (paragraph 3.2.7), which continuously contributes to maintaining the electrochemical gradients between the inside and outside of the cell. In addition, the resting permeabilities of each ion able to cross the membrane must be sufficiently low so that the gradients do not disperse too quickly across the membrane.

Active ion transport (paragraph 3.2.5 and 3.2.6) such as the electrogenic work of the Na^+/K^+ ATPase, contributes to the increase in potential energy by maintaining the **chemical gradients** of sodium and potassium and the electrical **gradients**, to stabilize the membrane potential. Membrane potential, in each cell, changes according to **the activity of specific ionic permeabilities**.

If it is assumed that the only ions involved are sodium and potassium and that both have an equal probability of moving freely through the phospholipid bilayer, the potential across the membrane approximates the mathematical average of the equilibrium potentials of the two ions. The potential difference would be at -16 mV, an intermediate value between the equilibrium potentials of potassium, -87.6 mV, and sodium, +57.1 mV (tab. 3.2).

If, on the other hand, the permeability for the sodium ion is very low, i.e., the probability of the specific pores being open is practically zero, the potential difference across the membrane matches the value of the potassium equilibrium potential, i.e., around -80 mV. In practice, potassium ions continue to move through their specific pores until the membrane potential coincides with the potassium equilibrium potential. At this value of membrane potential, even if the specific pores are open, the net flux of potassium is null (paragraph 3.2.8). Similarly, but with an opposite situation, i.e., a high probability that the specific pores for the sodium ion are open and a high probability that those for potassium are closed, the membrane potential becomes stable at the sodium equilibrium potential, i.e., around +50 mV. Supremacy in the decision of the membrane potential value belongs to the most permeable ion, which forces the membrane potential to be close to its equilibrium potential. However, all other permeable ions contribute to set the membrane potential value of a cell. The influence of each ion is directly proportional to its permeability level.

It therefore can be stated that in the presence of specific pores for sodium and potassium ions and a membrane dividing two environments at different solute concentrations (Table 3.2), there will be different values for the membrane potential, but in any case within the range delimited by E_{Na} and E_K, the **equilibrium potentials of the ions involved**.

The membrane potential of a cell depends on the equilibrium potential of all the ions capable of crossing the plasma membrane through specific pores. The contribution of each ion is the tendency to bring this potential towards the value of its own equilibrium potential, taking into account the extent of the permeability and the electrochemical gradient of each individual ion. This behavior is fully described by the **Goldmann equation** (paragraph 3.5.1) which takes into account the Nernst equation for the different ions involved in the determination of the membrane potential and their respective membrane permeabilities.

3.3 The calcium homeostasis

The ions most responsible for creating and maintaining the electrochemical potential difference across the membrane are sodium and potassium (paragraph 3.2.9). There are other ions in biological solutions, such as calcium and chloride (Table 3.2), which influence, modulate or change the potential difference of specific cell types. Ions such as calcium, magnesium, copper, iron and zinc are particularly important for cell function as activators of enzymes, as co-factors in protein synthesis or as stabilizers of nucleic acids.

Of all the ions mentioned, the only one that can perform a variety of functions in the cell, each with a particularly high level of effectiveness, is the calcium ion. Indeed, it has an **electrogenic action** on the membrane potential as a permeant ion and a cytoplasmic function as a **second messenger**. Second messengers are molecules that transfer information within the cell and are activated by an appropriate primary stimulus, which may be electrical or come from the activation of a membrane receptor. The primary stimulus rapidly increases the cytoplasmic concentration of the calcium ion, which transfers the message, amplifying it, within the cell to other molecules and activates specific functions, such as the activity of enzymes or the regulation of numerous intracellular enzyme cascades, or even the stimulation of the translation of several genes. Calcium is also solely responsible for the variation of the membrane potential in many types in the central nervous system, e.g., the lower olive, and in the hear cell t muscle (paragraph 5.14).

The possibility of infinite modulation means that, as a second messenger, calcium also plays a key role in skeletal and cardiac muscle contraction, synaptic communication and the modulation of membrane channels (Figure 3.16).

The primary stimulus, usually a molecule that binds to a specific calcium channel, rapidly increases the cytoplasmic concentration of calcium ions. Calcium transfers the message within the cell to other molecules: the message is amplified and activates specific functions, such as the opening of other ion channels like potassium or chloride.

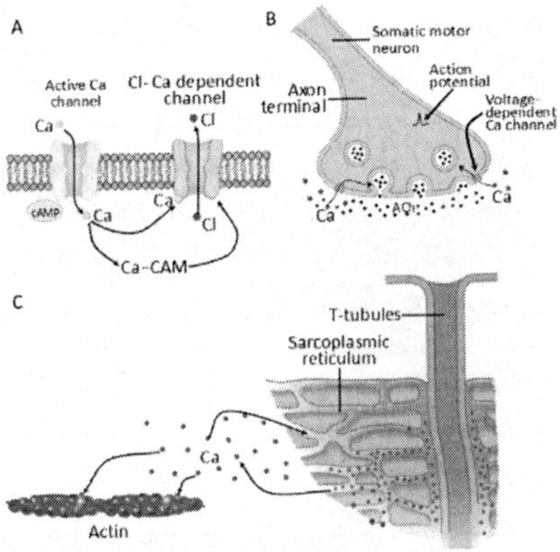

Figure 3.16 Examples of the role of the calcium ion (Ca) in cell physiology. The accumulation of intracellular calcium is essential (A) as a second messenger in numerous intracellular enzymatic cascades involving calmodulin (CAM) and cyclic adenosine monophosphate (cAMP) to activate, for example, calcium-dependent chloride channels, (B) for synaptic neurotransmitter release in the presence of the primary action potential stimulus and (C) to activate the contractile protein actin in the excitation-contraction coupling of skeletal and cardiac muscle.

In synaptic release, there is first an entry of calcium from the external environment very rapidly. The increase in calcium concentration activates, with relatively long timescales, its release from the intracellular stores.

In the muscular system, the primary stimulus consists of a change in the membrane potential due to the activity of sodium and potassium channels in skeletal muscle and calcium channels in cardiac muscle (paragraph 5.14). Calcium is released from the sarcoplasmic reticulum stores and activates the contractile machinery (paragraphs 7.2 and 7.4.1).

Calcium's function as a second messenger can be a slow process. Most of the time, enzyme synthesis is required. By permeating the plasma membrane through specific channels, the calcium ion can change its cytoplasmic concentration, especially in the micro-environment close to the membrane, contributing to calcium-dependent functions, such as regulating the function of membrane proteins, including calcium channels themselves.

At rest, the intracellular concentration of calcium ions is kept extremely low, in the order of 10^{-8} - 10^{-9} M (to be compared with 10^{-5} mM in Table 3.2). Calcium ions are highly reactive and concentrations between 10^{-6} M and 10^{-3} M, if maintained for long periods, are toxic to various cellular functions, as they can induce protein precipitation, block ATP synthesis, alter the cytoskeleton and activate calcium-dependent hydrolytic enzymes.

To ensure that the cell has a sufficient supply of calcium ions, they are actively sequestered in intracellular stores, a specialization of the endoplasmic reticulum, present in all cells. Calcium stores are able to accumulate the divalent ion by means of a calcium-dependent ATPase. The **calcium pump** or **Ca^{2+}-dependent ATPase** is located on the membrane of the stores, is always active and increases its calcium sequestration function in relation to the cytoplasmic concentration of the ion. Calcium concentration inside the stores can reach values of 10^{-2} to 10^{-3} M. Calcium is usually bound with low affinity to a protein, **calsequestrin**, to decrease its apparent concentration and thus facilitate the work of the pump.

When the cell receives an external or internal signal, which may be electrical, mechanical, chemical or hormonal, it is able, in a few tens of milliseconds, to increase the concentration of calcium ions, either diffusely throughout the cytoplasm or in subcellular micro-environments, to a value of 10^{-3} to 10^{-4} M. This increase in concentration can occur in two precise ways: either by inward flux from outside through specific channels in the plasma membrane, or by release from intracellular stores, again through specific channels in the membrane of the endoplasmic reticulum. The external environment and intracellular stores can therefore be considered reserves of calcium ions and thus environments with a high level of chemical potential energy.

To prevent the high concentration of calcium ions from damaging the cell, there is usually an increase in cytoplasmic concentration following a stimulus in the form of calcium transients. The dynamic process involves the massive and instantaneous release of calcium from stores and its re-capture by the Ca^{2+}-dependent ATPase in an equally short time of a few tens of milliseconds. However, there are actions promoted not only by different concentrations of cytoplasmic calcium, but also by the time that this divalent cation spends in the cytoplasm. Intracellular actions promoted by calcium release can last a few milliseconds, like in muscle contraction, or tens of minutes, for example during the fertilization process.

Because calcium cannot remain elevated in the cell for such a long time without being toxic, the solution used by the cell is to operate with very large repeated calcium transients, but with a duration of only a few milliseconds, which increase the intracellular concentration up to 10^{-3} M. Calcium-dependent reactions, which have longer timescales for their activation, 'sense' a cytoplasmic calcium concentration, which is the time integral of the transients. In this way, the concentration becomes dependent on the frequency with which calcium is released from the stores. Different frequencies at which calcium transients are produced are used to control the concentration in a systematic way over time. The frequencies often coincide with the frequency of stimulation of the target cell.

3.4 Ionic reserves

The average composition of the internal and external solution of a typical cell is given in Table 3.2. There are hydrophilic solutes including protein aggregates negatively charged at physiological pH, positive potassium, sodium, calcium and magnesium ions, anions such as chloride, carbonate and phosphate and hydrophobic molecules such as glucose and ATP. The plasma membrane has a hydrophilic outer and inner layer, formed by the water-soluble polar heads of the phospholipids. In addition, there is a hydrophobic lipid bilayer in between, formed by the phospholipid tails.

Although hydrophilic molecules in aqueous solution can move freely, and therefore have a high probability of being close to the plasma membrane, they cannot cross the lipid barrier. On the other hand, hydrophobic neutral molecules are able to permeate into the lipid bilayer, but they need a "carrier" to move in the aqueous solution. Because of its chemical and physical structure, the plasma membrane would therefore be an impermeable barrier to all types of molecules. However, the cell needs to exchange both hydrophilic and hydrophobic substances with the outside world. To do this, it can use specific integral membrane proteins for polar molecules or protein carriers for neutral compounds.

3.4.1 The chemical gradient

In order to perform a task, such as moving molecules from the inside to the outside of the cell, energy is required. If we look at Table 3.2, we see that the concentration of all the apolar molecules, amino acids, glucose and ATP, is different between inside and outside. If, for example, the plasma membrane has integral proteins capable of passing glucose, the molecule is more likely to pass from the external environment, where the concentration is higher, to the internal environment, where the concentration is lower, because it is continuously consumed by the mitochondria to form ATP. This transition occurs due to the presence of a concentration difference between the outside and inside: the chemical gradient, which provides the necessary energy. In cells, there is also a concentration difference for amino acids, proteins, ATP and other molecules. For each of these molecules, therefore, there is a chemical gradient independent of that of the others.

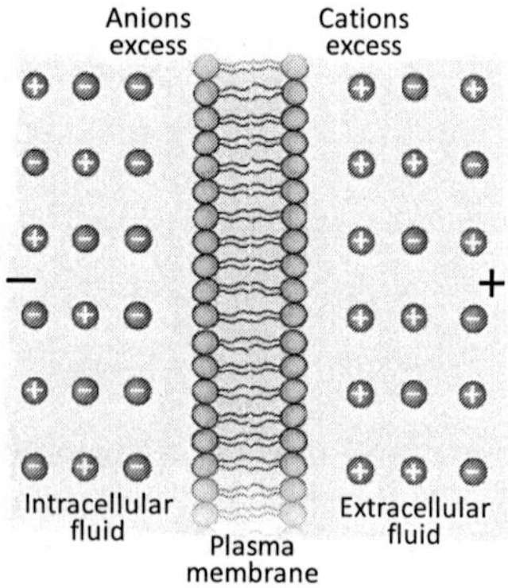

Figure 3.17 A model of a biological membrane capacitor. Phospholipid heads are the capacitor plates. Lipid tails work as a dielectric. Excess of cations or anions on one side of the membrane induces an equal number of counter-ions on the other side. This sets an electrically balanced system. However, in the close proximity of the membrane, there is a physical separation of charges, representing the ability of this structure to accumulate potential energy.

Looking again at Table 3.2, it can be seen that about 95% of the extracellular solution and about 75% of the intracellular solution consists of polar molecules, the ions derived from the dissolution of sodium, potassium, calcium, magnesium chlorides, various carbonates and phosphates in water. The two environments, like the whole organism, are electrically neutral. In fact, there are as many positively charged cations as there are negatively charged anions. Looking at the table, it would seem that there is an excess of potassium ions in the intracellular solution compared to the anions. This is due to the fact that at the physiological pH of about 7.4, proteins are present in the form of protein anions with a high affinity for potassium, which can thus accumulate in the cell.

3.4.2 The membrane capacity

An additional fundamental role played by the lipid bilayers is charge buffer capacitance. Unbalanced charges (see for example Figure 3.15) inside or outside the biological membrane are immediately re-balanced by the membrane capacitor. A transient capacitive current coordinates the accumulation of opposite charges across the membrane to re-establish electro-neutrality in the micro-environment formed by the membrane and the solutions in its immediate vicinity. This generates charge accumulation and thus the possibility to accumulate potential energy (Figure 3.17).

There is a logic to follow for a clear understanding of the membrane capacitor structure/function relationship. The environment has an important role in shaping membrane structure. In terms of capacitor characteristics, an important element is the distance between the opposite polar heads (capacitor plates) created by the lipid tails (dielectric). This is in the order of 8-10 nm and is a universal measurement for all biological cell membranes. This distance between the "plates" is instrumental to the forces of attraction for ions of opposite signs. With all of the parameters of Equation S2.3 constants, the only variable remaining is the cell surface. Thus, the time in which the membrane potential changes after stimulation (τ, Equation S2.7) is directly proportional to the cell size.

3.4.3 Conductance or resistance

Ions immersed in a restricted environment move due to thermal agitation. For ions like Na, K and Cl, there is a high probability of colliding with the membrane. If the membrane contains integral proteins forming aqueous pores, it is possible for the charged particles to cross the membrane (Figure 3.18). The parameters regulating ion movement through a channel protein are the resistance to passage (R) and the conductance (G) of the pore. These parameters being reciprocal (G = 1/R), the pore diameter is directly proportional to G and inversely proportional to R.

Similarly, but oppositely, the relationship is valid for channel length: this is directly proportional to R and inversely proportional to G. The two extremes for R are: (1) the membrane is sealed to ion movement, where the resistance is infinite and the conductance is zero and; (2) ionic permeabilities are fully open, where the resistance is the same as the physiological solution (100 Ω·cm) and the conductance is maximum.

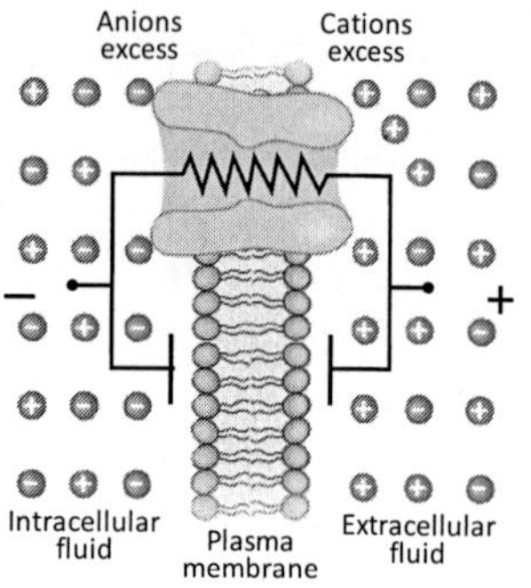

Figure 3.18 In the electrical model of a plasma membrane, the capacitor and resistor are connected in parallel.

3.5 The membrane potential

The plasma membrane has different types of integral proteins (Figure 2.1), defined as ion channels (paragraph 4.4), which allow the passage of charged particles. When the cell is at rest, it is in a dynamic equilibrium. The value of the **membrane potential** is determined by averaging the Nernst potentials for each permeable ion multiplied by its permeability. Generally speaking, the main basic conductance that influences the resting membrane potential is due to potassium ions. However, often the membrane potential values reflect the function that a particular cell is called on to accomplish. Because potassium permeability is essentially ubiquitous, the range of membrane potentials spans from the potassium equilibrium potential and the average between potassium and sodium reversal potentials. This value is between -15 and -20 millivolts. There is actually continuous leakage of potassium, which tends to bring the difference in electrical potential across the membrane, the **resting membrane potential**, towards its equilibrium potential. In fact, we have seen how the escape of an ion by the chemical gradient proceeds until the

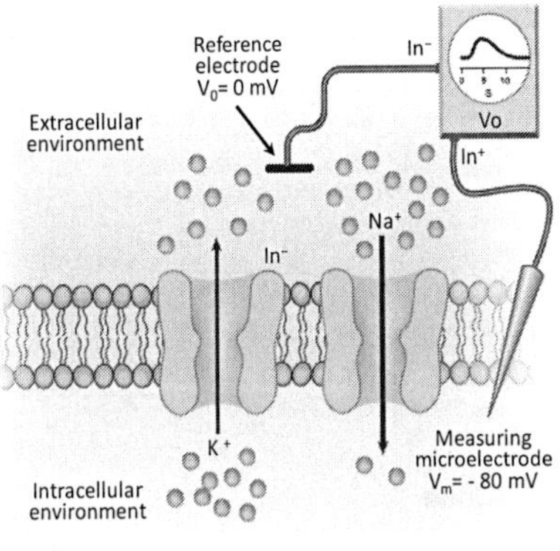

Figure 3.19 Using a glass microelectrode filled with a concentrated KCl solution and a voltmeter (Vo), the membrane potential can be measured.

electric potential generated reaches the value of the ion's equilibrium potential.

To measure the membrane potential of a cell, an experimental apparatus used is the voltmeter. A microelectrode (Data Sheet 3.4) is moved with a micromanipulator until it touches the plasma membrane and perforates it. Due to its fluidity, the plasma membrane closes and seals itself on the capillary glass. The microelectrode (Figure 3.19) is connected to the positive input of the voltmeter (In^+), while the electrode placed in the extracellular solution, or reference electrode, is connected to the negative input of the same voltmeter (In^-). The voltmeter (Vo) measures the differences in potential between the inside of the cell (V_m) and the outside (V_O). It is convenient to set the external potential equal to zero by connecting the input In- of the voltmeter to ground (reference electrode). To identify the electrical energy content of a cell, the correct term is "the potential difference across the membrane". Conventionally, the external reference electrode is connected to ground at zero voltage. This allows discussion of "cell membrane potential" instead of "difference in potential between the inside and outside of the membranes".

The microelectrode is unlikely to be positioned close to the membrane, as shown in Figure 3.19, but more probably towards the center of the cell. Nevertheless, it **senses** the potential in the immediate proximity of the membrane where is actually generated (Figures 3.17-3.18).

To set the resting membrane potential, the main contribution comes from potassium and sodium permeability. However, all of the ions present on both sides of the membrane, chloride, calcium, etc., can influence the resting membrane potential if they can move across the membrane. Lastly, the mechanism of maintaining stability in the resting membrane potential is recharging through the continuous work of the electrogenic active Na/K ATPase (Figure 3.15).

A complete mathematical description of the cell membrane potential is given by the Goldmann equation.

3.5.1 The Goldmann equation

The **Goldmann equation**, which is derived directly from the Nernst equation (Equation 3.1), relates the membrane potential V_m to the concentration and permeability of the ions exchanged by the cell at rest (potassium, sodium and chloride):

$$V_m = \frac{RT}{F} \ln \frac{P_K[K^+]_e + P_{Na}[Na^+]_e + P_{Cl}[Cl^-]_i}{P_K[K^+]_i + P_{Na}[Na^+]_i + P_{Cl}[Cl^-]_e}. \quad (3.4)$$

Given a temperature of 18 °C (291 K), known R and F values (Data Sheet 3.2) and when the cell is not undergoing any kind of stimulation, the sodium and chloride to potassium permeability ratios are according to the proportion

$$P_K : \frac{P_{Na}}{P_K} : \frac{P_{Cl}}{P_K} = 1 : 0.014 : 0.02 \quad (3.5)$$

It follows that the membrane potential of the cell at rest with an ion distribution such as that in Table 3.2, or **resting potential**, will be:

$$V_r = 0.058 \cdot \log \frac{4 + (0.014 \cdot 145) + (0.02 \cdot 4)}{140 + (0.014 \cdot 15) + (0.02 \cdot 115)} = -0.0793\text{V} = -79.3\text{mV} \quad (3.6)$$

like that measured experimentally (paragraph 3.5).

3.5.2 Ohm's law

The electric circuit in Figure 3.18 represents the plasma membrane. A charge passing through a resistor is a current which depends on the potential difference according to **Ohm's law**:

$$I = \frac{1}{R} \cdot V = G \cdot V, \quad (3.7)$$

in which the current I expressed in amperes (A) is proportional to the voltage V expressed in volts (V) and the proportionality constant is the reciprocal of the resistance R expressed in ohms (Ω) or the conductance G expressed in Siemens (S). If we consider the cell, the current (given by the ion I crossing the plasma membrane using its ion channels) is proportional to the difference between the membrane potential V_m and the equilibrium potential of the ion I E_I, since this influences, as we have seen, the movement of the ions across the membrane, according to the relationship that is normally called **generalized Ohm's law**

$$I_I = G_I \cdot (V_m - E_I) \quad (3.8)$$

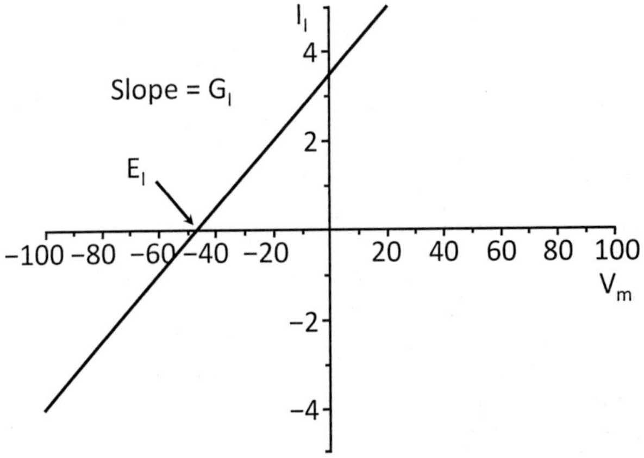

Figure 3.20 Equation 3.8 represents a straight line with slope G_I and zero current when V_m=E_I. V_m is expressed in mV ($V_I \cdot 10^{-3}$) and I_I in pA ($A \cdot 10^{-12}$).

where the constant of proportionality is the conductance G_I for ion I. The difference between the membrane potential V_m and the equilibrium potential E_I of ion I is generally called the **electromotive force** (e.m.f.) or driving force. From Equation 3.8, which in a current-voltage graph represents a straight line with slope G_I crossing the x-axis when $V_m = E_I$ (Figure 3.20), it can be deduced that I_I is constant if V_m is constant and varies when V_m varies, but in turn varies V_m. Moreover, if V_m becomes equal to E_l, the current becomes, as observed above, equal to zero and the channels must be open, i.e., the conductance must be greater than zero, since a current has passed that has made V_m vary. It will be seen later that the conductance, in many cases, also depends on V_m (paragraph 5.8).

Bibliography

Nutman A, Bennett V, Friend C, Van Kranendonk M, Chivas A. Rapid emergence of life shown by discovery of 3,700-million-year-old microbial structures. Nature (2016) 537(7621): 535–538. Neil A Campbell NA and Reece JB., Meccanismi dell'evoluzione ed origini della diversità. Zanichelli 2004.

Data Sheet 3.1

Donnan equilibrium (Gibbs-Donnan equilibrium)

The English chemist Frederick G. Donnan (Colombo 1870 - Canterbury 1956) demonstrated that in an experimental system consisting of two environments separated by a semi-permeable membrane, where the membrane is permeable to water and a salt but impermeable to a proteinate, the diffusible ions of the solute become unevenly distributed

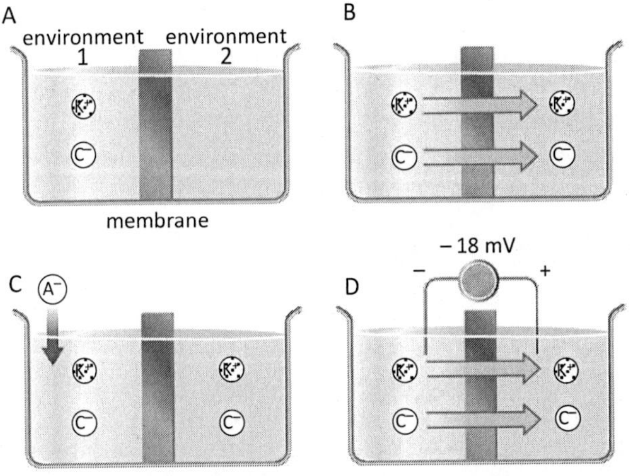

Figure S3.1 A salt dissolved in one environment (A) diffuses into the other environment (B) through the semi-permeable membrane and the concentrations become equal, but a non-diffusible anion (A^-) placed in environment 1 (C) causes redistribution of diffusible ions (D).

between the two compartments and generate, at equilibrium, a difference in potential at the sides of the membrane. This phenomenon is known as Donnan equilibrium, or Gibbs-Donnan equilibrium.

Let us assume an initial situation in which potassium chloride has been dissolved in room 1 at a certain concentration. The potassium chloride dissociates into potassium K^+ ions and chloride Cl^- ions (Figure S3.1A), which by diffusion, according to a gradient, pass from room 1 to room 2. At equilibrium, electro-neutrality will be restored and the ions will have the same concentration in the two rooms (Figure S3.1B). At this point, a potassium salt of a non-diffusible A^- anion, such as a protein, is added to environment 1 (Figure S3.1C), resulting in an increase in the concentration of the potassium ion in environment 1. The potassium ion tends to diffuse into environment 2 via a concentration gradient, also bringing along the chloride ion, the only anion that can move across the membrane. At equilibrium, the potassium ion is more concentrated in environment 1, where it has to neutralize the charges of the chloride ion and A^-, than in environment 2, while the chloride ion is more concentrated in environment 2 (Figure S3.1D). The two environments again reach electro-neutrality, i.e., the total number of positive charges is equal to the number of negative charges.

The equation can be proved

$$\frac{[K^+]_1}{[K^+]_2} = \frac{[[Cl^-]_2}{[Cl^-]_1} \tag{S3.1}$$

with unequal distribution of diffusible ions, characteristic of Donnan equilibrium.

The diagram in Figure S3.1C could represent the situation of a cell where environment 1 represents the inside and environment 2, the outside. The presence of the non-diffusible proteinate A^- (Figure S3.1C) decreases the concentration of chloride and increases the concentration of potassium inside the cell compared to outside. The more concentrated (by the same amount) chloride outside creates: a) a potential difference between inside and outside (Figure S3. 1D), b) a higher concentration of diffusible ions inside the cell, c) an increase in osmotic pressure with an inward flow of water that causes the cell to swell, a situation incompatible with its proper function, d) a certain hydrostatic pressure that opposes the osmotic pressure.

An equilibrium similar to Donnan's equilibrium, more correctly known as colloidosmotic pressure or oncotic pressure, is partly responsible for the osmotic pressure due to the presence of non-diffusible proteins. This equilibrium is established, for example, between the blood contained in the capillaries and the interstitial fluid, since the blood contains the plasma proteins to which the capillary membrane is impermeable, while the interstitial fluid is very poor in proteins.

More generally, the cell must be considered as a system similar to that represented by a Donnan equilibrium, but only as a simple introductory model for studying the mechanisms of ion distribution between two environments separated by a plasma membrane. Compared with a functional situation such as that of the Donnan equilibrium, a) the cell maintains an almost constant volume; b) in the cell, active transport plays a decisive role in ion distribution; c) the plasma membrane is not perfectly semi-permeable; d) in reality, the cytoplasm is not a homogeneous aqueous solution; d) the activity of diffusible ions is generally much lower than their concentration because the salts are not completely dissociated.

Data Sheet 3.2

Calculation of the equilibrium potential of ions according to the concentrations given in Table 3.2.

In the Nernst equation

$$E_I = \frac{RT}{zF} \cdot \ln \frac{[I]_e}{[I]_i} \tag{S3.2}$$

the term RT/zF calculated at 18 °C, for z=+1 converting to decimal logarithms, becomes:

$$\frac{RT}{zF} = \frac{8.314 \frac{J}{°K \cdot mole} \cdot 291\,°K}{1 \cdot 96.487 \frac{coulomb}{mole}} \cdot \ln 10 = 0.058 \frac{J}{Coulomb} = 0.058\ V \tag{S3.3}$$

while for z=+2, it becomes:

$$\frac{RT}{zF} = \frac{8.314 \frac{J}{°K \cdot mole} \cdot 291\,°K}{2 \cdot 96.487 \frac{coulomb}{mole}} \cdot \ln 10 = 0.029 \frac{J}{Coulomb} = 0.029\ V \tag{S3.4}$$

Potassium (z = +1)

$$E_K = \frac{RT}{1 \cdot F} \cdot \ln \frac{[K^+]_e}{[K^+]_i} = 0.058 \cdot \log \frac{4.0}{140.0} = -0.0896\ V = -89.6\ mV \tag{S3.5}$$

Sodium (z = +1)

$$E_{Na} = \frac{RT}{1 \cdot F} \cdot \ln \frac{[Na^+]_e}{[Na^+]_i} = 0.058 \cdot \log \frac{145.0}{15.0} = +0.0571\ V = +57.1\ mV \tag{S3.6}$$

Chloride (z = -1). Knowing that ln A/B = -ln B/A

$$E_{Cl} = \frac{RT}{-1 \cdot F} \cdot \ln \frac{[Cl^-]_e}{[Cl^-]_i} = \frac{RT}{1 \cdot F} \cdot \ln \frac{[Cl^-]_i}{[Cl^-]_e} = 0.058 \cdot \log \frac{4,0}{115.0} = -0.0846 - = -84.6\ mV \tag{S3.7}$$

Calcium (z = +2)

$$E_{Ca} = \frac{RT}{2 \cdot F} \cdot \ln \frac{[Ca^{2+}]_e}{[Ca^{2+}]_i} = 0.029 \cdot \log \frac{1.8}{10^{-7}} = +\infty \tag{S3.8}$$

Note that the value of +∞ for the equilibrium potential of the calcium ion is the theoretical value, due to the many interactions it has within the cell. Its measured equilibrium potential is typically between +70 and +150 mV.

Data Sheet 3.3

The membrane potential in a neuron

In order to measure the membrane potential in relatively small cells such as neurons or muscle fibers, an experimental apparatus such as that shown in Figure S3.2 is used.

The experimental dish, filled with an external physiological solution in which the cell under examination is placed, is placed on the table of an optical microscope with an average magnification of 200x-400x, not shown in the Figure. The microelectrode (tab. 3.4) is attached to a manipulator capable of making movements of the order of a micrometer and positioned in the bath, electrically regulated so that it registers a potential difference equal to zero, then brought close to the cell membrane and pushed until it

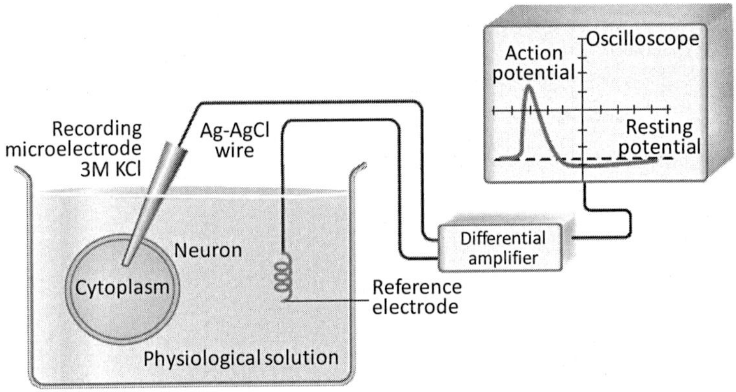

Figure S3.2 Schematic of an apparatus for measuring membrane potentials in a neuron.

perforates it.

When the tip of the microelectrode pierces the membrane, the circuit formed between the differential amplifier, the microelectrode, the inside of the cell, the outside of the cell and the differential amplifier allows the differential amplifier (Figure S3.2) to compare the potential inside the cell with that of the reference electrode. When the electrode is immersed in the bath outside the cell, it measures 0 mV because it is in the reference environment. At the instant the cell membrane is pierced, the signal that appears on the oscilloscope screen (tab 5.2) will be the difference in potential between the inside and outside of the cell (section 3.5). It should be borne in mind that, if for convenience the term *potential* is used for cellular electrical phenomena, in reality reference is always made to the *potential difference* between inside and outside: this simplification can be justified by the fact that the external reference potential is always 0 mV.

Under these conditions, it is possible to manipulate the external solution to change the equilibrium potentials of the ions, for example by increasing the external potassium concentration at the expense of the sodium concentration. This substitution causes, according to the Goldmann equation (Equation 3.4), a depolarization which, if sufficiently large, can lead to the stimulation of an action potential visible on the oscilloscope screen (Figure S3.2).

Data Sheet 3.4

The microelectrode

In order to carry out electrophysiology experiments on relatively small cells such as neurons or muscle cells, it is essential to use the

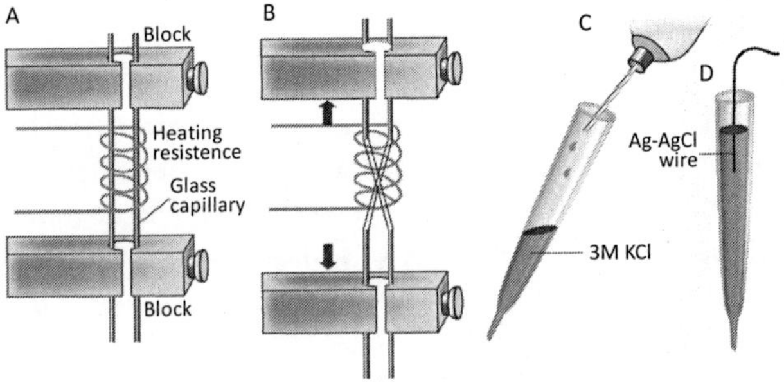

Figure S3.3 Schematic of an apparatus for measuring membrane potentials in a neuron. A and B are magnification of a microelectrode puller in the region of the heating resistance. Once current passes through the resistance the filament become incandescent. The glass capillary is simultaneously melted and pulled apart (black arrows). The result is the production of two microelectrode with a tip diameter directly proportional to the heat level and to the force of the pulling

microelectrode, a signal transducer which enables the charges distributed on the sides of the membrane to be transformed into measurable electrical signals without damaging the membrane.

The microelectrode is obtained using a puller (Figure S3.3A), an apparatus essentially consisting of: 1) two blocks to which a pervious glass capillary with an external diameter of 1 mm is attached with a very thin glass wool thread fixed inside; 2) a puller system by gravity or with electromagnets; 3) a spiral resistance surrounding the capillary; 4) a system to regulate the heating current of the resistance.

Heating the glass with the resistor, it becomes malleable, and the capillary is stretched by moving the two blocks apart (arrows in Figure S3.3B). It breaks, forming two microelectrodes, still intact and with a tip with a diameter varying from 100 µm to 0.1 µm (Figure S3.3C). The microelectrodes are filled with a highly conductive solution of 3 M potassium chloride (KCl), which diffuses up to the tip due to the presence of the glass wool wire, which considerably lowers the surface tension. Potassium chloride is preferentially used, because any small amount of diffusion inside the cell, which already has a high concentration of potassium ions, would not significantly alter its ionic equilibrium.

A silver wire coated with silver chloride (Ag-AgCl of Figure S3.3) is inserted into the microelectrode to obtain the connection with the measuring instrument (Figure S3.3D).

When making electrical measurements in a system consisting of metallic conductors, conductors of the first type that move electrons, and electrolyte solutions, conductors of the second type that move ions, a silver wire is used, onto which a layer of silver chloride is deposited by a

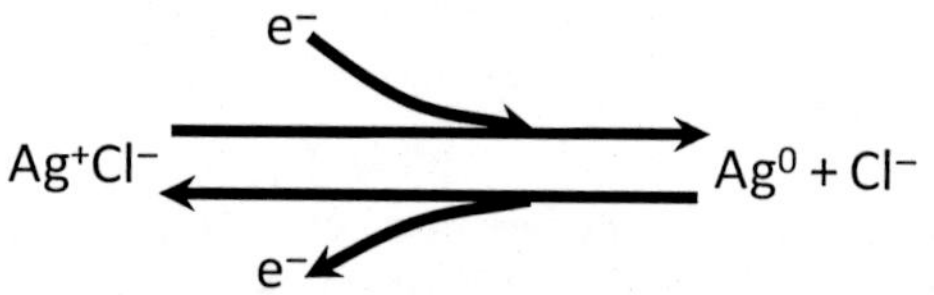

S3.3

chemical reaction with a suitable supplier of chloride ions. In this way, the equilibrium reaction takes place:

If an electron (e-) is brought from the metal circuit, a silver ion (Ag+) of the silver chloride is transformed into metallic silver (Ag0), the reaction goes to the right and an excess chloride ion (Cl-) goes into solution, making the solution negative. Vice versa, if an electron is subtracted, the metallic silver loses an electron, becomes a silver ion and forms, with a chloride ion subtracted from the solution, the salt that precipitates, so the reaction goes to the left. In the first case, a negative charge consisting of the electron passes from the metal to the solution as a chloride ion; in the second case, a negative charge of a chloride ion is subtracted from the solution and passes to the m metal as an electron.

CHAPTER 4

Transporters

To exchange the wide variety of substances needed by the cell, the plasma membrane can use a number of specific or non-specific passive mechanisms, without energy consumption, or specific active mechanisms with ATP consumption. Passive mechanisms include simple, non-specific diffusion and facilitated transport, which uses specific integral membrane proteins; active transport also uses integral membrane proteins but with ATP consumption. Facilitated transport and active transport allow exchanges to take place in a controlled and highly selective manner, to a greater extent than diffusion. Gaseous molecules, such as oxygen and carbon dioxide, on the other hand, are able to cross the membrane without the need for special mechanisms, but only thanks to the thermal agitation of the molecules themselves.

4.1 Diffusion

Diffusion is a mechanism which, by exploiting the mobility of the molecules due to thermal agitation, allows the movement of substances without the consumption of chemical energy. Diffusion can take place between two points in the same environment, such as a solution. Aqueous environment is the physical substrate from every "fast" chemical reaction because indeed medium is dynamic and allows mobility of molecules that a solid state would not. It also occurs between two environments separated by a biological membrane, provided there are aqueous pores.

Diffusion always occurs according to a gradient, i.e., from the environment with the highest concentration to that with the lowest concentration.

4.1.1 Simple diffusion

The mechanism by which a molecule can move from one point to another in an aqueous environment is that of **simple diffusion**, which, for example, allows dye deposited in a well-defined area of a container to invade the entire environment.

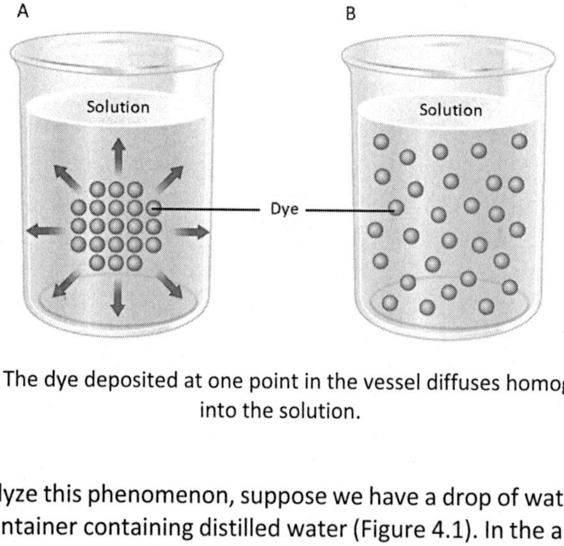

Figure 4.1 The dye deposited at one point in the vessel diffuses homogeneously into the solution.

To analyze this phenomenon, suppose we have a drop of water-soluble dye in a container containing distilled water (Figure 4.1). In the area where the drop is deposited, there is a high concentration of dye. In the surrounding area, it is the water with a high concentration (Figure 4.1A). Each molecule has kinetic energy proportional to its temperature. As long as it is above absolute zero (-273 °C), molecules continuously move by thermal agitation in uniform rectilinear motion, in a random direction in space, in a temperature-dependent manner. A change in direction occurs because one molecule hits another. Initially, water and dye are at high concentrations: it is highly probable that a molecule of dye will move from the drop of dye to the water, as well as a molecule of water moving from the water to the drop of dye, but the probability of opposite movements

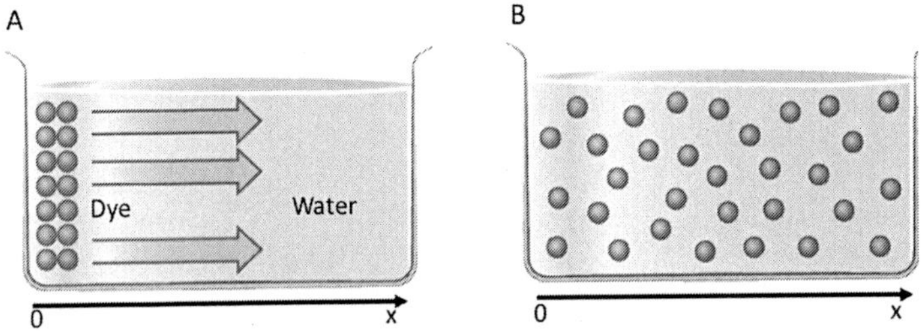

Figure 4.2 The dye diffuses into the solution (A). Only at the final time t_4 (Figure S4.1) is the concentration homogeneous (B).

is practically null. If we define by Φ_{1-2} the unidirectional flow of the dye through the entire outer surface of the drop towards the water and by Φ_{2-1} the unidirectional flow in the opposite direction, the net flow Φ_n will be given by the equation:

$$\Phi_n = \Phi_{1-2} - \Phi_{2-1} \tag{4.1}$$

At the beginning of the process, the Φ_{1-2} flow of dye from the drop to the water will be maximum due to the large difference in concentration between the drop and the water. The Φ_{2-1} flow of dye from the water to the drop will be zero as the concentration of dye in the water is zero. Therefore, according to equation 4.1 the net flow Φ_n will be equal to Φ_{1-2}.

A time course of the bidirectional flux shows a decrease of intensity in the Φ_{1-2} flow and a related increase in the flow Φ_{2-1} as the concentration of the dye increases in the water and decreases in the drop. Consequently, the probability that a new dye molecule can pass into the water and that a water molecule can pass into the drop of dye decreases. This phenomenon affects all the molecules of dye and water present in the container, until an equilibrium is reached so that the unidirectional flow Φ_{1-2} will be equal to the unidirectional flow Φ_{2-1} (Figure 4.1B).

Diffusion is thus proportional to the difference in concentration of the dye but also of the solvent, in this case water. A mechanism similar to the one just described can be assumed in a situation such as that depicted in Figure 4.2A. Due to the diffusion phenomenon, the dye, deposited at the point close to x=0 and with the same concentration in the whole area, will move until it is homogeneously distributed throughout the cylinder (Figure 4.2B). The flux Φ, i.e., the number of moles dn moving in the time unit dt, dn/dt, across the region A is given by Fick's law of simple diffusion of an apolar molecule (Data sheet 4.1):

$$\Phi = \frac{dn}{dt} = -AD \cdot \frac{dc}{dx}, \tag{4.2}$$

and is proportional to the concentration gradient according to the constant D, which is the diffusion coefficient and has the dimensions of a velocity x length.

Fick's law describes the flow of any apolar molecule in any solution because D is typical of each substance. The driving force for the diffusion process is exclusively the concentration gradient both of the solute and of the solvent.

4.1.2 Diffusion in the presence of a membrane

In a cylinder (Figure 4.3), a membrane of thickness X and area A_m separates environment 1, in which the water-soluble substance S is present at concentration C_{S1}, from environment 2, in which the concentration of S, C_{S2}, is initially zero (Figure 4.3A). The passage of S across the plasma membrane from environment 1, e.g., the outside of a cell, to environment 2, the inside of a cell, is indispensable for the cell. If S can diffuse freely across the membrane and if we assume that its diffusion

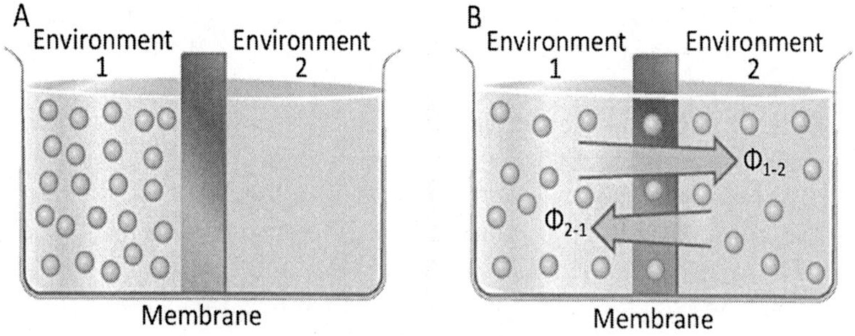

Figure 4.3 Diffusion in the presence of a membrane.

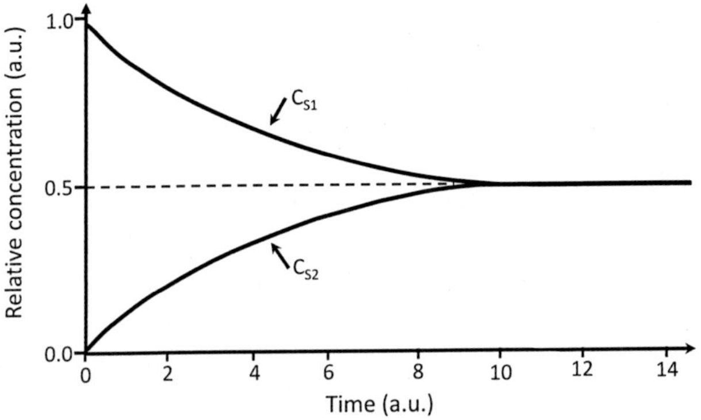

Figure 4.4 Relative concentration trend at time t=0 in environment 1 (C_{S1}) and environment 2 (C_{S2}).

coefficient D_S is sufficiently large and is equal in the two environments, we will have a net flux Φ_{nS} according to Equation 4.1 (Figure 4.3B). This flux tends to decrease in time until it is exhausted because each unidirectional flow is proportional to the difference in concentration between the two environments C_{S1} and C_{S2} and this varies in time. As C_{S1} decreases, C_{S2} increases, reaching an equilibrium in concentration between the two environments (Figure 4.4).

The equation describing the net flux of substance S per unit area A, J_{nS}, is derived from Fick's law, in which the difference in concentration between environments 1 and 2 is considered (Figure 4.2):

$$\frac{\Phi_{nS}}{A} = J_{nS} = P_m (C_{S1} - C_{S2}) \qquad (4.3)$$

P_m is the permeability coefficient of S in the membrane, is proportional to the membrane diffusion coefficient D_m and is the size of a velocity.

By measuring J_{nS} with flow measurement experiments and knowing C_{S1} and C_{S2}, one can derive P_m in different biological membranes for various substances of physiological interest.

P_m varies greatly for different molecules crossing membranes of different cells. For example, in erythrocytes, it can vary from $1 \cdot 10^{-12}$ to $1 \cdot 10^{-2}$ cm/s. In addition, this parameter can be modified by the action of various substances through specific membrane receptors. An example is the case of antidiuretic hormone, which can make the nephron's collecting ducts up to 10 times more permeable to water. Other examples are several neurotransmitters that can increase the membrane ionic permeability of the target cell.

4.2 The movement of water

Water must be able to move passively through cell membranes to compensate for the flow of salts and other substances, in order to maintain constant osmotic pressure. Water, like any other molecule, flows according to the concentration gradient. Water is a solvent, but from a chemical point of view, is like any other component present in the solution. The concentration of water is less in the presence of a solute compared to pure water. Thus, if the solute at time = 0 retains the higher probability to diffuse towards pure water, water moves in the opposite direction with the aim of diluting the solution.

4.2.1 Aqueous pores

The plasma membrane can allow apolar molecules to pass through the phospholipid bilayer or through large **aqueous protein pores** or aqueous

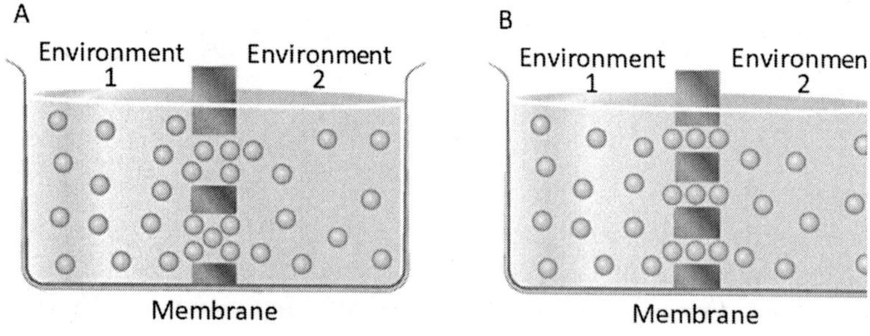

Figure 4.5 The flow through the large-pore (A) and small-pore (B) porous membrane.

pores with a diameter comparable to the diameter of the molecule to diffuse. For large pores, Equation 4.3 is valid. The diffusion coefficient D is that of free diffusion, the partition coefficient is equal to 1 and the area involved in diffusion is the total area of the pores A_p (Figure 4.5A). Thus, the permeability of the substance is

$$P_p = \frac{A_p D}{A_m \Delta X}, \quad (4.4)$$

where A_m and X are the area and thickness of the membrane, respectively.

In the case of **small-diameter pores**, on the other hand, the net flux J_n can reach saturation, as within the pore, molecules can interact with the pore walls and with each other as they cross the membrane in a single row, so the two unidirectional fluxes are not independent (Figure 4.5B). The permeability coefficient is variable and can only be determined experimentally. Aqueous pores of small diameter include, for example, ion channels.

4.2.2 Water flows

Water molecules move if there are two phases with different concentrations of a given substance S. Such movements are always simple diffusion but can occur in different physiological conditions.

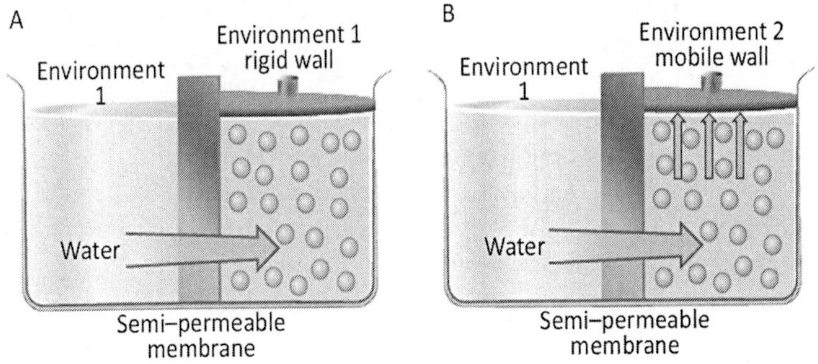

Figure 4.6 The diffusion of water in the presence of a semi-permeable membrane with rigid (A) and mobile (B) walls.

Suppose we have a cylinder similar to the one in Figure 4.3. Environment 1 contains distilled water and environment 2, distilled water with a solute, e.g., sucrose at a given concentration. The two environments are separated by a **semi-permeable membrane**, i.e., permeable to water but not to the solute (Figure 4.6A). The concentration of water in environment 1, C_{a1}, is greater than the concentration of water in environment 2, C_{a2}, because in the latter, each unit volume contains both water and solute molecules.

The individual water molecules pass through the phospholipid layer by a mechanism similar to the diffusion of an anelectrolyte. Equation 4.3 becomes:

$$J_a = P_a \cdot (C_{a1} - C_{a2}). \qquad (4.5)$$

J_a is the net flux of water per unit area of the membrane. P_a is the permeability coefficient for water. P_a takes into account the mobility and the partition coefficient r of water in the membrane.

The passage of water from environment 1 to environment 2 is called **osmosis**, from the Greek *osmos*, to push. If environment 2 has rigid walls (Figure 4.6A), following the physical principle of liquid incompressibility, there is an increase in the water pressure at the membrane without a physical passage of water. If the top of environment 2 is mobile (Figure 4.6B), water flows from 1 to 2. Under these conditions, the water actually passes through the membrane, increasing the volume in 2. Water flows until the difference in hydrostatic pressure between 1 and 2 (due to the greater weight of the water column in 2) counterbalances the net flow of water between 1 and 2. The difference in hydrostatic pressure that counterbalances the flow of water is the **osmotic pressure**. The osmotic pressure is therefore the pressure that can be measured as a net flow of water from an environment with a high concentration of water to one with a low concentration due to the presence of a solute.

This is referred to as the volumetric flow of water J_V:

$$J_V = P_{osm} \cdot (\Delta\pi - \Delta p). \qquad (4.6)$$

$\Delta\pi$ and Δp are the difference in osmotic pressure and hydraulic pressure, respectively, between environment 1 and environment 2. P_{osm} is the osmotic permeability coefficient, which depends on the mobility of water and the partition coefficient for water between the membrane and environment 1. In steady state conditions, J_V is = 0 in both cases: the osmotic pressures and hydraulic pressure are equal with an opposite sign.

4.2.3 Osmolarity

The osmolarity is the sum of the concentrations of all molecules and ions in solution regardless of whether they can cross the plasma membrane or not. A solution containing sucrose at a concentration of 1 mol/L (1 M solution) has an osmolarity of 1 Osm (Osmole), whereas a 1 M NaCl solution will have an osmolarity of 2 Osm, as the salt dissociates into ions and both the cations and anions contribute to the osmolarity. The higher the osmolarity of a solution, the greater the osmotic pressure it exerts on water to diffuse through a semi-permeable membrane. The higher osmolarity means a higher concentration of particles and therefore a lower concentration of water. This leads to a greater osmotic flow of water and therefore a greater hydrostatic pressure, i.e., a greater osmotic pressure to balance the system. If a solution is **isosmotic** with respect to another solution separated by a semipermeable membrane, the water flow is null; if it is **hyperosmotic**, water diffuses from the other solution; if it is **hypoosmotic**, water diffuses towards the other solution.

The osmolarity, therefore, although referring to the concentration of solutes, acts on water flows. In most living organisms, the solution inside the cell and the solution outside the cell are isosmotic with an osmolarity of about 300 mOsm ($300 \cdot 10^{-3}$ Osm).

4.2.4 Tonicity

Let us consider a cell containing a solute S_i that is not diffusible through the membrane, with a concentration that gives an osmolarity of 300 mOsm, immersed in a solution with an osmolarity equal to that internally and containing a molecule S_e that diffuses freely (Figure 4.7A). In the two solutions, the water concentration is equal and therefore there is no flux, i.e., the two solutions are isosmotic. The S_e molecule, however, diffuses inside the cell, due to the presence of a concentration gradient, according

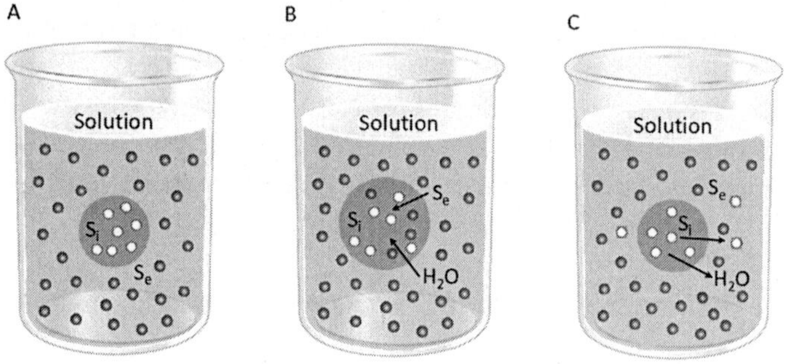

Figure 4.7 If S_i (light dots) is non-diffusible and S_e (dark dots) is diffusible (A), the cell becomes loaded with Se and water (B). If S_i is diffusible and S_e non-diffusible (A), the cell loses S_i and water (C). In the first case, the external solution is hypotonic and in the second case, hypertonic.

to Equation 4.3; the osmolarity inside the cell increases because the water concentration decreases, so an inward flow water is established according to Equation 4.5 (Figure 4.7B). The cell swells until it bursts or until the elasticity of the membrane generates a level of hydrostatic pressure that blocks the inflow of water. The solution containing the S_e molecule is **hypotonic** compared to that inside the cell. On the other hand, the solute inside the cell S_i diffuses through the membrane, but the S_e molecule does not, and the two solutions are still isosmotic: there will be flow of the solute S_i towards the external environment, accompanied by water escaping from the cell. The cell reduces its volume and shrivels (Figure 4.7C). The external solution in this situation is **hypertonic**. Of course, in the situation in which, even in the presence of non-dispersible substances, there are no water flows, we would consider this an isotonic solution. Therefore, while the osmolarity of a solution depends on the total concentration of diffusible and non-diffusible solutes and directly activates water flow, the tonicity is related to the concentration of non-diffusible solutes and the water flow is secondary to the flow of the solutes according to their concentration gradient.

4.2.5 Water transport

The transport of water, for example through the intestinal epithelium, occurs by osmosis, but is regulated by the mechanism outlined in Figure

4.8. The Na^+/K^+ ATPases present on the basal membranes forming the interstices between cells create a strong sodium gradient. This gradient increases from the interstitial fluid in the serosal zone of the intestinal epithelium towards the apex, near the intestinal lumen. The sodium gradient draws water by osmosis: water passes through both the apical membranes and the gap junctions between one cell and the next. The continuous passage of sodium through the epithelial cell therefore produces, by osmosis, a continuous passage of water in the same direction. Water and sodium continuously diffuse through the basal membrane, which is not a barrier, to the blood capillaries. The water uptake model in Figure 4.8 is also known as the **Curran and McIntosh model** with three environments: the lumen, the interstitium and the serosal side of the epithelium.

It should be taken into account that whenever passage of sodium occurs, it is accompanied by the flow of chloride, which follows it passively to maintain the electroneutrality of the environment.

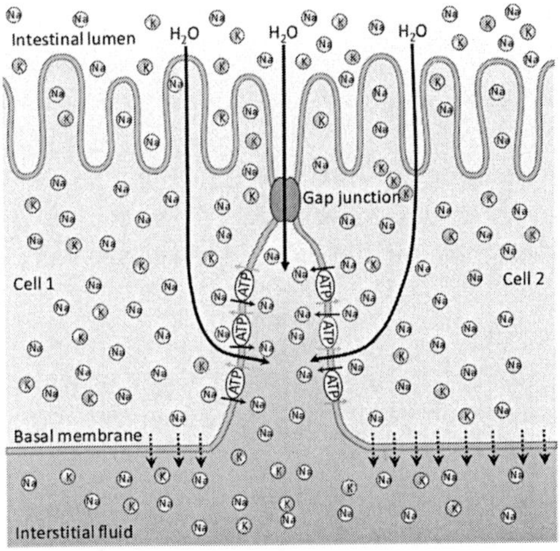

Figure 4.8 The passage of water (large black arrows) through the intestinal epithelium and gap junctions is regulated by the accumulation of sodium ions (Na) in the interstitium between cells due to the activity of a Na^+/K^+ ATPase (ATP) which exchanges sodium (small black arrows) and potassium (K and grey arrows) ions.

Water diffusion and flow occurring between two environments can be constant, establishing a dynamic equilibrium that is essentially stationary, coupled with the continuous cycling of the Na^+/K^+ ATPase. A different situation occurs for the absorption of sodium (Figure 4.12) and glucose (Figure 4.13) which occurs between the lumen and the blood vessels through the epithelial cells or, in some epithelia, the gap junctions (Data sheet 6.1). In this case, there is a continuous supply of molecules, ions and water into the intestinal lumen as a result of digestion and into the lumen of the renal tubules as a result of ultrafiltration (Figure S4.2). Continuous passage of these substances through the epithelium are constantly moved into the bloodstream (Figures 4.8, 4.12, 4.13). The draining action by the blood vessels, which keeps the concentration of sodium, glucose and water low in the deeper areas of the epithelium, allows a constantly high concentration gradient to be established from the intestinal lumen to the blood capillary. The continuous absorption of these molecules and ions produces a **stable imbalance**. Ions, glucose and water absorbed by the intestinal epithelium must be transferred to tissue sites. Here, there is again a stable imbalance between the capillary and the cells that make up tissues. Cell metabolism drains nutrients continuously. The low cellular concentration of solutes induces a flow of substances from the capillaries to the tissues.

A third situation occurs when diffusion and flow are established between two environments in which fluid flows continuously, as happens, for example, with sea water and blood in the gills of fish (Figure 4.15).

4.3 Transporters

Epithelia use three mechanisms for transporting substances: facilitated transport, active transport and counter-current systems. In detail, facilitated transport uses specific membrane proteins for each substance to be transported according to a concentration gradient. Active transport uses chemical energy in the form of ATP and transfers molecules against the gradient. The counter-current system, also according to the gradient, exploits the properties of two parallel flows occurring in opposite directions (section 4.3.5).

4.3.1 Facilitated transport

Some molecules, such as monosaccharides or amino acids, cannot diffuse through the phospholipid bilayer, because they are hydrophilic, nor through the membrane pores, because they are large, with a molecular weight greater than 120 Da. In addition, the cell is strongly selective for imported molecules. Among the integral membrane proteins, there are particular types that are generally referred to as **transporter** or **carrier proteins** (Figure 4.9).

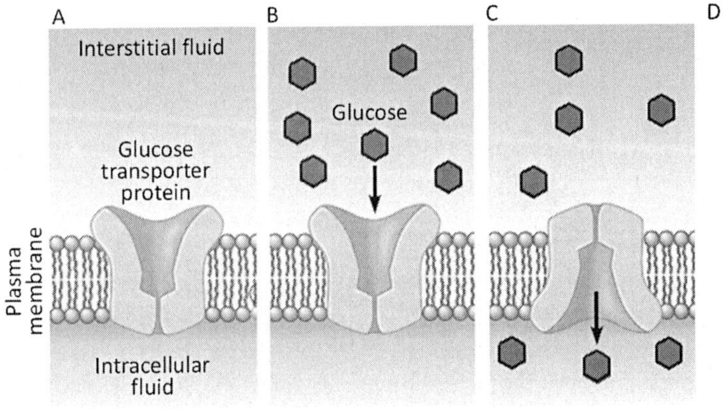

Figure 4.9 The carrier can change conformation randomly by thermal agitation, exposing the binding for glucose outwardly, empty in low glucose concentration (A), ready to bind glucose once it increases in concentration (B). Glucose is more likely to bind to the carrier on the side with higher concentration and more likely to be released towards the side with lower concentration (C).

The carrier has a specific binding site for different molecules, e.g., glucose, and due to thermal agitation, this site is randomly exposed to the outer (Figure 4.9A) or cytoplasmic side of the plasma membrane (Figure 4.9D) by conformational change. If the concentration of glucose in the external solution is high and its concentration in the cytoplasm is low because it is consumed by the cell, the probability is high that glucose will bind to the carrier on the external side (Figure 4.9B). The carrier randomly changes conformation by thermal agitation, independently of the presence of bound glucose and exposes the specific site with glucose towards the cytoplasm. The probability that the glucose molecule will detach from the site is high (Figure 4.9C), since its concentration is low in the cytoplasm. At this point, the carrier randomly returns to its initial conformation by thermal agitation and is ready to bind another molecule (Figure 4.9A).

It can be shown that if substance S is transported by a carrier across a plasma membrane and its cytoplasmic concentration is negligible, J_S flux follows Michaelis-Menten kinetics for enzymatic reactions:

$$J_S = \frac{Jmax \cdot [S]}{[S] + Km}. \tag{4.7}$$

K_m is the dissociation constant between carrier and substance S, has the dimensions of mol/cm^3 and corresponds to the concentration of S that produces a flux equal to $J_{max}/2$. To demonstrate this, it is sufficient to put in Equation 4.7, $[S]=K_m$.

From Equation 4.7, it can be deduced that for very large concentrations of S with respect to K_m, $[S]+K\cong[S]$. Then we have that the flux of S is equal to J_{max}, i.e., facilitated transport is characterized by saturation kinetics due to the finite number of carrier molecules available in the membrane. For small concentrations of S relative to K_m, on the other hand, we have $[S]+K\cong K_m$. The flux becomes approximately linear and proportional to the concentration of S, as in diffusion according to Equation 4.3 (Figure 4.10, S-curve).

A **competitive inhibitor** S_{IC} (a compound that can compete for the same site as S on the carrier), and a **non-competitive inhibitor** S_{IN} can interfere with the flux of S. Assuming a constant concentration of the inhibitors, in the presence of S_{IC}, the same maximum flux J_{maxIC} will be reached for S, but at higher concentrations, i.e., when there is a greater probability of S binding than S_{IC} (Figure 4.10, S_{IC} curve) and K_{mIC} is greater than K_m. Conversely, in the presence of a non-competitive inhibitor S_{IN}, which binds to the carrier at a different site than S, a maximum J_{maxIN} flux will be reached at the same concentration of S. However, this maximum will be lower than the flux without the inhibitor, with a K_{mIN} dissociation constant unchanged with respect to K_m (Figure 4.10, S_{IN} curve), since S_{IN} interferes

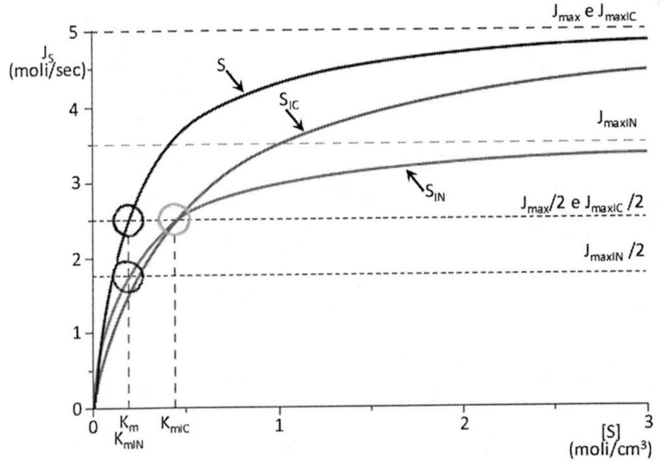

Figure 4.10 Development of J_s flux as a function of S concentration under normal conditions (S curve), in the presence of a competitive inhibitor (S_{IC} curve) and a non-competitive inhibitor (S_{IN} curve). The numerical values of the Cartesian axes are purely indicative.

with conformational changes in the carrier, but not with the binding of S to its site.

4.3.2 Active transport

With sodium and potassium distributed differently, there is a strong concentration gradient for the diffusion of potassium ions outwards and sodium ions inwards. These two fluxes can be measured and occur due to the presence of specific and non-specific pores in the plasma membrane.

Over time, these flows would lower the internal potassium concentration and increase the sodium concentration. It is therefore necessary to generate an opposite flow for sodium and potassium, which would be against the concentration gradient and would therefore need energy in the form of ATP. This task is carried out by integral membrane proteins that are generally called **pumps**, since they generate a flow against a concentration gradient, or ATPases, since they are able to hydrolyze ATP. We will see the example of the Na^+/K^+ ATPase, which serves to maintain concentration gradients for sodium and potassium (section 3.2.7 and Data sheet 4.3).

The Na^+/K^+ ATPase is a highly conserved integral membrane protein expressed in virtually all cells of higher organisms. In resting humans, it is estimated that about 25% of all ATP is hydrolyzed by the Na^+/K^+ ATPase, up to 70% of which occurs in nerve cells. Depending on the cell type, there are between 0.8 and $30 \cdot 10^6$ Na^+/K^+ ATPase molecules on the plasma membrane, distributed across the surface, clustered in specific areas or stored in the cytoplasm, ready to be inserted into the membrane. The Na^+/K^+ ATPase consists of two α subunits of approximately 70 kDa each, which bind sodium and potassium ions and have a phosphorylation site, and two β subunits of approximately 35 kDa each. β subunits are glycoproteins that are essential for the activity of the entire molecule and also appear to facilitate membrane insertion of the α subunits. The Na^+/K^+ ATPase appears to consist of 8-10 transmembrane domains, although its membrane arrangement is still uncertain.

Flow measurements have shown that there is a ratio of 3 to 2 in the expulsion of sodium ions compared to the introduction of potassium ions into the cell. As there is no known structure of the Na^+/K^+ ATPase, the description of the kinetics of the asymmetric double transport is based on models. In the most accepted model, of the infinite number of conformations that Na^+/K^+ ATPase can assume through thermal agitation, two extremely stable conformations are very likely. The first conformation is with three high-affinity binding sites for sodium ions (Sn, Figure 4.11A) and two low-affinity binding sites for potassium ions (Sk, Figure 4.11A) facing the cytoplasm (Figure 4.11A); the second is with the sodium and potassium sites facing inwards (Figure 4.11D). The structure of the molecule itself makes it stable and determines the different affinities for the two ions.

Despite being in a low concentration environment, the sodium ion has a high probability of binding to its site, due to its high affinity, while the potassium ion, present at high concentration, has zero probability (Figure 4.11B). The molecule with the three sites for sodium ions occupied has ATPase action and is auto phosphorylated if sodium is present, hydrolyzing an ATP molecule, present at a high concentration inside the cell and therefore with a high probability of being near the membrane (Figure

4.11C). The Na^+/K^+ ATPase becomes unstable and changes conformation, bringing outwards the sites for sodium, which become low affinity, and those for potassium, which become high affinity (Figure 4.11D).

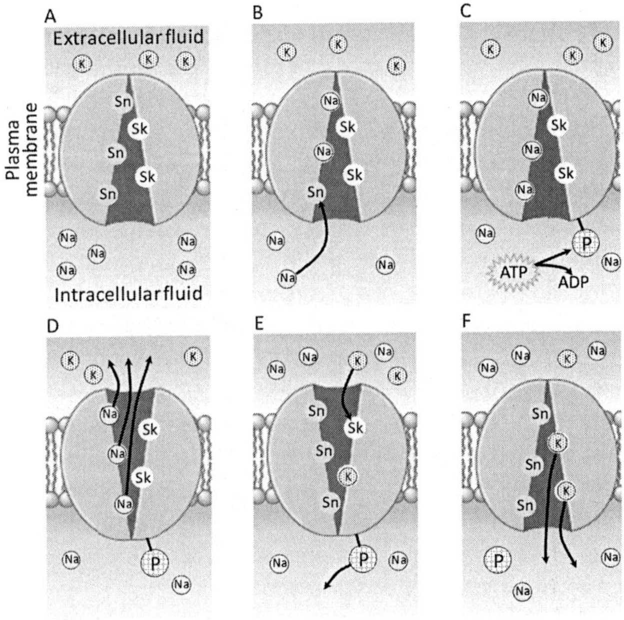

Figure 4.11 Functional model of Na^+/K^+ ATPase.

The sodium, due to the low affinity with respect to its binding site, detaches and goes into solution. At this point, the probability of potassium binding to its own high-affinity sites in the Na^+/K^+ ATPase becomes high (Figure 4.11E). The binding of potassium decreases the affinity for the phosphate ion, so the Na^+/K^+ ATPase dephosphorizes and returns to the energetic condition where the stable conformation is that with the sites facing the cytoplasm with high affinity for sodium and low affinity for potassium (Figure 4.11F). Potassium is released and the molecule is ready to perform another cycle (Figure 4.11A).

The Na^+/K^+ ATPase works continuously as long as there are sodium ions in the cytoplasm and potassium in the external solution, even at low concentrations, and ATP in the cytoplasm at sufficiently high concentrations.

The result of each cycle is the transport of three sodium ions outside the cell and two potassium ions inside at the same time, with the net loss of a positive charge by the cell. The Na^+/K^+ ATPase is thus said to be electrogenic and is the third element (after the charge separation realized by the plasma membrane and the flux of potassium and sodium ions, discussed above) that maintains the resting potential of the cell at negative values.

There are a number of ATPases that operate in similar ways while exchanging other ion pairs, even with different stoichiometric ratios. The most important of these is the **Ca^{2+}-dependent ATPase**, which is present in the membranes of many cell types, such as cardiac myocytes, and on the membranes of the sarcoplasmic reticulum of skeletal muscle cells. It is activated by intracellular calcium in the presence of ATP, has one phosphorylation site and two calcium-binding sites, and probably functions with conformational changes and changes in ion-binding affinity similar to those of the Na^+/K^+ ATPase.

The phenomena described, schematized in Figures 4.3, 4.4, 4.5, 4.6 and 4.7, relate to fluxes or transport of molecules and ions between an environment 1, which may represent the internal environment of the cell, and environment 2, the exterior of that cell. Both environments are characterized by the solutions being essentially stationary; under these conditions a situation of electrochemical and osmotic equilibrium can be achieved between the interior and exterior.

When epithelia are present, such as those lining the intestinal lumen, a different situation arises. In the epithelia, in fact, in addition to being involved in the phenomena of diffusion and transport, numerous flanking cells play a fundamental role in establishing equilibrium between different environments, the thick network of arterial and venous capillaries and sometimes the entire epithelium.

4.3.3 Sodium uptake

An example of specific absorption is that of sodium in Figure 4.12. Sodium introduced, for example, through the diet and present in the intestinal lumen at a high concentration, can enter the cell by diffusion according to a concentration gradient through specific ion channels present only on the apical membrane (long upper arrow) if its intracellular concentration is kept low. This is done by a Na^+/K^+ ATPase, present only on

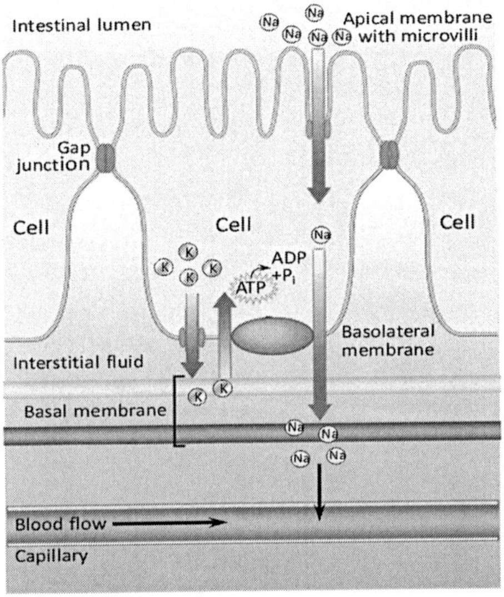

Figure 4.12 Schematic representation of the sodium absorption pattern in the intestine. See text for explanation.

the basolateral membrane, which releases sodium (long lower arrow) by exchanging it for potassium (short upper arrow).

Potassium leaves the cell by gradient diffusion across the basolateral membrane using specific ion channels present only on that side of the membrane (short down arrow) and sodium enters the bloodstream by gradient diffusion across the basal membrane (black down arrow). This complex mechanism keeps the potassium concentration constantly high and the sodium concentration constantly low in the cell, but produces a continuous gradient of sodium flow from the lumen to the circulatory system.

4.3.4 Glucose uptake

Glucose is not able to cross the plasma membrane freely. A probable functional model of glucose uptake at the level of the intestinal epithelium is shown in Figure 4.13; a similar situation is also present in the proximal convoluted tubule of the renal nephron for reabsorption of glucose from the ultrafiltrate. The cells that form the intestinal epithelium are polarized: they have an apical membrane facing the intestinal lumen and a basolateral membrane facing the basement membrane and blood capillaries (Figure 4.13).

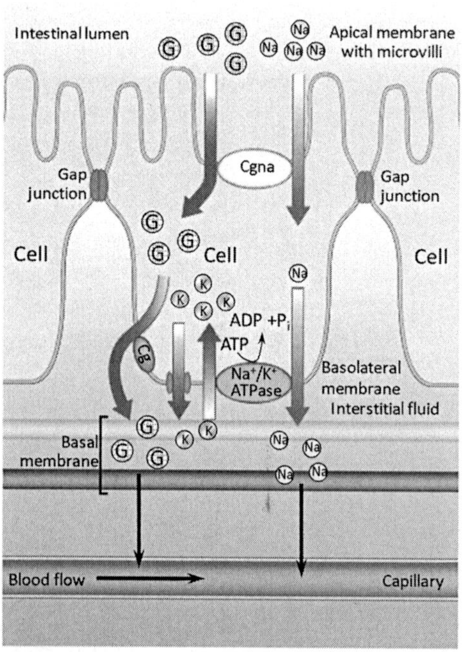

Figure 4.13 Schematic representation of the glucose absorption pattern in the intestine. See text for explanation.

In the apical membrane, there are carriers, that is, integral transporter proteins (Cgna) with a high-affinity binding site for glucose (G) and one for sodium (Na). The carrier continually changes conformation due to thermal agitation, but has a very high probability of exposing the two binding sites to the lumen. In the basolateral membrane, there are Na^+/K^+ ATPases and specific carriers for glucose (Cg).

If there is a high concentration of glucose in the lumen as a result of ingestion and digestion of sugars, there is a high probability of it binding to the carrier (Cgna in Figure 4.13). In the presence of carrier-bound glucose, the probability is also high that a sodium ion will bind to its specific site and induce a conformational change in the carrier that turns the two sites towards the interior of the cell. Its low concentration makes sodium release highly likely and induces glucose release.

Straddling the apical membrane of the epithelial cell, there is a strong electrochemical gradient for sodium produced by the Na^+/K^+ ATPase, which makes the concentration of intracellular sodium particularly low by exchanging a sodium ion (bottom long arrow) for a potassium ion (top short arrow). This results in a gradient-dependent flow of sodium (upper straight arrow), which also induces a gradient-dependent flow of glucose (upper curved arrow) from the lumen into the cell. Potassium entering the cell through the Na^+/K^+ ATPase and glucose entering the cell through the carrier pass through the basement membrane via specific potassium channels (short downward arrow) and another carrier (lower curved arrow), respectively, and then diffuse by concentration gradient to the capillaries. In the capillaries, continuous drainage makes concentrations low, and potassium and glucose enter the blood stream (lower black arrows).

There is sodium and glucose co-transport and secondary active glucose transport also against the gradient. The energy required for glucose flow is provided by the sodium gradient created by the Na^+/K^+ ATPase (Data sheet 4.3) present on the basolateral membrane, for gradient-facilitated transport (section 4.3.1) of glucose.The overall result is that the continuous entry of sodium according to the gradient is associated with a continuous entry of glucose against the gradient, regulated only by the glucose concentration in the lumen and the difference in sodium concentration between the lumen and the epithelial cell interior.

4.3.5 Counter-current transport

Consider the situation in which two tubes, T1 and T2, carry an aqueous solution of a salt (Figure 4.14). The solution of tube T1 reaches a constant, high concentration of 200 mM (Figures 4.14A and 4.14B, sector 1) because it comes from a large-volume reservoir, while the solution of tube T2 reaches a constant, low concentration of 80 mM (Figures 4.14A, sector 1; Figure 4.14B, sector 6) because it comes from a tissue that metabolizes the solute. The epithelia of T1 and T2 are semi-permeable, i.e., permeable to the solute but not to water.

If the two solutions flow in the same direction (Figure 4.14A), the solute diffuses from T1 to T2 and the process goes on as the two solutions proceed in their respective tubes. The high concentration gradient induces a large solute flow and the rapid attainment, in the example of Figure 4.14A already in sector 2, of a **stable equilibrium** of concentration between the solution in T1 and the solution in T2, without exploiting the whole length of the tubes for the exchanges. These exchanges are in limited quantity (the concentration of the solute in B increases by 60 mM), such that the solution in T1 exits the tube with a high concentration of solute (140 mM, Figure 4.14A, sector 6).

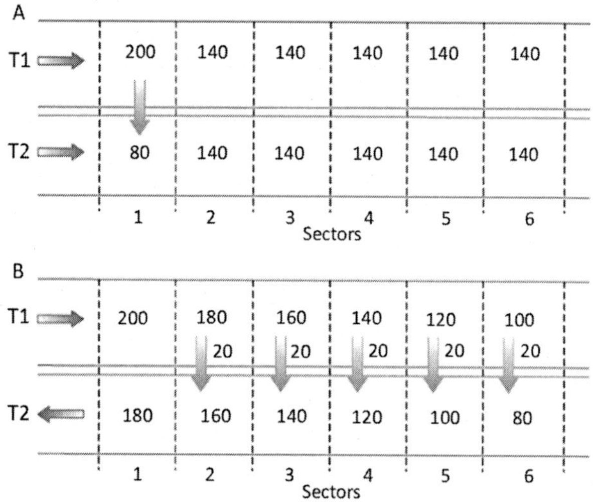

Figure 4.14 If the solutions in two parallel tubes flow in the same direction, there is a limited exchange of substances (A). If the two solutions flow in the opposite direction, the exchange of substances is much more effective (B). The numbers represent the solute concentrations in mol/L in the tubes and the quantities exchanged in moles at the downward arrows, which are purely illustrative.

If, on the other hand, the two solutions flow in the opposite direction (Figure 4.14B), the solution in T2 acquires 100 mM from the solution in T1 using the whole length of the tubes for the exchanges and the solution in T1 exits the tube with a low solute concentration (100 mM, Figure 4.14B, sector 6). The concentration of solute in T1, in fact, decreases steadily from sector 1 to sector 6; the concentration in T2 increases steadily from sector 6 to sector 1. This is because the solute passes in all sectors from T1 to T2 due to the establishment, along the whole length of the tubes, of a **stable disequilibrium** with a small constant concentration gradient at the level of each sector (in the example of Figure 4.14B of 20 mM), but sufficient to produce the necessary fluxes of solute from T1 to T2. There is a functional combination of solution flow rate and solute flow rate. For example, at T1 in sector 4, a solution arrives in which the solute concentration has changed from 160 mM to 140 mM due to solute flux from T1 to T2, and at T2, a solution arrives with a solute concentration of 120 mM because, by moving from sector 5 to sector 4, it has acquired solute from T1. Extending this mechanism to all sectors, and remembering that concentration changes along the tubes are continuous, it becomes clear how small single gradients can produce a large solute flux.

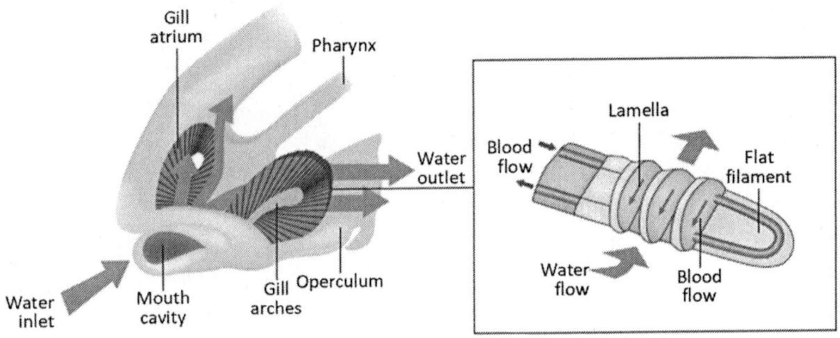

Figure 4.15 Diagram of the teleost gill apparatus (A) and structure of the gill arches (B).

Counter-current transport, a clear example of how structure and function are closely related, is present, as mentioned, in the gills of teleost fish. The gills of other species of fish and those of amphibian larvae are similar. Teleost gills are formed by 8 gill arches from which emerge, perpendicular to their longitudinal axis, two rows of flat filaments. These filaments in turn form, by eversion of the epithelium, a large number of lamellae arranged perpendicular to the surface of the flat filaments and transversal to their major axis (Figure 4.15B). Seawater with high pO_2, partial pressure of dissolved oxygen, enters through the mouth cavity, flows forcefully along the lamellae from outside to inside (Figure 4.15B) and exits, after passing through the pharynx, the branchial arches and the branchial atria, through the opercula (Figure 4.15A). The blood, with low pO_2, flows inside the gills in the opposite direction. The situation described is similar to that in Figure 4.14B, substituting the concentrations for the partial pressures of oxygen: the flow of seawater corresponds to that of the solution in tube T1, although in this case the pO_2 can be considered almost constant, given the large volume of solution and the high diffusion coefficient of gases in water, while the flow of blood corresponds to the solution in tube T2. The mechanism just described is so efficient that seawater can yield up to 80% of dissolved oxygen to the blood.

4.3.6 Counter-current exchange

A system exists that makes it possible not to disperse an excessive quantity of solute into an environment in which it is present at a low concentration and not to absorb an excessive quantity from an environment in which it is present at a high concentration. Such a system occurs when the mechanism of two solutions that flow in counter-current takes place in a single tube folded like a loop.

Consider if a solution at high concentration, for example 200 mM, enters a straight tube and arrives unaltered at the end of the tube where, in the external environment, the same solute is present at low concentration, for example 50 mM. The solution would lose solute, which would diffuse in great quantity towards the outside because of the high concentration gradient (in this example, 150 mM).

Consider if instead we have the situation of Figure 4.16, in which the tube T is bent as a loop and forms the tube T', which is placed parallel to the first. As the solution in T flows from sector 1 to sector 7, it yields a part of the solute, which is taken up by the solution in T', with a mechanism analogous that in Figure 4.14, and arrives in sector 7 at low concentration,

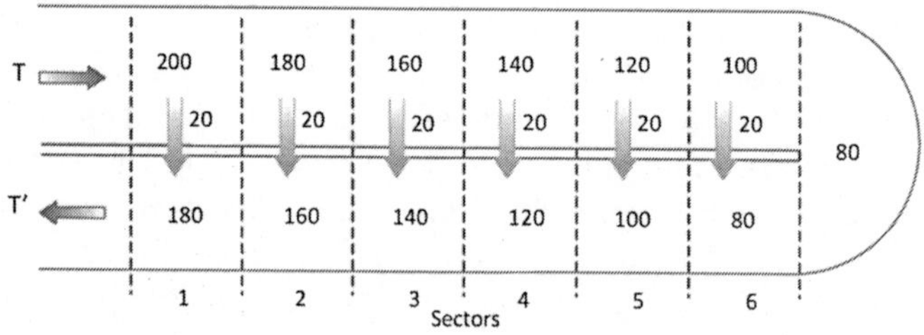

Figure 4.16 In a looped tube, it is possible to limit the loss of solute to the environment by recovering it from the inlet to the outlet tube. The numbers, representing the solute concentrations in mol/L in the tubes and the quantities exchanged in moles at the arrows, are to be considered examples only.

therefore with a small concentration gradient towards the outside. This gradient generates a small flux and a reduced loss of solute. Also, in this case a **stable disequilibrium** is established, because while the solutions flow continuously, a small concentration gradient is generated within each sector, which in turn generates the flow of solute from T to T'. Of course, if the aim of the counter-current exchange system is *not* to absorb an excessive quantity of solute from an environment where it is present at high concentration, in the tube T (Figure 4.16) the concentrations will increase from sector 1 to sector 7; in the tube T' they will decrease from sector 7 to sector 1; the flows will be from T' to T.

The counter-current exchange can involve solutes but also heat. For example, for feeding, river birds remain immersed for long periods with their long legs in water at low temperature. They have the arteries and veins of the legs arranged like the tubes T and T' in Figure 4.16: the arterial blood arrives at sector 1 of T at body temperature but it drops at the end of the leg at a lower temperature, reducing the loss of heat to the external environment due to a reduction in the gradient.

4.3.7 Counter-current multiplication

If specific ATPases are present in a loop system that actively transports the solute from the solution in T' to the solution in T, **counter-current multiplication** occurs (Figure 4.17). This system is used to create a hypertonic environment by increasing the concentration of a solute with

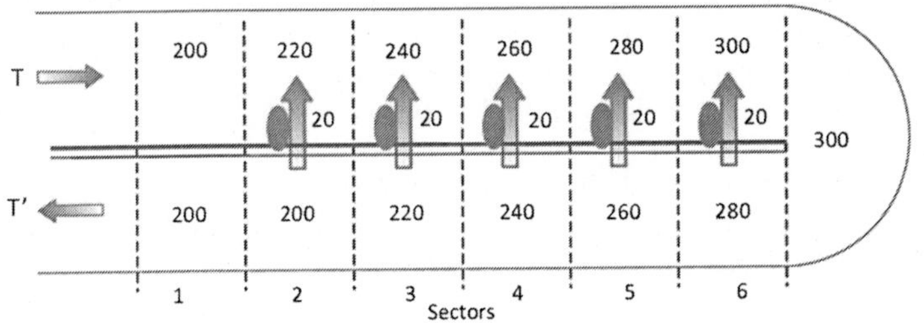

Figure 4.17 A solute can be concentrated in a looped tube if, at minimum in the innermost section, there is active transport. The numbers, representing the solute concentrations in mol/L in the tubes and the quantities exchanged in moles at the arrows, are purely illustrative.

low ATP consumption, while active transport works against small concentration gradients (20 mM in the example of Figure 4.17) within each sector of the loop system.

This is possible because the single transport effect within each sector is multiplied by the fact that the solution in T, which has been enriched in solute by flowing from sector 1 to sector 5, in turn becomes a supplier of solute for the solution that eventually arrives in sector 1 of T (by flowing from sector 5 to sector 1 of T').

A simple system that uses a counter-current multiplier is the one allows teleosts to increase the volume of the swim bladder, to decrease the specific weight of the animal and allow it to move towards the surface of the sea (Figure 4.18).

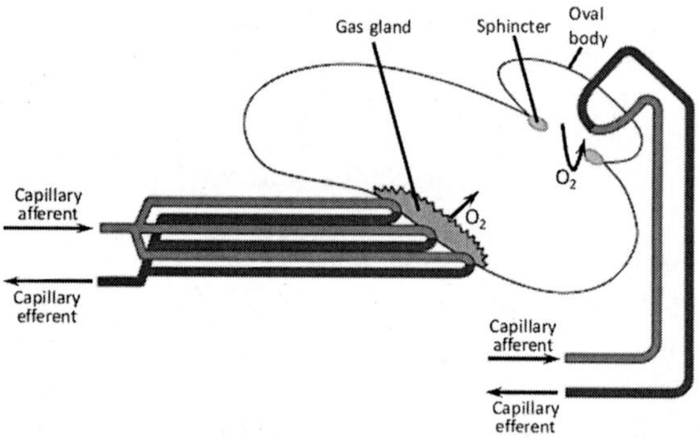

Figure 4.18 Increased metabolism in the afferent capillaries causes the swim bladder to fill with O_2, which increases its volume and lowers the specific gravity of the animal. The degree of sphincter openness regulates the flow of O_2 to the oval body from where it passes, by partial pressure gradient, to the circulatory system.

Each afferent capillary of the very dense system (about 100,000 vessels) that impacts the swim bladder bends into a loop and becomes an efferent capillary arranged in parallel to the first. Within the gas gland, located near the swim bladder and irrigated by the dense network of capillaries, the increase in metabolism and consequent ATP consumption produce a large amount of carbon dioxide and lactic acid that lower the pH of the blood: the consequence, similar to the increase in solute concentration in Figure 4.18, is a progressive increase of the pO_2 due to the release of oxygen by hemoglobin (a phenomenon called the **Bohr effect**).

Oxygen passes, by partial pressure gradient, into the swim bladder with an increase in volume and decrease in specific gravity.

Another use of the mechanism described is Henle's loop of the mammalian kidney nephron (Data sheet 4.4). The counter-current multiplier system, associated with a counter-current exchanger system used by the vasa recta running parallel to Henle's loop, is used to produce and maintain a high osmolarity in the renal interstitium, which can be as high as 2000 mOsm in the deep medullary zone, compared with 300 mOsm in the cortical zone. This high osmolarity is used to produce hypertonic urine by osmotic reabsorption of water, which passes from the collecting duct to the renal interstitium and vasa recta.

4.4 Ionic permeabilities as integral membrane proteins

In order to utilize the potential energy from the different distribution of ions on either side of the plasma membrane, integral proteins are required. Integral proteins consist of a hydrophobic portion that allows them to be inserted into the membrane and a hydrophilic portion that can form an aqueous pore that crosses the membrane throughout its thickness. Although water molecules are overall uncharged, they are **dipoles** due to the presence of a negative charge fraction towards the oxygen atom and a positive charge fraction towards the hydrogen atoms. Therefore, both a positive ion (Figure 4.19A) and a negative ion (Figure 4.19B) in solution are surrounded by **hydration water** molecules in numbers inversely proportional to their diameters. An ion in solution can cross the membrane only through the aqueous pores formed by the ion channels and can do so only when hydrated, having a very low probability of crossing the hydrophobic phospholipid layer. These integral membrane proteins are the ion channels which, depending on how they function, can be passive or active.

4.4.1 Aqueous pores as ion channels

These channels are generally open, always allowing ions to pass according to a gradient and are generally present in the cell in limited numbers. Their molecular structure is little-known because they have been little studied. They are inserted into the membrane in such a way as to form an aqueous pore, probably with a diameter and a distribution of charges that select, although not precisely, the ion passing through it. The flux of potassium and sodium ions through passive channels is small but constant and tends to reduce the potential energy due to the asymmetric distribution of the ions between the inside and outside. The Na^+/K^+ ATPase, which functions continuously in the presence of ATP and sodium in the cytoplasm and potassium in the external solution, maintains the electrochemical gradients of potassium and sodium that would otherwise be reduced by the passive flows described.

4.4.2 Ion channels

The maintenance of electrochemical gradients is essential to generate the electrical phenomena which, in muscle cells, activate the process of contraction and, in nerve cells, allow the generation and transmission of messages between different regions of an organism and the processing of the messages received in the central nervous system. These phenomena are temporally confined to a range of a few milliseconds to tens of milliseconds. Macroscopic mechanisms at the cellular level at this order of magnitude require individual elements that activate ion flows across the plasma membrane with much faster kinetics. These elements are the **ion channels**.

Ion channels are aqueous pores that have a high probability of taking on two conformations from among the infinite number of possibilities: they can be closed or open. Since these are conformational changes of amino acid residues that are part of the aqueous pore, the time to transition between one state and the other is practically null and the probability of the transition depends on changes in the potential difference across the membrane, the time the protein channel remains in that state, environmental factors and specific modulating molecules. All of these factors can alter the probability that the channel protein is in the open state, and therefore permissive to ion passage, or is in the closed state (Chapter 5).

Ion channels are widely used in the mechanisms of excitable cells. Sodium, calcium, potassium and chloride channels produce, control, modulate and re-establish the initial conditions of excitation-related phenomena in nerve and muscle cells. They are, for the most part, **voltage-dependent** and **time-dependent**. This means that appropriate changes in the membrane potential and the time the channel is active cause conformational changes in the channel protein, such that the probability of being in one state rather than the other is altered.

Functional, structural and molecular experimental approaches have enabled the intrinsic properties of ion channels to be studied in depth. There are **families** of sodium, potassium, calcium and chloride channels (Figure 4.20), composed of numerous elements expressed in different organisms or in different tissues of the same organism. Sometimes channels within a family have different functions, so that each channel must be distinguished with numbers, for example K_{IR} 1.1 or K_{IR} 2.1; the abbreviation identifies the type of channel, the first digit identifies the member of the family and the second the isoform of this member. The sodium, potassium and calcium channel families are encoded by three types of genes that are thought to have evolved from a single ancestral gene, probably a potassium channel with two transmembrane domains (Figure 4.20D); for the chloride channel family, the origin of the encoding genes is not yet clear.

The molecular structure of voltage-dependent and time-dependent ion channels is now known. The basic configuration is that of the potassium channel. This is a homotetramer consisting of four transmembrane domains with six elements each and with hydrophilic amino acid sequences linking the various elements (K_V, Figure 4.20D). In the case of potassium channels, the gene encodes for a protein that forms one of the four domains. The sodium (Data sheet 4.5) and calcium channels have the same quaternary structure, but unlike the potassium channel, they are heterotetramers. The gene for these channels encodes a protein whose secondary structure comprises four transmembrane domains.

The chloride channel differs from those seen so far. Recently, several models of anion channels have been presented, although there is as yet no definitively agreed structure.

Once activated, sodium channels usually cause an inward current with a depolarizing function. Calcium channels behave in the same way. The latter, in addition to acting on the value of the membrane potential, perform the fundamental function of supplying calcium ions that will be used inside the cell as second messengers (section 3.3). With regard to repolarization and hyperpolarization, the family of potassium channels plays an important role. The various potassium channels are used to re-establish rest conditions in excitability processes and modulation of various bioelectric phenomena.

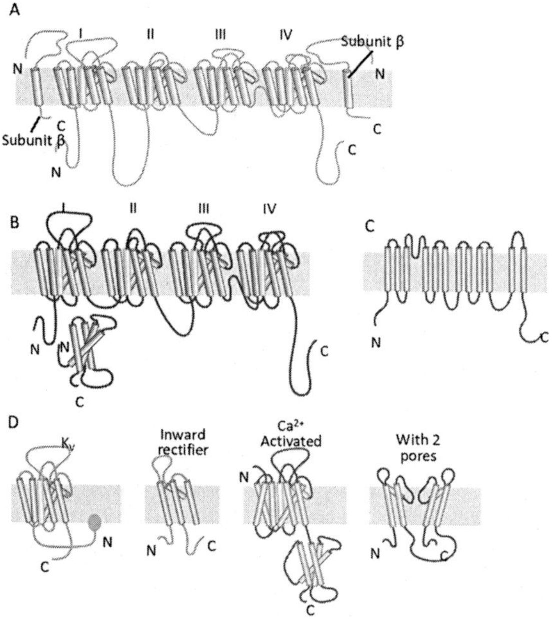

Figure 4.20 Schematic representation of a sodium channel (A), a calcium channel (B), a chloride channel (C) and some members of the potassium channel family (D).

Finally, chloride channels play a key role in modulating nerve excitability and are very important in maintaining the resting potential of skeletal muscle cells.

There are also ion channels in non-excitable cells, mainly calcium, potassium and chloride channels. In these cells, there are also non-specific ion channels, often with mixed selectivity for sodium and potassium. The function of these channels does not depend on voltage or time, but on other factors such as mechanical stress on membranes, cytoplasmic pH, oxidative stress or intracellular calcium concentration. Voltage- and time-independent channels are, for example, the sodium and chloride channels of olfactory receptors (Figure 8.2), the potassium and sodium channels of gustatory receptors (Figure 8.4), the sodium channels of the Pacini corpuscle (Figure 8.7), the potassium channels of inner ear Corti cells (Figure 8.13), and the sodium channels of retinal rods and cones (Figure 8.24).

A very important role in non-excitable tissues is played by chloride channels. Chloride permeabilities control cell volume, certain mechanisms involved in cell movement and transepithelial water transport. Point mutations in the protein that forms certain chloride channels are responsible for serious diseases that are still incurable today. A clear example is the ΔPhe508-CFTR mutation causing, among several other mutations, Cystic Fibrosis disease (Imberti et al. doi: 10.1186/s12931-018-0901-1).

Bibliography

Hille B. Ionic Selectivity, Saturation, and Block in Sodium Channels. A Four-Barrier Model. J Gen Physiol, (1975) 66: 535-560.

Data sheet 4.1

Fick's Law

In Figure 4.2A, we have a cylindrical container with a small diameter compared to its length and we consider only the dye, which can only move along the x-axis. The dye molecules are deposited at the point near x=0 and at the same concentration through the whole area. Due to the phenomenon of diffusion, the dye will move until it is homogeneously distributed throughout the cylinder (Figure 4.2B). The small number of moles dn diffusing by thermal agitation in the time unit dt will be proportional to:

a) the area A of the section of the cylinder at any point x, since every section x of the cylinder along dx has equal area and homogeneous concentration of the dye;

(b) the concentration gradient dc/dx, which is given by the difference between the concentration c_{x+dx} at point x+dx and the concentration c_x at point x of the cylinder considered;

(c) the absolute temperature T, which determines the amount of thermal agitation, making diffusion of the dye possible;

d) the gas constant R, which is 8.314 J/K · mol and represents the work that a mole of substance can do for each degree of temperature due to thermal agitation;

and inversely proportional to:

e) the coefficient of friction f, which according to Stokes' law is 6πηr, proportional to the viscosity η and the radius r, two properties that tend to oppose diffusion;

f) Avogadro's number N, which is $6.022 \cdot 10^{23}$ molecules per gram mole of substance and is the component that tends to limit the amount of displacement caused by thermal agitation due to collisions between molecules:

$$\frac{dn}{dt} = -A\frac{RT}{Nf} \cdot \frac{dc}{dx}. \qquad (S4.1)$$

In the cgs system, dn/dt is expressed in mol/s, A in cm^2 and dc/dx as $(mol/cm^3) \cdot (1/cm)$. Since simple diffusion occurs, as we have seen, from an area of higher concentration to an area of lower concentration, the difference $dc = c_{x+dx} - c_x$ is less than zero; but dn/dx cannot be negative since in the presence of a concentration gradient, and therefore of diffusion, the number of moles at time t+dt is greater than the number of moles at time t. Therefore, dn = n_{t+dt} n_t is greater than 0 and there is therefore a negative sign in the second part of the equation. If the temperature is constant, we can write:

$$D = \frac{RT}{Nf} . \qquad (S4.2)$$

D is the diffusion coefficient, which is proportional to the temperature, an increase in which increases the thermal agitation and the probability of a molecule diffusing, and inversely proportional to the friction coefficient f, which has been shown to be directly proportional to the viscosity and radius of the molecule. This means that a small molecule diffuses better than a molecule with lower viscosity.

Substituting Equation S4.2 into Equation S4.1, we obtain **Fick's law** of simple diffusion of an anelectrolyte, i.e., a molecule with no charge:

$$\Phi = \frac{dn}{dt} = -AD \cdot \frac{dc}{dx} . \qquad (S4.\ 3)$$

dn/dt, variation of the number of moles in the unit time, is the flow per unit time through area A.

In the cgs system, dn/dt is expressed in mol/s, concentration dc in mol/cm^3, area A in cm^2 and distance dx in cm. Therefore, considering the dimensions in equation S4.3 we have:

$$\frac{Moli}{sec} = cm^2 \cdot D \cdot \frac{Moli}{cm^3} \cdot \frac{1}{cm} \qquad (S4.4)$$

Simplifying and highlighting D we have:

$$D = \frac{cm^2}{sec} \qquad (S4.5)$$

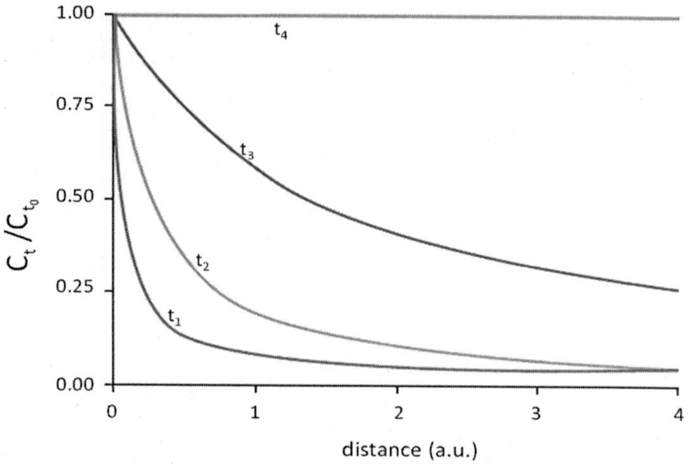

Figure S4.1 The dye concentration profile at times t_1, t_2, t_3, t_4.

From Equation S4.3, we deduce that the flux is constant if the concentration gradient is constant. However, in time dt, dn molecules are moved from point x to point x+dx; at x and x+dx, the concentrations will change and thus the gradient between the two points.

The concentration trend as a function of x at times t_1, t_2, t_3 and t_4 is shown in Figure S4.1. It can be seen that at a certain distance x, the concentration profile is different at different times because the molecules move towards the areas of lower concentration, causing it to increase. The flow proceeds until there is a homogeneous distribution of molecules, the concentration gradient becomes zero and the flow ceases according to Equation S4.3.

Data sheet 4.2

Diffusion through a membrane

Consider environment 1 with substance S at concentration C_{S1}, and environment 2 with S at concentration C_{S2}. The two environments are separated by a membrane of non-infinite thickness X and area A_m (Figure 4.3). The substance S is at a constant concentration in each of the two environments. If we assume that the concentration gradient within the membrane is constant, we can apply Fick's law (Equation 4.2) to the diffusion of n moles of S through the membrane in time t, so that dC/dX can be replaced by DC/DX. The concentration gradient ΔC that generates the diffusion is that given by the difference between the concentration of S immediately inside the membrane towards environment 1 and towards environment 2, taking into account the partition coefficient r, given by the ratio between the concentration of S in the membrane and that of S in water:

$$\Delta C = r \cdot C_{S2} - r \cdot C_{S1} = -r \cdot (C_{S1} - C_{S2}) \quad (S4.6)$$

Equation 4.2 becomes:

$$\Phi_{nS} = \frac{\Delta n}{\Delta t} = A_m \cdot D_m \cdot r \cdot \frac{(C_{S1} - C_{S2})}{\Delta x} \quad (S4.7)$$

where D_m is the diffusion coefficient in the membrane.

Considering the flux per unit area A_m

$$J_{nS} = \frac{\Phi_{nS}}{A_m} \quad (S4.8)$$

and assuming

$$P_m = \frac{D_m \cdot r}{\Delta x} \quad (S4.9)$$

we obtain, from equations S4.7, S4.8 and S4.9, the equation of flow of substance S through a membrane:

$$\frac{\Phi_{nS}}{A} = J_{nS} = P_m(C_{S1} - C_{S2}), \qquad \text{(S4.10)}$$

where P_m is the permeability coefficient of S in the membrane. From equation S4.9, we see that it is proportional to the membrane diffusion coefficient D_m and the partition coefficient r, and is inversely proportional to the membrane thickness ΔX. D_m has dimension cm^2/s (equation S4.5), r is dimensionless, since it is a ratio of concentrations and ΔX has dimension cm. Therefore, P_m has dimension

$$\frac{cm^2}{sec} \cdot \frac{1}{cm} = \frac{cm}{sec} \qquad \text{(S4.11)}$$

and represents the linear velocity with which S crosses the membrane.

Data sheet 4.3

Na^+/K^+ ATPase: functions and localization

1) Na^+/K^+ ATPase is distributed more or less uniformly in the plasma membrane, as is the case, for example, in nerve or muscle cells. It is essential for maintaining low sodium and high potassium concentrations intracellularly (Table 3.2) but is unable to regulate extracellular concentrations because of its large volume relative to that of the cell. This latter function is performed by the circulatory and lymphatic systems, either directly or by means of particular structures, such as the glia in the nervous system or the pigmented epithelium in the retina.
2) The Na^+/K^+ ATPase:
   (a) contributes to the formation and maintenance of resting potential, as it is electrogenic;
   b) prevents the accumulation of ions inside the cell that could occur either due to the passive flows that contribute to the formation of the resting potential or, as will be seen later, to the action potential. Intracellular accumulation of ions would lead to swelling of the cell by osmosis and destruction of the cell;
   c) maintains the electrochemical potential of sodium and potassium constant, making the flow of ions possible for the action potential.
3) In epithelia, Na^+/K^+ ATPase is generally located in specific areas of the cell.
   (a) Na^+/K^+ ATPase is localized at the basolateral membrane, it serves to keep the intracellular concentration of sodium low, so that it can enter the cell through the apical membrane according to a gradient (section 4.3.3). The sodium gradient also provides the energy required for the co-transport of glucose (section 4.3.4), other sugars and amino acids in the presence of specific transporter proteins on both the apical and basolateral membranes.
   b) When the Na^+/K^+ ATPase is located on the membrane areas delimiting the interstices, it is essential for water reabsorption in the epithelia (section 4.2.5).

4) Na+/K+ ATPase: properties

(a) For the Na^+/K^+ ATPase to function, simultaneously, sodium must be present inside the cell and potassium outside the cell, the latter even at low (but not zero) concentration.

b) ATP must be present, together with the activator ion magnesium, near the cytoplasmic side of the membrane and must be at a sufficiently high concentration.

c) Sodium and potassium transport is proportional to the intracellular sodium concentration and has saturation kinetics due to the finite number of Na^+/K^+ ATPase molecules present in the membrane. The semi-saturating concentration for sodium is 20 mM and for potassium 2 mM: this explains why the transport of sodium and potassium is proportional to the intracellular concentration of sodium.

(d) Its operation is strongly influenced by temperature ($Q_{10} \approx 3$) and pH (the optimum pH is 7.4) as chemical reactions take place.

(e) If the electrochemical gradients of sodium inwards or potassium outwards become too high, e.g., if there is no potassium in the extracellular solution, there is a flow of potassium outwards through the Na^+/K^+ ATPase according to its own gradient. This flow causes phosphorylation of the Na^+/K^+ ATPase and a flow of sodium inwards according to its own gradient, which displaces the phosphate ion bound to the Na^+/K^+ ATPase onto an ADP molecule with formation of ATP.

Data sheet 4.4

The nephron

The nephron shown in Figure S4.2 is the functional unit of the kidney. It represents one of the thousands of nephrons present in the mammalian kidney. In the renal cortex, there are the afferent arteriole, the glomerulus, part of the vasa recta and efferent arteriole, Bowman's capsule, the proximal convoluted tubule, part of the distal convoluted tubule and part of the collector. The vasa recta, loop of Henle, distal convoluted tubule and collector penetrate more or less deeply into the renal medulla; they are practically straight, parallel and close to each other to facilitate exchanges with the interstitial tissue.

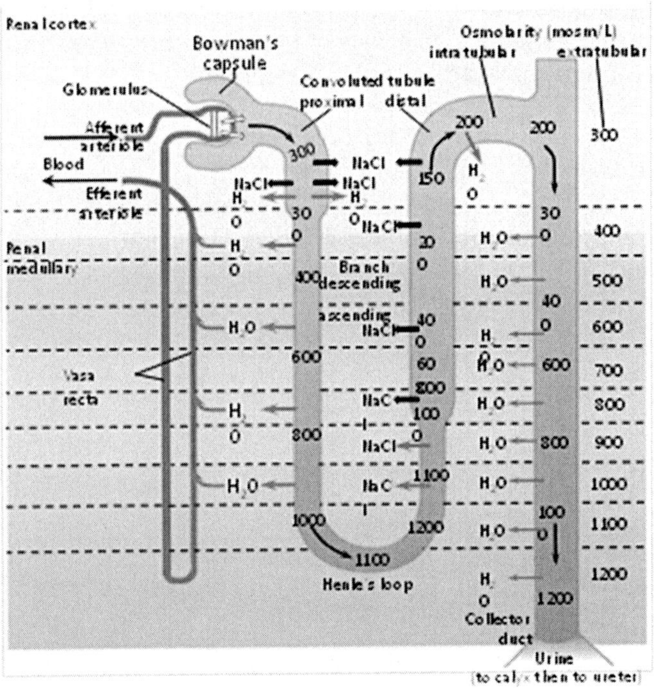

Figure S4.2 Schematic representation of a nephron, the functional unit of the

The proximal convoluted tubule and the distal convoluted tubule are formed by an epithelium often similar to that in Figure 4.13, as the site of active transport and co-transport (section 4.3).
The afferent arteriole carries the arterial blood into the glomerulus, from where the non-corpuscular components of the blood pass, according to a concentration gradient, by ultrafiltration through a size filter (white arrows) to Bowman's capsule. An isotonic ultrafiltrate is formed with plasma, with an osmolarity of 300 mOsm/L, which travels throughout the nephron to the collecting duct and goblet.
In the proximal convoluted tubule, there is reabsorption of glucose by co-transport and secondary active transport (section 4.3.4), of amino acids and other solutes by specific mechanisms (section 4.3).

The counter-current multiplier system (section 4.3.7), consisting of the proximal convoluted tubule, loop of Henle, distal convoluted tubule and vasa recta, uses the active sodium transport of the Na^+/K^+ ATPase (thick black arrows) and the gradient diffusion of water and sodium (thick grey arrows) to create an increasingly hypertonic environment in the nephron interstitium. Water passes, by osmotic gradient, from the collector to the interstitial tissue (thin grey arrows). Urine with a final osmolarity of up to 1200 mOsm/L is formed in the collector from an initial osmolarity of 300 mOsm/L. Water and sodium are finally reabsorbed by the vasa recta and returned to the circulation.

Data sheet 4.5

The sodium channel

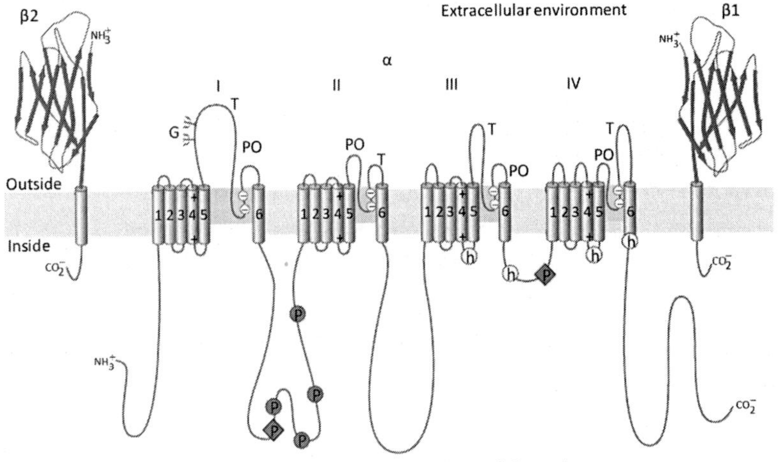

Figure S4.3 Schematic representation of the secondary structure of the α, β1 and β2 subunits of the sodium channel. The 4 domains forming the channel (I, II, III, IV), the 6 segments of each domain S1, S2, S3, S4, S5, S6 (1, 2, 3, 4, 5, 6), the phosphorylation sites (P), the glycosylation sites (G), the pore forming loops (PO), the inactivation loop (h) and the tetrodotoxin binding sites (T) are highlighted.

The sodium channel is a molecular complex weighing about 320 kDa, consisting of an α subunit, a β1 subunit and a β2 subunit (Figure S4.3). The α subunit, weighing about 260 kDa and consisting of about 1800 amino acids, forms the channel. The β1 subunit, weighing about 36 kDa, is weakly bound to the α subunit. The β2 subunit, weighing about 32 kDa, is bound to the α subunit by disulphide bridges. Together, β1 and β2 function to modulate both the expression and the activity of the channel itself.

Biochemistry and molecular biology studies, which have also made it possible to determine the amino acid sequence, have shown that the α subunit consists of a single polypeptide chain made up of four homologous domains of approximately 250-300 amino acids each, called I, II, III and IV. In addition, the use of specific antibodies made it possible to demonstrate that the N-terminal, C-terminal and phosphorylation sites are intracellular, while the glycosylation sites (important, as mentioned, for the protein's membrane insertion) and the binding sites for specific toxins are extracellular (Figure S4.3).

A hydrophobicity profile highlights the parts of the protein that are more likely to be inserted into the phospholipid bilayer, because hydrophobic properties prevail among its component amino acid residues, and the parts that are more likely to be intracellular or extracellular, because they are predominantly hydrophilic. The hydrophobicity profile of the sodium channel has shown that each domain consists of six predominantly hydrophobic transmembrane segments, denoted S1, S2, S3, S4, S5 and S6, joined together by predominantly hydrophilic protein tracts, the intracellular loops and the extracellular loops.

The four domains are joined together by other intracellular loops: the one between domains I and II has numerous phosphorylation sites, which are important for regulating ion flow, while the one between domains III and IV, called the h loop, is responsible for channel inactivation.

The tertiary and quaternary structure model of the α-subunits of the sodium channel considered most valid is shown in Figure S4.4.

The four loops between the S5 and S6 segments of each domain form the pore, which determines both conductance and ion selectivity. When a hydrated sodium ion diffuses into the aqueous pore, it reversibly binds with weak electrostatic bonds to an active site, consisting of an amino acid residue with a negatively charged fraction. The active site, together with one of the hydration water molecules interacting with another negatively charged residue, stabilizes the charge of the sodium, which would otherwise be in an unfavourable energy state due to dehydration. After less than a microsecond, electrostatic and diffusion forces cause the ion to detach from the active site: the ion rehydrates and diffuses along the aqueous pore to the exit.

The conformation of the active sites, their relative position and the size of the hydrated and dehydrated ion, which are typical of each type of channel, determine its conductance and selectivity, i.e., they make a single ion species highly likely to permeate.

The four S4 segments, placed immediately behind S5 and S6 (Figure S4.5A), are the voltage sensors. This property derives from the fact that the stretch of protein that forms S4 has six to eight positive charges, since

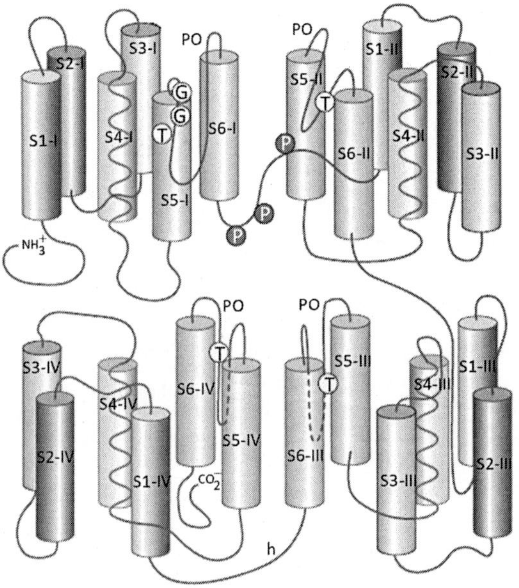

Figure S4.4 Tertiary and quaternary structure model of the α-subunit of the voltage-dependent sodium channel. Phosphorylation sites (P), glycosylation sites (G), pore forming loops (PO), inactivation loop (h) and tetrodotoxin binding sites (T) are highlighted.

every two amino acids always include a lysine or an arginine, both with a fraction of positive charge. The four intracellular loops between the S5 and S6 segments also contain the binding site for the best-known channel blocker, tetrodotoxin (TTX), a toxin found in the intestines of pufferfish (Sphoeroides rubripes) and other related species. When the membrane potential is equal to the resting potential, e.g., -80 mV, the S4 segments and the h-loop have a conformation such that the channel is closed and not inactivated (Figure S4.5A). The accumulation of positive charges on the inner face of the plasma membrane, caused by depolarization with the consequent variation of the electric field between the two sides of the plasma membrane, determines a conformational change of the α subunit. This change consists of a shift of about 0.5 nm of the four S4 segments outwards, with a simultaneous rotation of about 60° (curved arrows in

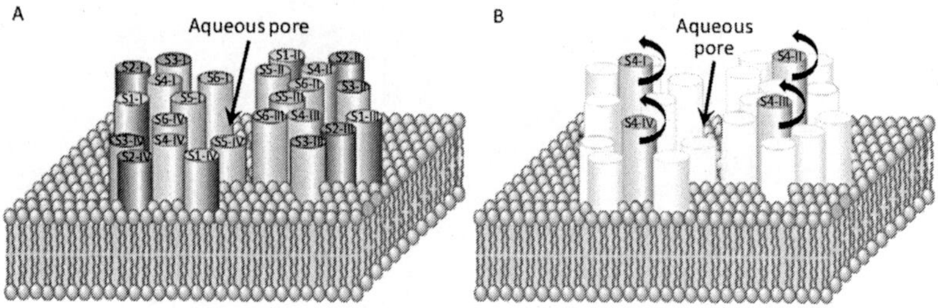

Figure S4.5 (A) Model of the membrane arrangement of the α subunit of the sodium channel. (B) Depolarization causes a conformational change. The four S4 subunits make an upward shift of approximately 0.5 nm and a counterclockwise rotation of approximately 60° (curved arrows) and the channel switches from the closed to the open state.

Figure S4.5B). This movement is transmitted to the four S5-S6 loops and the channel switches to the open state. The same change in the electric field also causes a conformational change in loop h and the channel switches to the inactivated state. When the membrane potential is returned to its resting value, the S4 segments and loop h return to their original position (Figure S4.5A) and the channel switches to the closed, non-inactivated state. The conformational changes are virtually instantaneous.

CHAPTER 5

Communicators 1: the primary signal

5.1 Excitable cells
5.2 The action potential
5.3 Ionic membrane currents are voltage-dependent
5.4 Ionic membrane currents are time-dependent
5.5 The ionic nature of the action potential
5.6 The ionic basis of the genesis of the action potential
5.7 The ionic nature of membrane currents
5.8 Voltage dependence of membrane ionic currents
5.9 Time dependence of membrane ionic currents
5.10 The Hodgkin and Huxley equation
5.11 Inactivation of the sodium current
5.12 From membrane current to channel population dynamics
5.13 New equations for voltage and time dependent currents
5.14 Shaping the action potential

In the course of evolution, in parallel with their increased size and complexity, biological organisms have developed a highly refined network of communication within the cell. They have also developed more complex communications between different cells that are near and far from the source of the primary signal. The construction of an information exchange system was necessary to be able to have precise control over the various parts of the organism. In addition, the ability to control the relationship with the surrounding environment was equally important. The possibility to monitor all the spatial/temporal information coming from outside the organism was instrumental to develop a diffuse system to acquire, elaborate and respond to a vast variety of external stimuli.

The ways in which communication takes place between the inside and the outside of the organism or between the various parts of the organism have evolved with its size and with the specialization of its different regions.

In general, the transfer of information within cells and between cells is essentially chemical, electrical or both, in appropriate combinations. However, chemical and electrical transmission are not interchangeable, as they have their own particular characteristics with respect to the different ways communication takes place in the biological sphere. If, for example, the speed of information transfer is essential, the electrical mode is the most suitable, operating with times in the order of micro- and milliseconds. If, on the other hand, the time span of the message covers minutes, hours or days, it is clear that the potential of chemical modes of transmission is unlimited. The latter considerations do not exclude a slow electrical signal and a fast chemical information transfer These ideas will be explored in more detail in the following chapters.

In the communication system between cells, the **primary electrical signal** is generally the event known as the **action potential**, an electrochemical phenomenon involving charges carried by ions, which essentially affects the difference of potential across the membranes. The action potential is the basis for the transfer of information both in and out of the biological organism and is mainly characterized by the fact that it is generated in a few milliseconds; there are a few examples of long-lasting action potentials (seconds and also minutes).

The action potential is a passive phenomenon, in that its generation does not require the simultaneous production of energy, but instead exploits the **potential energy** needed to maintain the potential difference across the membranes of excitable cells (Paragraph 3.5). These cells have resting membrane potential values approximately between -60 and -90 mV (Chapter 3), a potential close to the equilibrium potential of the potassium ion (Data Sheet 3.2).

The potential difference is maintained by the presence of constraints, namely a very low and extremely selective membrane electrical permeability. These constraints, in the presence of an appropriate stimulus, are momentarily removed, allowing the biological system to move spontaneously and rapidly towards a stable equilibrium. The complete cycle includes the ability to re-establish the original conditions at a high level of potential energy. In practice, the mechanism consists of a depolarization of the membrane lasting one to two milliseconds from the resting potential to a positive value of around +30 to +40 mV, which is close to the equilibrium potential of the sodium ion (Data Sheet 3.2); the depolarization activates a mechanism that, at the end of the upstroke, is able to return the membrane potential to its initial value. The result of this process is the generation of a transient signal of approximately 100 mV in amplitude, which varies in duration, depending on the type of cell, from a few milliseconds to several minutes.

Four characteristics of the action potential should be emphasized: transient occurrence, low energy consumption, defined amplitude, and the digital nature of the signal.

The **transient occurrence** makes the action potential a communication system that does not produce permanent changes in the physiology of the cell that generates the signal. Quantification of inward and outward ionic flows demonstrated a negligible amount of ions crossing the membrane during the action potential (10^5-10^6 ions), which is several orders of magnitude less than the ion content in the intracellular and extracellular fluids (10^{22} ions). The **consumption of energy** is practically zero, as the single action potential wave generated uses a very small part of the potential energy accumulated in the form of electrochemical potential. A dissected sciatic nerve placed in an experimental chamber without nutrients is able to produce action potentials for several hours without apparent signs of fatigue. As mentioned above, the **amplitude** of the signal of about 100 mV fluctuates between the potassium and sodium equilibrium potentials. It is important to notice that, once triggered, the action potential wave cannot be modulated in amplitude. Finally, being an **all-or-none** event, the action potential can be defined as a **digital signal.** The code used by the excitable cells to communicate is the frequency of action potentials (firing frequency).

What has just been described represents a generic wave of excitation of nerve cells (Figure 5.1): it will be seen that its duration can vary, depending on the type of neuronal cell, from one to tens of milliseconds, while its amplitude strictly varies between the equilibrium potential of potassium and that of sodium and is therefore immutable regardless of cell type.

Due to its digital nature, the single action potential, which goes from a value = 0, close to the potassium equilibrium potential, to a value = 1, close to the sodium equilibrium potential, does not have functional significance. It is not uncommon for nerve cells to have irregular spontaneous activity in which single action potentials are randomly produced that do not, in fact, constitute a message for other cells. While a digital signal cannot be **modulated in amplitude**, the signal for communication between different cells can be **modulated in frequency** with the generation of spike trains of action potentials or **neuronal firings** (Paragraph 6.4). Neuronal firing consists of a large number of individual events that are generated at regular time intervals and in such a way that different frequencies can encode different functions.

For the sake of completeness regarding the communication process between cells, it is necessary to point out that the phenomenon of excitability does not only apply to nerve cells, but also to muscle cells, oocytes and certain glandular cells, all of which are capable of responding to particular stimuli with a sudden change of the membrane potential. In these cases, the generation of an action potential is closely linked to a specific function. In the cardiac action potential, at the ventricular level, the action potential lasts several hundred milliseconds. The depolarization is maintained by additional membrane calcium channels (Paragraph 7.4).

In order to study the principles of cellular excitability, we will consider a model of a cell whose characteristics are as similar as possible to those of a neuron, the predominant cell type involved in cellular excitability in the mammalian nervous system. Action potential generation in the neuron can be used as an illustration of the basic mechanism of cell excitability. Specific aspects of excitability in different tissues and specialized cells can be evaluated on the basis of the neuronal action potential. For example, during fertilization, the oocyte produces an action potential lasting 30-40 minutes to prevent polyspermy. This action potential is basically equal to a nervous action potential lasting a few milliseconds, with the addition of a calcium conductance able to maintain a long-lasting depolarization.

5.1 Excitable cells

The model of the excitable cell that will be considered must have certain characteristics as described in previous chapters:

1. *The shape of the cell is spherical to allow a homogeneous distribution of both the membrane potential and its variations.* This is not always the case, since in an organism there are excitable cells with different shapes that certainly have transitory potential differences between one point and another in the cytoplasm.
2. *The lipid composition of the membrane is homogeneous.* There are many exceptions to this statement: for example, a membrane rich in cholesterol also influences the function of the ion channels by making the membrane in which the channels are embedded less fluid.

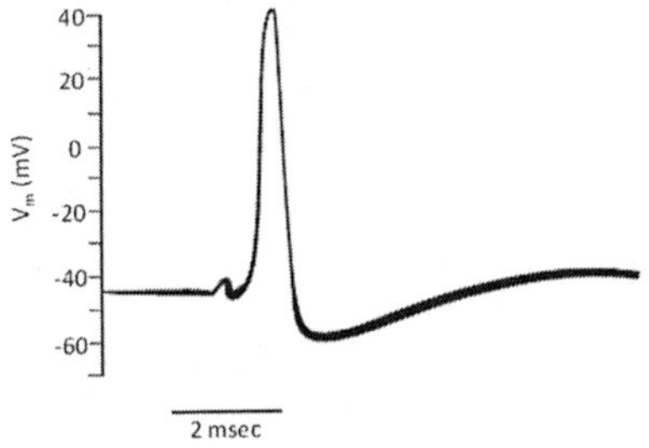

Figure 5.1 Recording of an action potential in a giant squid axon. Modified from Hodgkin and Huxley, 1945.

3. *Ions respond to the laws of free diffusion as they move from one environment to another across the membranes using passive* micro environments where the movement of ions and molecules may not follow the laws of free diffusion. Furthermore, passive ionic permeabilities do not always have the same diffusion coefficient as the one calculated in a solution because of the restrictions presented by the ion channel.
4. *The cell must have at least two types of potassium channels, one type of sodium channel, a Na^+/K^+ ATPase, and a glucose transport system in its membrane and have potential energy in the form of a membrane potential.*

The model cell under consideration has all of the minimal characteristics to be considered an excitable cell. It is able to generate and conduct functionally efficient potassium and sodium currents across the membrane using channels that have been suitably selected over the course of evolution with distinctive biophysical properties. The efficacy consists of

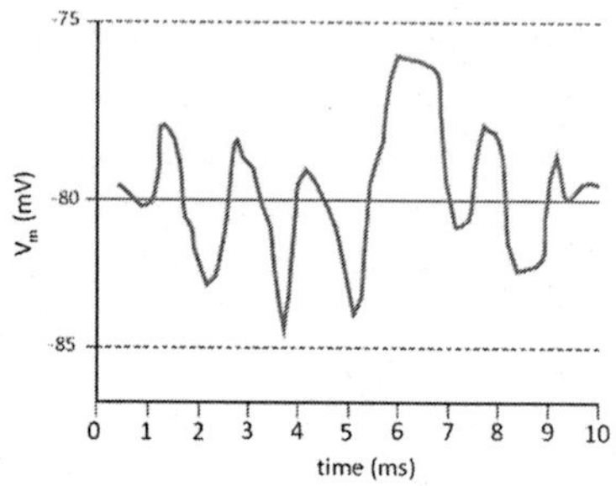

Figure 5.2 In a cell, the recorded potential randomly oscillates around the value of the resting potential.

the production of a far stronger response, from an energetic point of view, with respect to the stimulus received.

5.2 The action potential

The action potential is defined as the phenomenon of a transitory depolarization of the membrane potential from -65/-70 mV up to +35/+40 mV. The duration of an action potential varies from a few hundred microseconds, to a few milliseconds, up to several minutes, depending on the cell type. The membrane potential then returns, to the resting values within a time span typical of each excitable cell type (Figure 5.1). From the beginning to the end of the voltage wave, the sum of the inward and outward currents must be zero to maintain electroneutrality of the solution. The energy required to perform this work comes from the potential energy stored across the membrane. The Na^+/K^+ ATPase, which works slowly and constantly to maintain a high level of potential energy in excitable cells, is not directly involved in the generation of the action potential.

The action potential is usually generated by an excitable cell once it is hit by a mechanical, chemical, electrical, light or thermal stimulus. A stimulus can be considered adequate if it is able to depolarize the cell membrane to the point of causing the all-or-none change in membrane potential. A stimulus is not necessary in cells where the action potential is produced spontaneously in a rhythmic pattern. The cardiac pacemaker and intestinal peristaltic smooth muscle, for example, have automatic mechanisms which, by exploiting specific phenomena of membrane ionic conductance, are able to generate spontaneous action potentials that are rigidly regulated in frequency (Data Sheets 6.2 and 7.3).

The **resting membrane potential** in a generic nerve cell is not stable over time (Figure 5.2), but can be correctly defined as **dynamically stable.** Membrane voltage oscillations are physiological events due to the actions of intrinsic cellular mechanisms and to external factors. The cell's response to compensate for spontaneous hyperpolarization or depolarization consists of the automatic activation of mechanisms which tend to bring the potential value back towards its resting value by means of positive or negative currents.

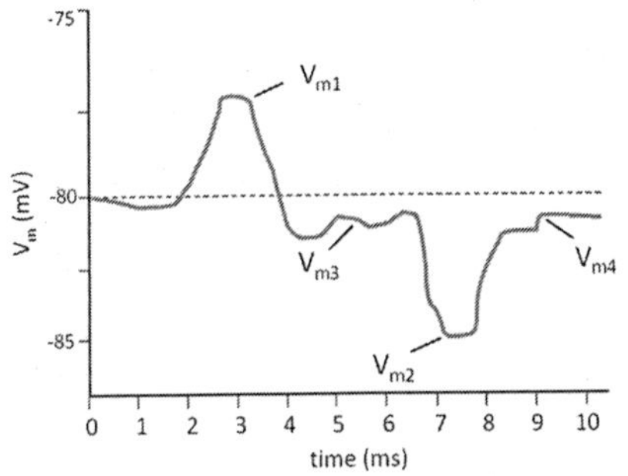

Figure 5.3 The cell has membrane structures, often consisting of voltage-dependent potassium channels, that can compensate for potential oscillations. The dashed line indicates the equilibrium potential.

According to the convention of biological systems, **negative currents** are defined as those currents due to positive charges entering the cell or to negative charges leaving the cell through membrane channels, causing depolarization. **Positive currents** are defined as those currents generated by positive charges leaving the cell or by negative charges entering the cell through membrane channels, causing hyperpolarization. The average membrane potential measured over a particular interval of time during which the number of charges generating a negative current are equal to the number of charges generating a positive current will be considered as the resting membrane potential. Therefore, if $\bar{I}_m$ is the average membrane current, and I_e and I_u are the inward and outward currents, respectively, generated by the active and passive ionic fluxes present at the plasma membrane at equilibrium, the relationship is as follows:

$$\bar{I}_m = I_e + I_u = 0 \tag{5.1}$$

The spherical excitable cell model under consideration is a very simple system from a dynamic point of view. It consists of an active component with ATP consumption, which creates and restores the electrochemical gradient exploited by passive components over time, mainly the conductances for sodium and potassium. At negative potentials such as the resting membrane potential, sodium has a large driving force, that is the difference between the membrane voltage and the equilibrium potential for a particular ion. However, the sodium membrane permeability is close to zero, so no significant current is measured. On the contrary, potassium has low or null driving force because the resting membrane potential is close to potassium equilibrium potential (Data Sheet 3.2), but a high membrane permeability. However, if the membrane potential moves away from its equilibrium and becomes more positive (V_{m1} of Figure 5.3) or more negative (V_{m2} of Figure 5.3), an outward or inward potassium ion flux generates a hyperpolarizing (outward) or a depolarizing (inward) current, respectively, to re-establish the equilibrium (V_{m3} and V_{m4} in Figure 5.3). In both cases, the current responsible to maintain the resting membrane potential stable (IK_1 or inward rectifying potassium current) is directly proportional to the voltage displacement. In the proximity of the cell average resting membrane potential, this potassium current allows outward-flowing or inward-flowing ions to maintain the resting membrane potential stable (V_{m3} and V_{m4} in Figure 5.3). This is an important mechanism in excitable tissues which are activated only after they have been stimulated. The cardiac ventricular contraction is the clearest example. Systole occurs only following the command from the pacemaker cells. The IK1 potassium current is less important in nervous cells where the message is coded by the frequency of action potentials.

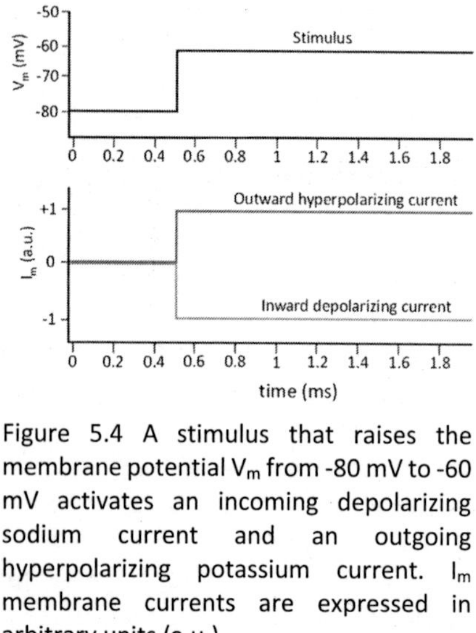

Figure 5.4 A stimulus that raises the membrane potential V_m from -80 mV to -60 mV activates an incoming depolarizing sodium current and an outgoing hyperpolarizing potassium current. I_m membrane currents are expressed in arbitrary units (a.u.).

In the presence of the potassium current alone, which can quickly compensate for variations in the membrane potential, the cell is not able to generate a rapid depolarization. No matter how great the stimulus, the potassium current will always prevail. It is therefore necessary to activate a **depolarizing current**, which in the proposed model of the cell is carried out by sodium ions. The current must be voltage- and time-dependent. The voltage dependency is evident since at -80 mV, the current is not appreciable, but a depolarization reaching -55 to -60 mV reaches the threshold for maximal activation of an action potential. At this potential, all the membrane currents follow Ohm's law and the result is activation at the precise moment of the voltage variation (Figure 5.4). If this occurs, it is clear that the net current though the membrane is null without a membrane potential variation. Even if the hyperpolarizing current were greater than the depolarizing current, there would be an automatic adjustment of the latter and the establishment of a new membrane potential equilibrium between the inward (i.e., sodium) and the outward (i.e., potassium) current.

In order for the cell to generate a rapid depolarization, it is necessary that the membrane currents do not follow Ohm's law and that they have specific characteristics which take into account the **voltage** and **time** parameters which change at each instant.

In the following paragraphs, particular attention is devoted to the **voltage dependence** and **time dependence** of the membrane currents. The idea is to find a logical way in the model where, following a depolarizing stimulus, membrane currents retain the characteristics able to produce an action potential.

The model should show that in a generic excitable cell:

1. at the end of the membrane potential variation (action potential), the net current though the membrane is zero;
2. the stimulus causes a depolarization by inducing membrane capacitor current;
3. membrane ionic currents must be voltage- and time-dependent;
4. activation of ionic currents occurs upon membrane depolarization both for inward sodium and outward potassium ion flows;
5. activation of depolarizing sodium current allows the membrane potential to move towards the sodium equilibrium potential;
6. there are two types of potassium ionic permeability, one devoted to stabilizing the membrane potential and the second which

becomes active during the excitation wave. The two currents have different voltage- and time-dependent characteristics. Both currents participate in the membrane repolarization process.

5.3 Ionic membrane currents are voltage-dependent

In a current/voltage graph (I/V plot), the currents described by Ohm's law (equation 3.7) for potassium and sodium are linear (Figure 5.5) and the slope of the lines is the conductance or the reciprocal of the resistance, which is constant at each voltage value. In the case of Figure 5.5, the conductance for potassium is greater than that for sodium. Such a system, if left uncontrolled, stabilizes the membrane potential at an intermediate value of about -20 mV, a value between the equilibrium potential for potassium (E_K) and that for sodium (E_{Na}).

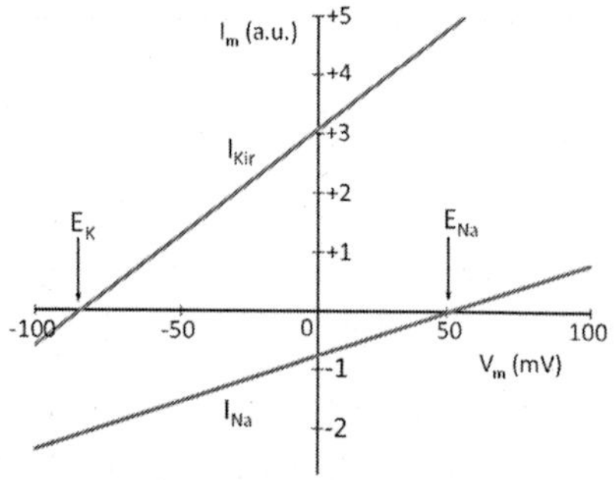

Figure 5.5 Linear trend of sodium currents (I_{Na}) and potassium currents (I_{Kir}), expressed in arbitrary units (a.u.), described by Equations 5.2 and 5.3.

In order to describe the sodium and potassium currents in the cell model under consideration, Ohm's law needs to be supplemented with the ion driving force, that is the difference between the membrane voltage and the equilibrium potential for a particular ion. Therefore, we must use the generalized Ohm's law, derived from Ohm's law (equation 3.8), applied to sodium

$$I_{Na}=G_{Na}\cdot\left(V_m-E_{Na}\right) \tag{5.2}$$

and potassium

$$I_K=G_{Kir}\cdot\left(V_m-E_K\right), \tag{5.3}$$

where V_m is the membrane potential and I_{Na}, E_{Na}, G_{Na} and I_{Kir}, E_K, G_{Kir} are current, equilibrium potential and conductance for sodium and potassium, respectively. In the cell model under consideration, E_{Na} is equal to +50 mV and E_K is equal to -90 mV (Data Sheet 3.2). For the membrane potential to stabilize at a potential close to the equilibrium potential of potassium, the sodium current must not be allowed to be active.

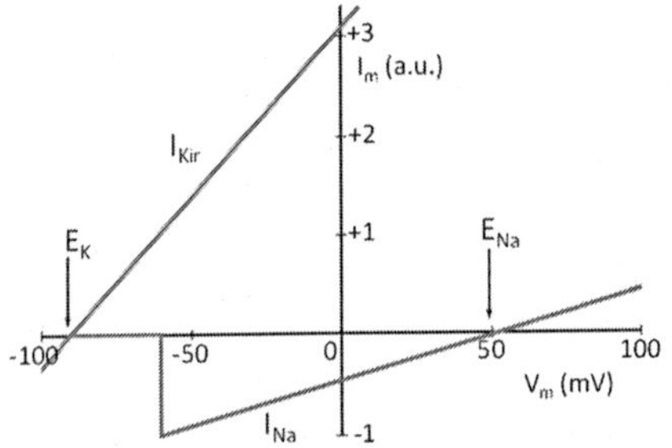

Figure 5.6 Linear trend of sodium (I_{Na}) and potassium (I_K) currents, expressed in arbitrary units (a.u.), described by Equations 5.2 and 5.3.

This phenomenon, shown in Figure 5.6 in the case of the sodium current, introduces a characteristic that belongs to all ionic currents in excitable cells: **voltage dependence.** To maintain the membrane potential at a negative value, the prevailing current must be the hyperpolarizing current of the potassium ion. At potentials between the equilibrium potential of potassium and –55 mV, the sodium current is practically absent. Thus, the membrane potential stabilizes at a value slightly more positive than E_K. If a depolarizing stimulus from -70 mV to -55 mV is applied to the system described in Figure 5.6, the result will be instantaneous activation of the sodium current. Under these conditions, the inward sodium current finds a hyperpolarizing potassium current already operating. Again, the membrane potential

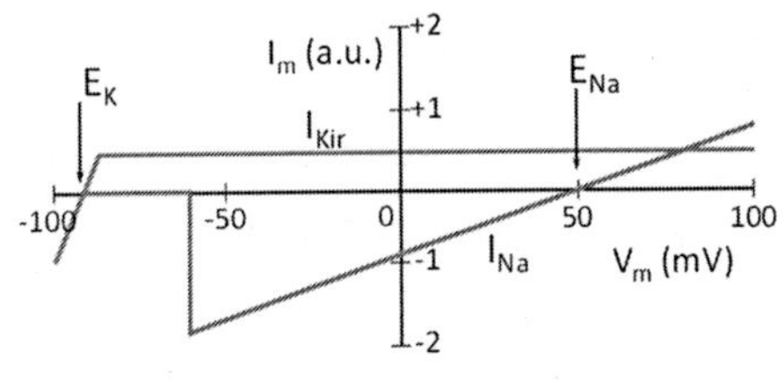

Figure 5.7 The potassium current is active at any membrane potential, but above a certain potential value it is constant and small.

will stabilize at an intermediate value between the potassium equilibrium potential and the sodium equilibrium potential, at about -20 mV.

In order for the membrane to depolarize to positive potentials, the potassium current, which is decisive in maintaining the negative potential, must also be voltage-dependent. In practice, as shown in Figure 5.7, this particular current is active around E_K, but at a more depolarized potential following an appropriate stimulus, it reduces to a minimal flow. This behavior of the potassium current allows depolarization of the membrane to positive values, promoted by the entry of sodium ions, which shift the membrane potential towards the E_{Na} value. In this case, the membrane potential governed by the sodium equilibrium potential would be stabilized at a positive potential close to E_{Na}.

To return the membrane voltage to its resting state, it is necessary to activate a mechanism to repolarize the membrane. Essentially, the repolarization must be carried out by potassium current. This current has different kinetics from that described above, even if the overall characteristics are similar, such as the reversal potential and voltage dependency. Its function must be to counteract the sodium current and therefore its activation must occur during depolarization, when the potential is greater than -55 mV. Its task is to repolarize the cell after a positive potential has been reached by the sodium current (Figure 5.8). Despite the activity and modulation of currents that we have assumed in the excitable cell model, voltage dependence is necessary, but alone is not sufficient to determine the action potential of the nerve cell. The second potassium current fails to conclude the depolarization/repolarization cycle and re-establish the negative potential after the transient change in membrane potential. Due solely to the voltage dependence of the currents (Figure 5.8), after an initial depolarization, the membrane potential would stabilize once again

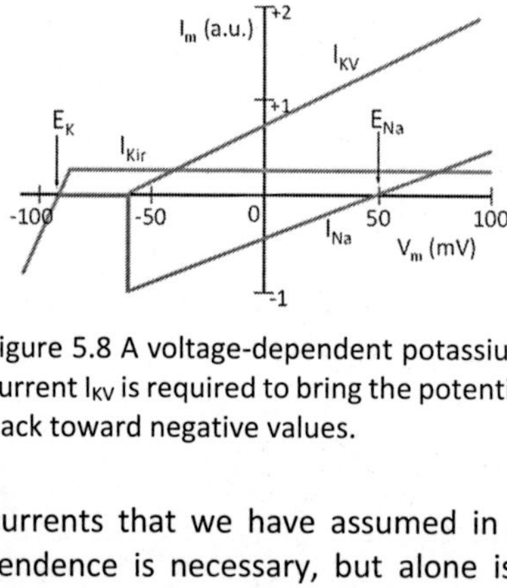

Figure 5.8 A voltage-dependent potassium current I_{KV} is required to bring the potential back toward negative values.

at an intermediate value between E_K and E_{Na}. At potential values more positive than -55 mV, both sodium and potassium currents become active. Depolarization occurs because the sodium driving force is greater (V_m - E_{Na}) than the potassium driving force (V_m - E_K). However, the sodium current decreases as the membrane potential approaches E_{Na}. In parallel, the potassium current increases as the membrane potential moves away from E_K, and increases the driving force for the potassium ions to leave the cell. Again, at the end of the process, the membrane potential will stabilize at an intermediate value between E_{Na} and E_K, at about -20 mV.

In conclusion, we can say that the voltage dependence of the membrane currents is a necessary but not sufficient condition to guarantee the correct generation of the action potential in excitable cells. For correct development of the excitatory wave of the membrane potential, the membrane currents, in addition to being voltage-dependent, must be also **time-dependent**.

5.4 Time-dependent membrane ionic currents

The appropriate depolarizing stimulus activates the capacitive currents (I_C, Equation S2.2). Accumulating charges on the membrane capacitor (armatures with kinetics that depend on the biophysical characteristics of the membrane itself) effect an initial depolarization up to -50/-55 mV (between 0.5 and 0.8 ms in Figure 5.9). Capacitive depolarization increases both sodium and potassium currents. If the increase were synchronous, i.e., if the currents were only voltage-dependent, the membrane potential would behave as described in Paragraph 5.3. What enables a change towards positive potential values is the difference between the kinetics of the membrane currents, defined as the **time dependence** of the ionic conductances. In the description of membrane currents using Ohm's law (Equations 5.2 and 5.3), time is not taken into account. Both currents flow immediately following a stimulus, but only when voltage- and time-dependent currents are involved is it possible to have an effective transient excitation wave.

The sodium current is activated at a potential of -50/-55 mV with fast kinetics. The potassium current also activates at the same potential, but its kinetics are much slower (Figure 5.9). In this way, the membrane potential reaches positive values very quickly. The sodium current increases with a positive feedback mechanism, generally called the **Hodgkin cycle** (Figure 5.10), according to which the same sodium current, by depolarizing the membrane, exponentially increases its permeability. If time dependence were only a characteristic of the activation of membrane currents, after an initial peak of the potential up to positive values, the membrane potential would stabilize at an intermediate value between E_K and E_{Na}. In order for the potassium current to re-establish the negative membrane potential, the sodium current must be impaired. Because depolarization activates the

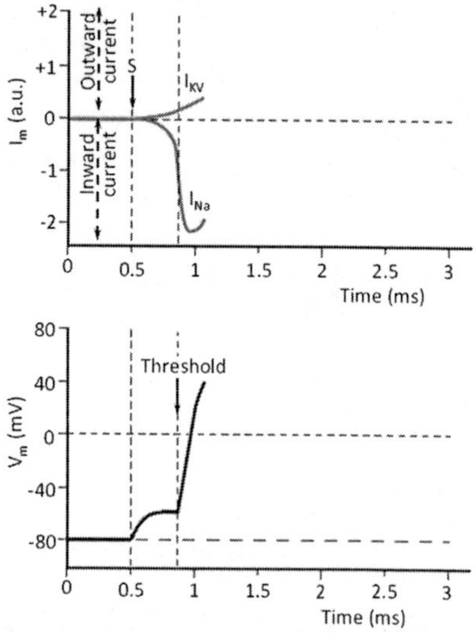

Figure 5.9 The stimulus (S) applied to the cell causes the membrane capacitance to charge. The membrane depolarizes, and activating the sodium and potassium currents (I_{Na} and I_K, top), reaches the peak of the action potential (bottom). I_m currents are expressed in arbitrary units (a.u.).

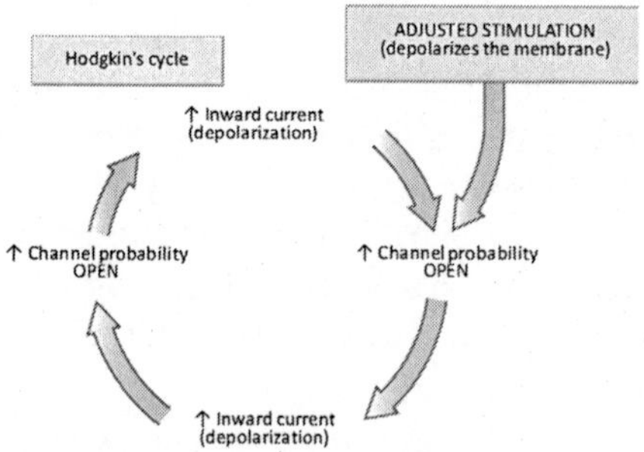

Figure 5.10 A positive feedback (Hodgkin cycle) is involved in the generation of the sodium current of the action potential.

sodium current, it is not possible that the same process promotes its deactivation. Rather, the mechanism is based on a time-dependent process. Immediately following activation, the sodium current declines using a separate process from that responsible for activation. The presence of **time-dependent inactivation** of the fast sodium current is an essential condition to allow the potential to repolarize (Figure 5.11).

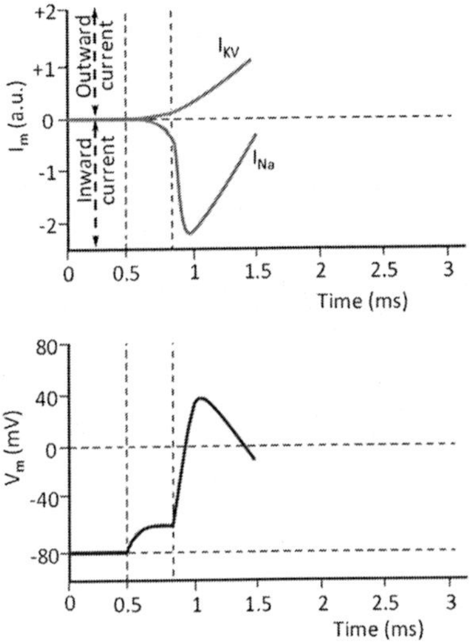

Figure 5.11 Inactivation of the sodium current (I_{Na}) and prevalence of the potassium current (I_K) allow repolarization of the membrane potential. I_m currents are expressed in arbitrary units (a.u.).

As the sodium current decreases, the outward potassium current takes over, driving the membrane potential back towards E_K. To repolarize the membrane, the potassium current decreases, but not in a time-dependent manner, as in the case of the sodium current, but by a decrease in the driving force for the potassium ions. This decrease is called **deactivation** (Figure 5.12). When the sodium current inactivates and potassium remains the sole current present, the membrane potential temporarily hyperpolarizes

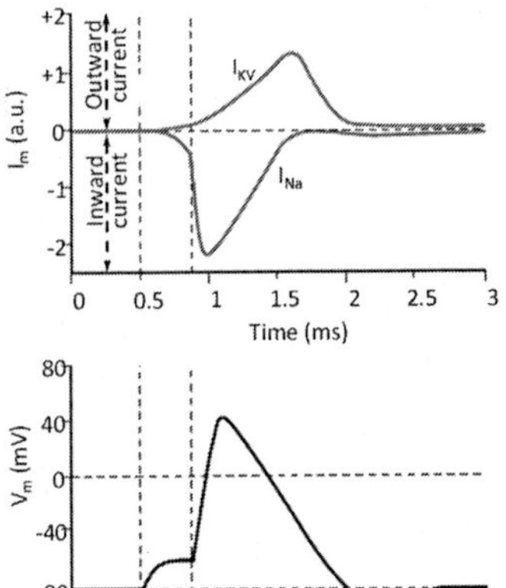

Figure 5.12 During repolarization, the potassium current (I_K) decreases but with a delay relative to the potential, which becomes more negative than the resting potential. I_m currents are expressed in arbitrary units (a.u.).

due to the persistency of sodium current inactivation. Inactivation is removed in a voltage-dependent manner by hyperpolarization. This concludes the action potential cycle by re-establishing the original cell **resting potential**.

5.5 The ionic nature of the action potential

The proposed model of an excitable cell can produce a transient oscillation of the membrane potential that is rapid and large enough to invert the polarity of the membrane potential, but with characteristics that allow resting conditions to be restored in a short time.

The fact that a nerve cell responds to an external stimulus by causing an oscillation of the membrane potential was already known in the 1930s. Kenneth S. Cole and Howard J. Curtis (1939) were the first to show that in a giant squid axon, prepared in a suitable experimental dish (Data Sheet 5.1), the shift in potential in a stimulated cell was associated with a change in membrane impedance, i.e. membrane resistance.

This was achieved by adapting the Wheatstone bridge, which was used to exactly calculate an unknown resistance or impedance. The Wheatstone bridge (Figure 5.13) is a circuit formed by two known resistors (R_1 and R_3) placed in parallel, a variable resistance R_2 in series with the first known resistor, and an unknown resistance R_X (constituted by the axon under examination) in series with the second known resistor and in parallel with the variable resistance. The circuit is completed by an alternating current generator (oscillator) and an oscilloscope (Data Sheet 5.2), which allows electrical phenomena lasting up to a few microseconds to be measured. In the presence of the high-frequency signal of the oscillator, R_2 is adjusted until the measured signal is minimal (0 ms in Figure

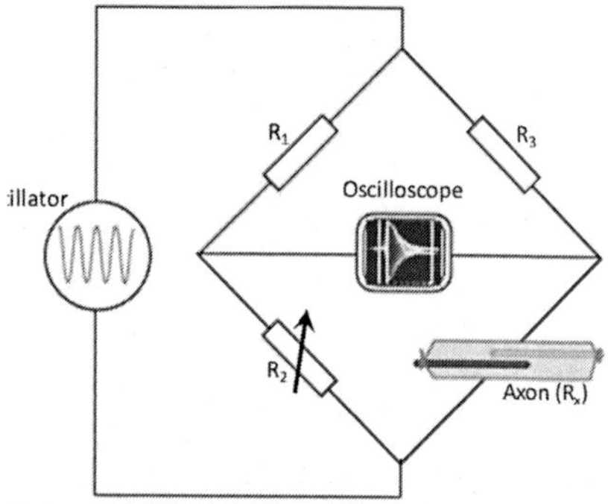

Figure 5.13 The Wheatstone bridge is used to measure impedance changes in a stimulated squid axon.

5.14). The axon is then stimulated appropriately and the response is shown as a large potential difference produced by the alternating current due to the imbalance of the system caused by the decrease of R_X (from 1 to 10 ms in Figure 5.14). Also observing the variation in potential of the axon superimposed on the oscillation (white trace), corresponding with the peak, there is a significant decrease of the impedance, which progressively attenuates until it disappears, when the system again becomes balanced.

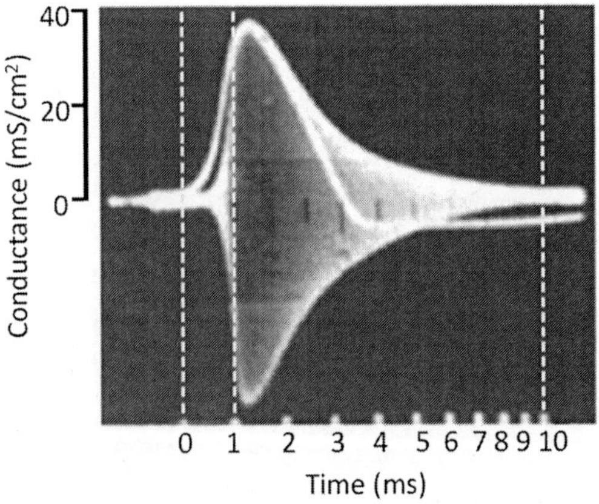

Figure 5.14. A Wheatstone bridge device used to capture the transient membrane impedance during squid axon excitability. Superimposed to a decrease in membrane impedance, indicated by an increase in oscillation amplitude, there is a typical action nervous potential to show two different ways to show cell excitability.

The experiment described above confirms, on one hand, that when an excitable cell is stimulated, it changes its resting potential, producing an action potential, and on the other hand, that the action potential is associated with a decrease in membrane resistance. To understand the ionic permeabilities involved in this very important phenomenon that characterizes excitable cells, it is necessary to carry out voltage measurements across the membrane. Experiments to study the action potential from various points of view were carried out by Allan L. Hodgkin, Andrew F. Huxley and Bernard Katz between 1940 and 1950. These researchers obtained very useful results on the ionic nature of excitability. Hodgkin and Katz (1949), in particular, demonstrated that in a squid axon prepared in a suitable experimental dish (Data Sheet 5. 3), lowering the external sodium concentration produces a decrease of the action potential amplitude (Figure 5.15). The experiment consists of substituting sodium with an impermeable solute, here dextrose. A lower sodium concentration modifies the shape of the action potential. With external sodium less concentrated, not only is the voltage variation of the action potential reduced proportionally but there is also a reduction in the variation of the potential in time (dV/dt) (track 2 of Figure 5.15). This phenomenon is not an artefact due, for example, to a plasma membrane damage. Reintroducing 100% artificial seawater re-establishes the original action potential waveform (track 3 of Figure 5.15).

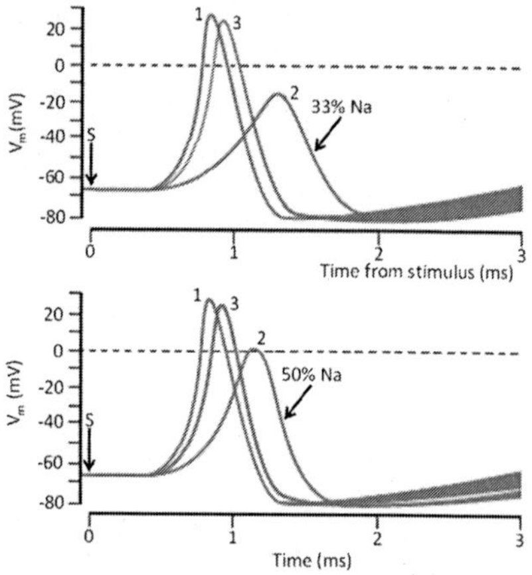

Figure 5.15 Effect of low sodium concentration on an action potential. Trace 1 is in seawater, trace 2 in 33% (top) and 50% (bottom) seawater and 100% supplementation with isotonic dextrose, trace 3 recovery. E_M, membrane potential.

The research of Hodgkin, Huxley and Katz made it possible to explain the ionic basis of excitability, beginning from the simple recording of the action potential. However, measurement of changes in the membrane voltage during excitability did not answer several questions about membrane current kinetics and the interplay between the different ionic membrane conductances. Research made a huge advance when, due to the need for new investigatory instruments, Hodgkin and Huxley invented the **voltage-clamp**. The idea was to visualize the kinetics of the different ionic species that are able to cross the membrane during excitability.

5.6 The ionic bases of the action potential genesis

Hodgkin and Huxley's contribution to explain the mechanisms at the basis of cell excitability has been fundamental. Their work was not limited to cell physiology, and was a theoretical as well as technical step forward towards understanding how dynamic biological processes function. After almost one hundred years from their important work, we are still discussing their concepts and using their revolutionary technical improvements. The main achievment of their research was the identification of the starting point, i.e., the practical demonstration of their theoretical idea about the basis of the action potential mechanism. The existence of ionic membrane currents was accepted by the scientific world. The problem was how to demonstrate the currents experimentally, what produces the current and what allows the flow (pore versus carrier). Most of all, determine what investigative procedure would enable the study of not only their voltage dependence but also and especially their time dependence. Ionic currents can theoretically be described by a modification of Ohm's law, introducing a time factor according to the equation:

$$\Delta I_{(t)} = \Delta V_{(t)} \cdot \Delta g_{(t)}. \tag{5.4}$$

Thus, the variation of the total current flowing through the membrane as a function of time $\Delta I_{(t)}$ is proportional to the variation of the membrane potential as a function of time $\Delta V_{(t)}$ and to the variation of the membrane conductance as a function of time $\Delta g_{(t)}$. The variables are $\Delta I_{(t)}$, $\Delta V_{(t)}$, $\Delta g_{(t)}$, so it was not feasible to solve the problem analytically. It was necessary to introduce an experimental procedure that would make it possible to keep one of the parameters constant and measure the second experimentally, in order to calculate the third.

Hodgkin, Huxley and Katz (1952) solved the problem by appropriately modifying the squid axon preparation techniques, designing completely new measurement circuits and introducing the voltage-clamp technique (Data Sheet 5.4). This technique allows one to change the membrane potential to a chosen value essentially immediately. The stability of the membrane potential is guaranteed by the current supplied from outside.

In fact, what is recorded with the voltage clamp is the current supplied to counterbalance the current flowing through the membrane. In this way, the individual currents active during the change in membrane potential and their ionic nature can be studied. Furthermore, by imposing an experimental protocol on the cell at different potentials, Hodgkin, Huxley and Katz were able to describe the voltage and time dependence of the different ionic currents involved in the action potential.

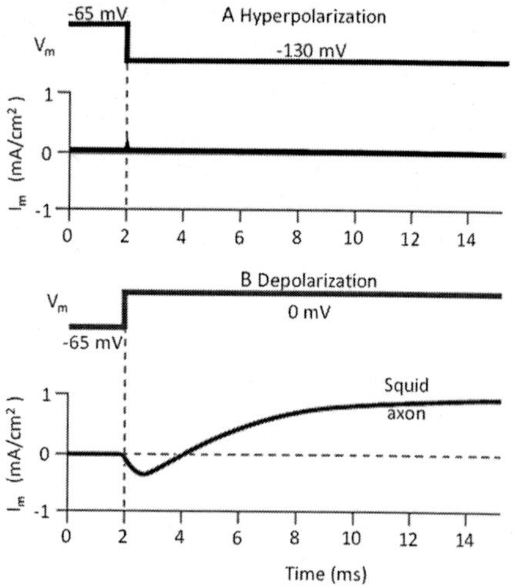

Figure 5.16 Time-dependent trend of the current density of a squid axon measured in voltage-clamp. At time zero, the membrane potential V_m is hyperpolarized (A) or depolarized (B) by 65 mV and held constant over time.

Using the voltage-clamp technique (Data Sheet 5.4), a current step was applied to a giant squid axon. The current step instantaneously hyperpolarized the membrane potential V_m from the resting value of -65 mV to the value of -130 mV, maintained constantly throughout the experiment without observing appreciable membrane currents I_m (Figure 5.16A). Vice versa, the application of a depolarization from -65 mV to 0 mV (Figure 5.16B) first causes an inward current, followed by a positive current which then remains constant throughout the duration of the stimulus.

Note that to be able to compare the currents of cells of different types and sizes, the values on the graphs do not represent the current in absolute values, but the current density expressed as the amount of current measured per square centimetre of membrane surface. A good approximation of this variable is the value of the capacity measured during the experiment and by the specific capacity, considered equal for all plasma membranes.

This fundamental voltage-clamp depolarization experiment first demonstrated the **voltage dependence** and ionic nature of the charge flow across the membrane given by an inward current of sodium ions and an outward current of potassium ions. In addition, it established that the membrane current changes its kinetics during a stimulus maintained at a constant voltage value over time. This demonstrated the **time dependence** of membrane ionic permeabilities.

In light of these results, it is clear that the action potential dynamics cannot be simply described by Ohm's law. Excitability, to be quantified, needs a more complex mathematical formulation, in which voltage and time dependence are considered as main features.

5.7 The ionic nature of membrane currents

Hodgkin and Huxley (1952a) were able to characterize the different membrane currents active during the action potential taking advantage of the voltage-clamp technique. For their experiments, they used the giant squid axon because was a unicellular preparation that was large enough to hold two transmembrane electrodes (Figure 5.16). Experimental configurations of this kind allow the recording of membrane currents elicited by voltage steps at different experimental membrane potentials. The squid axon is immersed in seawater and depolarized from -65 mV to -9 mV by a voltage step. The current trace in 100% external sodium shows a rapid and transient inward current followed by a sustain outward current (Figure 5.17, track A). The hypothesis is that we are in the presence of two separate transmembrane currents. It is impossible that a single ion, in the

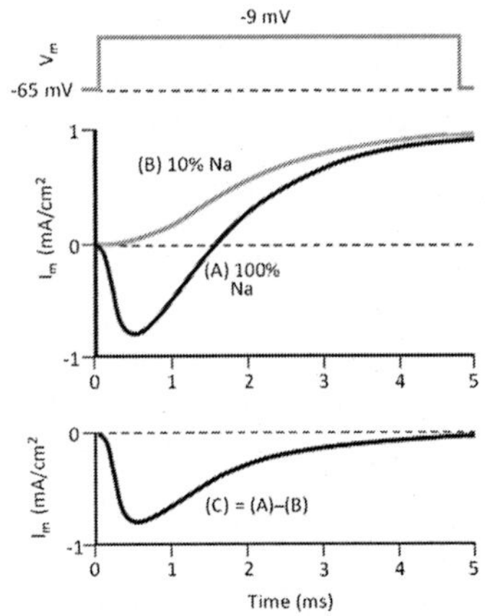

Figure 5.17 Voltage-clamp recording of I_{Na} currents in 100% seawater (trace A), in 10% seawater and 90% choline chloride solution (trace B) and the difference (trace C).

presence of a constant voltage, could change direction. In low sodium (10% seawater and 90% choline chloride) and using the same stimulation protocol, the inward current disappears and only the outward component of the membrane current remains (Figure 5.17 track B). The pure sodium component (Figure 5.17, track C) can be obtained by subtracting point by point from the total current (track A), that obtained in 10% sodium (track B). Regarding the outward component, the use of a cell model such as the squid axon, which has a diameter of up to 1 mm, also made it possible to change the internal solution by replacing the potassium ion and demonstrating its involvement in the outward current (Figure 5.17, track B).

With these "simple and innovative" experiments, Hodgkin and Huxley demonstrated that new ideas combined with technical advancements can solve very complex biological mechanisms.

Further information on the behavior of membrane currents in excitable cells has been obtained by applying voltage-clamp to a giant squid axon using voltage steps that increase the value of the membrane potential from -50 mV to +90 mV at 20 mV intervals. The initial potential was maintained at the holding potential of -65 mV (Figure 5.18A) in order to have comparable experimental conditions before each stimulus. The current induced by the different depolarizing stimuli delivered sequentially changes in amplitude and kinetics, depending on the imposed membrane potential. A stimulus from -60 mV to -50 mV failed to activate membrane permeabilities (Figure 5.18B). -30 mV induced a small and slow inward current. Additional

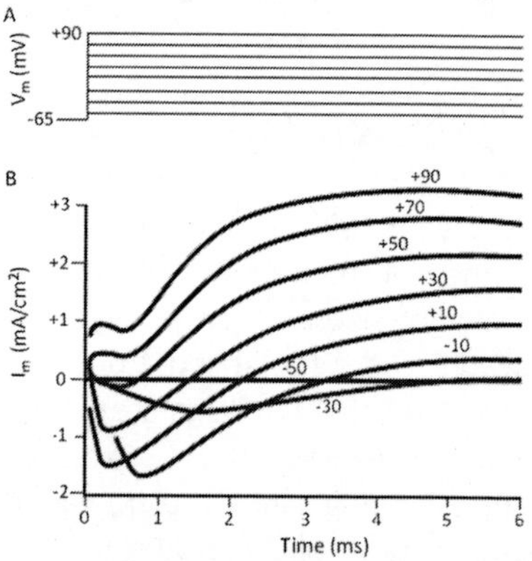

Figure 5.18 Stimulating the axon in voltage-clamp with a protocol that imposes potentials from -60 mV to +90 mV allows (A) recording of different currents at different potential values (B).

depolarization showed a family of currents where the inward current from a maximum value (-10 mV test) zeroed the current at +50 mV (E_{rev} for Na ions), and the reverse for further depolarizing steps. The outward current appears with the -10 mV voltage step and becomes predominant in +50 to +90 mV tests.

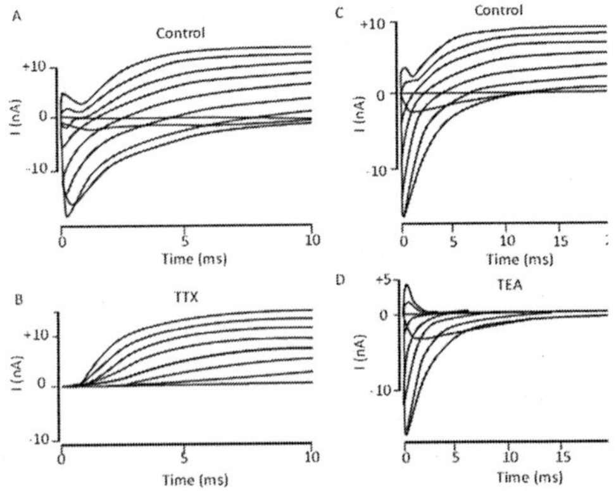

Figure 5.19 Comparison between membrane currents in controls (A and C) and after pharmacological blockade (B and D). TTX completely eliminates the inward current at nanomolar concentrations (B). TEA completely eliminates the outward current (D).

The definitive identification of the ions responsible for the membrane currents involved in the action potential was achieved with the use of specific blockers (Takata et al., 1966). **Tetrodotoxin (TTX)** was demonstrated to be a potent inhibitor that is very selective for the sodium current at nanomolar concentrations. Potassium current is inhibited by **tetraethylammonium (TEA)** at millimolar concentrations. TTX can be obtained from the spines and gonads of pufferfish while TEA is an inorganic component toxic to cells. Application of a depolarizing voltage-clamp protocol (Figure 5.18A; Hille, 1967) to the axon perfused with seawater triggers a family of membrane currents (Figures 5.19A and C). Applying the same stimulation protocol in the presence of 100 nM TTX causes a total disappearance of the inward current (Figure 5.19 B). The inward component of the membrane current can be isolated by perfusing seawater containing 100 μM TEA (Figure 5.19 D). The current reverses for positive voltage steps, showing that the impose potential is more depolarize than the current equilibrium potential (Figure 5.19 D). The possibility of pharmacologically separating the two main currents responsible for the excitatory process in nerve cells has made it possible to study their voltage dependence and time dependence in depth.

5.8 Voltage dependence of ionic membrane currents

Voltage dependence of membrane currents can be studied by creating the current/voltage relationship (I/V, Figure 5.20; Cole and Moore, 1960), which can be obtained from the giant squid axon voltage-clamp experiments (Data Sheet 5.4). The axon is first held at a holding potential (E_H) of -55 mV, then depolarized at various potential values from -50 mV to +85 mV, with increases of 5 mV for the first pulses, 10 mV for intermediate pulses and 20 mV for the largest pulses (inset of Figure 5.20). Currents are reported as steady-state values for TEA-sensitive current (Figure 5.19 B) and as peak values for TTX-sensitive current (Figure 5.19D). Current values are then plotted in a graph with the membrane potential on the x-axis and the current densities on the y-axis, where the positive current values indicate the outward currents and the negative values the inward currents (Figure 5.20). Both the positive outward current and the negative inward current start to assume a value other than zero at the same potential of ≅ -50 mV. The positive outward current shows a constant increase as a function of voltage. The negative inward current, in a few millivolts of depolarization, increases to a maximum current value (between -50 and -25). Following this increase the inward current shows a constant decrease as a function of voltage. The current reverses at ≅ +50 mV. The experiments done with ion substitution and with the current blockers previously suggested that inward currents are mainly carried by sodium ions and outward current is mainly potassium ion flow.

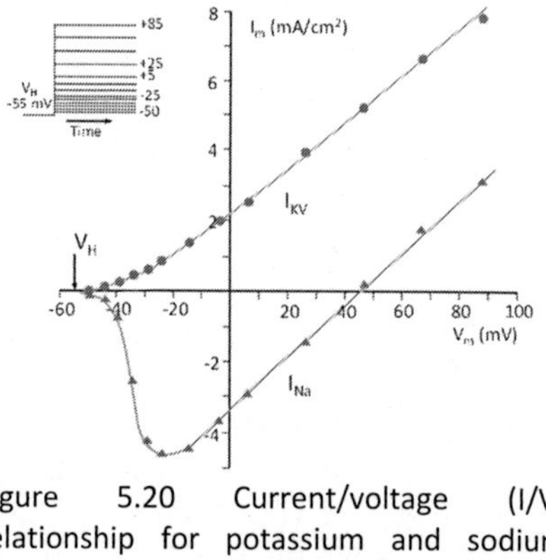

Figure 5.20 Current/voltage (I/V) relationship for potassium and sodium currents obtained from voltage-clamp recorded currents. Inset, the stimulation protocol. V_H = holding potential. Currents are expressed as current density to compare measurements on cells of different sizes.

In the I/V plot of Figure 5.20, inward current reverses the direction at ≅ +50 mV in agreement with the sodium equilibrium potential calculated with the Nernst equation (Paragraphs 3.2.8). Concerning the outward current, cytoplasmic ion substitution (Data Sheet 5.4) confers the ability to carry the transmembrane current with potassium ions. In addition, TEA is a specific blocker for potassium permeability. Current reversal potential for potassium should be ≅ -70 mV (Paragraphs 3.2.8). The current shows activation around ≅ -50 mV (like the sodium current; Figure 5.20). As in the case of sodium current, this is due to the voltage-dependent properties of the current.

From the current distribution of Figure 5.20 and using Equations 5.2 and 5.3 in the form

$$G_{Na}(V_m) = \frac{I_{Na}}{(V_m - E_{Na})} \quad (5.5)$$

and

$$G_K(V_m) = \frac{I_K}{(V_m - E_K)} \quad (5.6)$$

it is possible to determine the voltage-dependent conductances for sodium $G_{Na(Vm)}$ and potassium $G_{K(Vm)}$. The potential V_m imposed on the cell by the

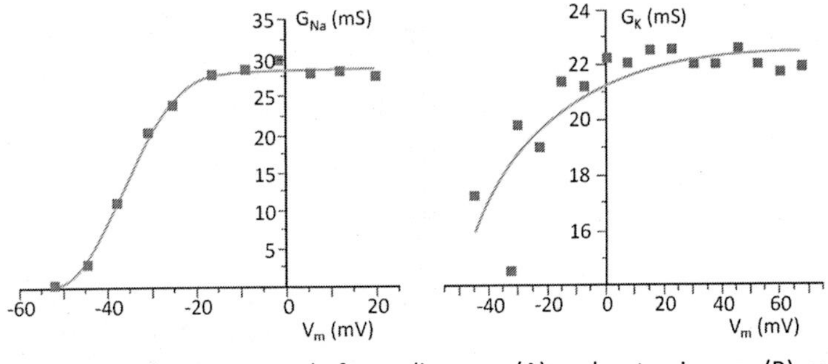

Figure 5.21 Conductance trends for sodium g_{Na} (A) and potassium g_K (B) as a function of membrane potential.

voltage-clamp is known and E_{Na} and E_K are constant, with the same solutions inside and outside the axon, and can be calculated using the Nernst equation.

It can be observed that the two conductances increase with voltage, even if they show different trends, up to about -10 mV for G_{Na} and +10 mV for G_K. Both reach saturation, even if the currents change beyond these values (Figure 5.21), but only due to the effect of the electromotive forces $V_m - E_{Na}$ and $V_m - E_K$, for sodium and potassium, respectively.

Taking into account the experiments in Figures 5.17 and 5.19, it can be stated that the sodium ions carry the inward current, with a reversal potential of about ≅ +50 mV, and that the potassium ions carry the outward current in Figure 5.20. Below -50 mV, no potassium current is recorded, so it is not possible to determine the reversal potential experimentally, which should be ≅ -70 mV.

Many membrane ionic currents are only outward, and this characteristic is due to the fact that some conductances have the peculiar property of behaving like diodes, i.e., they are more likely to let a current pass in one direction than the other (Data Sheet 2.1). A current that exhibits this behavior is the **inward rectifier** I_{KIR} (called I_K in Figure 5.8), which is very important, for example, in ventricular cardiac cells, where it is essential for the stability of the resting potential. I_{KIR} for values of potentials more negative than E_K is inward, depolarizing and very intense. This current is able to rapidly bring the membrane potential back towards the resting value and restore the degree of excitability of the cell. For more positive potential values of E_K, the I_{KIR} current is outward and counteracts any unwanted depolarization (Figure 5.3). However, the simple outflow of potassium ions is present only close to E_K. For more depolarizing membrane potentials, the outward current is inhibited (rectification). This is functional, to avoid competition with the inward sodium current once the ventricle cell receives the stimulus from the pacemaker cells (Figure 5.8). Similarly, for example in the giant squid axon, there is a potassium current, similar to that generically indicated as I_{KV} in Figure 5.8 or I_K in Figure 5.19 or 5.20, called **delayed rectifier** I_{KDR}. In this case, the activation of the current leads to a more positive membrane potential. At membrane potentials between E_K and the current activation (≅ -50 mV), the current is zero (Figure 5.20). The current is also absent in the inward direction for membrane potentials lower than E_K: a perfect diode.

The numerous experiments on the squid axon considered so far, carried out over a decade (Figure 5.15 to Figure 5.21), allow us to construct an electrical model of the membrane (Data Sheet 2.1) consistent with Equations 5.2 and 5.3 (Figure 5.22). The model consists of four branches in parallel: one branch with the capacitor C_m, two resistive branches for sodium G_{Na} and potassium G_K and the batteries E_{Na} and E_K, respectively, and the fourth branch with the conductance G_L and the battery E_L.

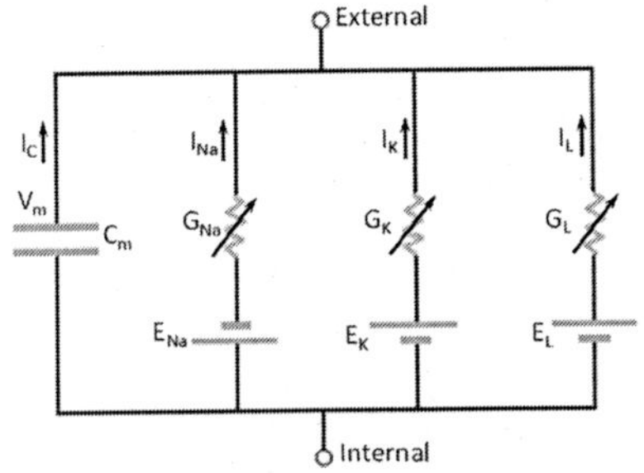

Figure 5.22 Electrical model of the axon membrane. C_m represents the membrane capacitance, G_{Na} and G_K represent the conductances, E_{Na} and E_K are the equilibrium potentials for sodium and potassium, respectively, and I_L, G_L and E_L are the leakage. G_{Na} and G_K are time- and voltage-dependent. All other parameters are constant.

The capacitor C_m represents the dielectric properties of the phospholipid bilayer and conducts the capacitive current I_C. The two resistive branches for sodium G_{Na} and potassium G_K are the variable, voltage-dependent conductances. E_{Na} and E_K represent the respective equilibrium potentials that provide the energy required for the passage of I_{Na} and I_K currents. Finally, the conductance G_L and the battery E_L pass the leakage current I_L. The latter follows the generalized Ohm's law (equation 3.8), is due to non-specific membrane conductances that are always active and is generally neglected if irrelevant to the currents studied or, if not, is determined during the experiment and subtracted from the total measured current. Only G_{Na} and G_K are variable, whereas all other parameters are determined by the biophysical characteristics of the plasma membrane and intrinsic proteins or by the difference in concentration of solutes between the inside and outside of the cell and are therefore constants.

The total current through the plasma membrane of an excitable cell is given by the following equation:

$$I_{TOT} = I_C + I_{Na} + I_K + I_L \quad (5.7)$$

To understand at this point how the excitable cell produces the action potential, it is necessary to study the time dependence of ionic currents, i.e., the behavior of membrane currents with respect to time, once again using the voltage-clamp technique in a squid axon.

5.9 Time dependence of membrane ionic currents

The voltage-clamp depolarization of a squid axon from -65 to -9 mV causes an ionic membrane current that for the first 1.5 ms is inward and then is outward for the whole duration of the depolarization (track A of Figure 5.17). Comparing the experiment carried out in artificial seawater at 100% with that carried out in artificial seawater at 10%, this ionic membrane current is formed by sodium ions for the inward part (track C of Figure 5.17) and by potassium ions for the outward part (track B of Figure 5.17). If we perform a sufficiently large number of measurements of the currents I_{Na} and I_K of Figure 5.17 at different times and use Equations 5.2 for the sodium current and 5.3 for the potassium current in the form

$$G_{Na}(t) = \frac{I_{Na}}{(V_m - E_{Na})} \quad (5.8)$$

and

$$G_K(t) = \frac{I_K}{(V_m - E_K)} \quad (5.9)$$

it is possible to determine the time course of the conductances for sodium and potassium (Figure 5.23), except for $V_m = E_{Na}$ and $V_m = E_K$, as this would have an indeterminate form. In fact, V_m is known, being the potential imposed on the cell by the voltage-clamp, and E_{Na} and E_K can be calculated using the Nernst equation with the known concentrations of the solutions inside and outside the axon (Data Sheet 3.2).

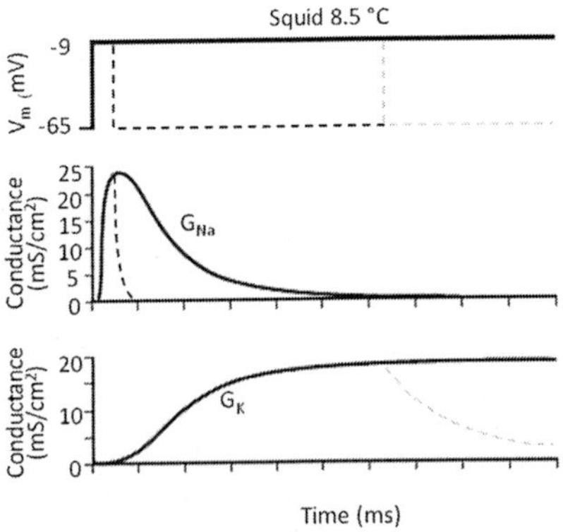

Figure 5.23 A depolarizing stimulus causes the conductances for sodium G_{Na} and potassium G_K to change over time with very different kinetics. Conductance is expressed per unit area in order to compare cells of different sizes.

It is evident that the graphs of the conductances for sodium G_{Na} and potassium G_K (Figure 5.23), even though they reach comparable maximum values, show very different temporal trends in each conductance, even in the presence of constant depolarization in time.

The conductance for sodium G_{Na} shows an increase to a maximum value in ≅ 0.3 ms, defined as **rapid activation**, which produces the initial increase of the sodium current to a peak. Peak current is followed by the current **inactivation** (Data Sheet 3.4) which reduces I_{Na} conductance to zero in ≅ 4 ms (Figure 5.17). The inactivation process is a time-dependent phenomenon. The experiment of Figure 5.23 shows the effect of interrupting the depolarization at the peak of the conductance. Sudden repolarization just before the current inactivation process starts causes a **rapid deactivation** (black dotted line of Figure 5.23). The conductance for sodium G_{Na} goes towards zero with an exponential trend due to the effect of voltage dependent closure of the permeability (Figure 5.20).

The depolarizing stimulus also causes the activation of the conductance for potassium G_K, at a rate ten times slower than G_{Na}, reaching the maximum value for the level of membrane depolarization about 2 ms after the peak of the conductance for sodium and with a sigmoid time trend. It then remains at this value for the entire duration of the stimulus without showing inactivation. At the cessation of depolarization and the restoration of the resting potential, for example after 6 ms, deactivation occurs up to the value of zero with a negative exponential time trend (grey dotted line of Figure 5.23). The potassium current follows the conductance trend in time: it starts to increase significantly only after about 1 ms from the stimulus and at about 0.5 ms from the peak of the sodium current, becoming significantly large only after about 2 ms, when the sodium current has reduced by about 70% (track B of Figure 5.17).

In conclusion, it can be stated that in excitable cells, the membrane currents responsible for generating the action potential have been directly measured. The actual properties of the membrane permeabilities are an accurate validation to the hypothesis formulated at the beginning of this Chapter concerning the characteristics of the currents.

The action potential is a transient variation of the resting potential of excitable cells which, in its various forms, is used by different cell types as a signal to ensure communication between cells of the same or different types. The excitation wave is caused by the activation, either spontaneously or as a result of a stimulus, of ionic currents that cross the plasma membrane.

The main characteristics of membrane currents are voltage dependence and time dependence, without which it would not be possible to generate the action potential. The activation of the TTX-sensitive sodium current and the TEA-sensitive potassium current are voltage-dependent, while the inactivation of the sodium current is voltage- and time-dependent, a property without which the membrane potential could not return to its resting value. The modulation of conductance parameters by voltage and time are intrinsic features of the membrane proteins responsible for ionic fluxes.

In contrast, current deactivation phenomena are extrinsic factors. However, the rapid deactivation of the sodium current is a different phenomenon from the deactivation of the potassium current. In the case of sodium, it is caused by an early end of the depolarizing stimulus that brings the membrane potential back to resting values, where the conductance of sodium is minimal. For the potassium current, deactivation occurs because repolarization brings the potential to values close to E_K and the current is exhausted due to the loss of the driving force for potassium.

At this point, it is possible to construct a mathematical model of the currents flowing through the plasma membrane that can describe each ionic current in a voltage- and time-dependent manner. The analogue formulation will be followed by the digital interpretation, supported by recent advances in the disciplines of cellular and molecular biology, based on direct measurements of the individual components responsible for the phenomenon of cellular excitability.

5.10 The Hodgkin and Huxley equation for sodium and potassium currents

Equations 5.2 for the sodium current and 5.3 for the potassium current, taking into account Equations 5.5, 5.6, 5.8 and 5.9, become:

$$I_{Na} = G_{Na(V_m,t)} \cdot (V_m - E_{Na}) \quad (5.10)$$

and

$$I_K = G_{K(V_m,t)} \cdot (V_m - E_K) \quad (5.11)$$

respectively, where $G_{Na(Vm,t)}$ and $G_{K(Vm,t)}$ are the conductances for sodium and potassium, respectively, depending simultaneously on the membrane potential V_m and on time t.

Performing voltage-clamp experiments with a stimulus protocol such as that in Figure 5.18 A results in a family of current curves recorded at various potential values, such as those in Figure 5.19 B for sodium and Figure 5.19 D for potassium. From these curves, it is possible to obtain the current values at different times and at different membrane voltages. Using these data in Equations 5.10 and 5.11, it is possible to calculate the corresponding conductance values for sodium and potassium. Finally, for each set of conductance values determined for each potential value, it is possible to draw the best fit, which graphically represents the time course of the conductances.

In order to characterize these conductances analytically, a few considerations are important:

1. even if the membrane potential increases, the conductances for sodium G_{Na} and potassium G_K reach a stationary maximum value (Figure 5.21);
2. for very long depolarizations at various potentials, the conductance for sodium reaches a peak (Figure 5.24 A). For potassium, the maximum is a stationary level (Figure 5.24 B) for potentials greater than +28 mV;
3. the currents that cross the membrane are relatively large compared to the size of the cells and are modulated by the membrane potential and

time (Figure 5.24);

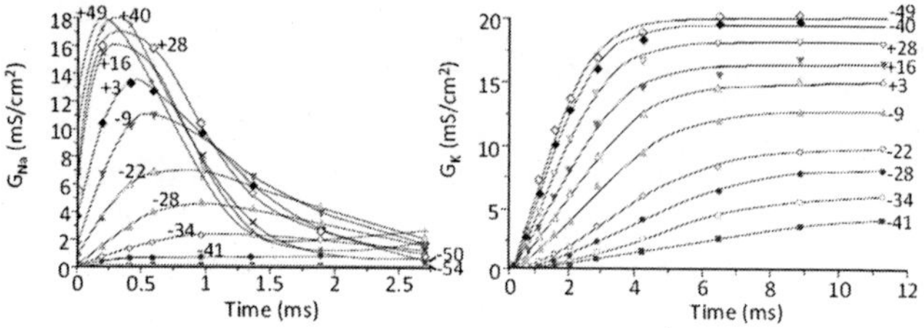

Figure 5.24 Time courses of conductance for sodium G_{Na} (A) and potassium G_K (B) for long depolarizations at various membrane potentials (the numbers on the curves) measured in two different axons. Symbols represent conductance values calculated with Equations 5.10 and 5.11 from families of current curves such as those in Figure 5.19B and 5.19D.

4. the conductance increases as the number of channels in the open state increases, but only up to a maximum stationary value, because the number of channels is finite;
5. the proteins constituting the channels are subject to spontaneous conformational changes due to thermal agitation of the molecules, each of which is instantaneous, all-or-nothing and random;
6. the dependence of conductances on the membrane potential (Figure 5.21) suggests that changes in ion permeability may be due to particular conformational changes induced by the variation of the electric field across the membrane.

On the basis of considerations of this type and the experimental data obtained up to that time, even without knowing the chemical and biophysical characteristics of plasma membranes, Hodgkin and Huxley (1952b) attributed the onset of currents and the voltage-dependent and time-dependent kinetics of membrane permeability to molecular mechanisms involving membrane channels. They proposed a mathematical model, universally known as the **Hodgkin and Huxley model** (**HH model**) for the action potential propagated in giant squid axons, which could describe with mathematical accuracy the course of conductances for sodium and potassium as obtained from the experiments in Figure 5. 24. This allows the theoretical reconstruction of the action potential.

If we consider the kinetics of the potassium conductance G_K during a depolarization, we observe a first phase of increase that is delayed with respect to the change of potential at t = 0, with a time trend that can be described with a sigmoid (continuous curve G_K of Figure 5.23). This is followed by a second phase in which it decreases, as soon as the stimulus ceases, with a time trend that can be described with a negative exponential (grey dotted line of Figure 5.23). Considering the interpolation of the activation and deactivation curves for the conductance of potassium, we realize that the single exponential function does not give a satisfactory result (n trace of Figure 5.25).

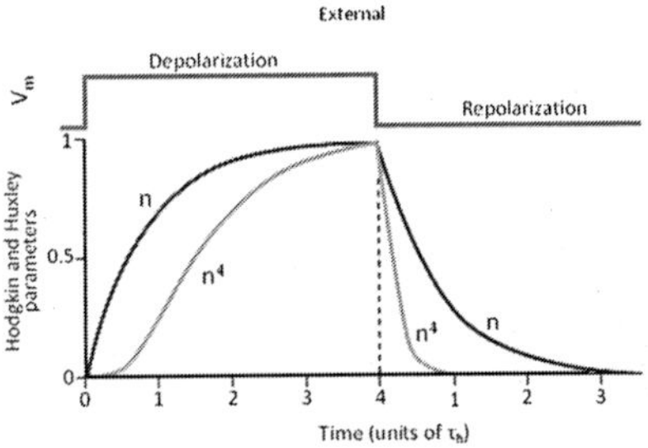

Figure 5.25 Hypothetical time course, according to the Hodgkin and Huxley model, of n and n^4 in the presence of depolarization followed by repolarization. τ_h is the time constant of the h-curve in Figure 5.26.

In order to reach a best fit, we need a function to the fourth power (n^4 trace of Figure 5.25). Therefore, we can write:

$$G_{K(V_m,t)} = \bar{G}_K \cdot n^4 \qquad (5.12)$$

which is graphically represented by the curve n^4 in Figure 5.25 and whose temporal trend is perfectly superimposable with that represented by the continuous lines in Figure 5.24B. $G_{k(Vm,t)}$ is the voltage- and time-dependent conductance for potassium and the proportionality constant G_K is the maximum conductance when all the channels for potassium available in the membrane are open. n is a percentage, dimensionless, varies from 0 to 1 and represents the fraction of channels actually open. It can be calculated, if its initial value is known, with the differential equation:

$$\frac{dn}{dt} = \alpha_n \cdot (1 - n) - \beta_n \cdot n, \qquad (5.13)$$

where n is time-dependent and indicates at each time interval dt the fraction of open channels, while 1 – n indicates the fraction of closed channels. n determines the delay in the increase and decrease of G_K in the time of Figure 5.24B due to the random latency time of each channel. It is close to 0 for negative potentials and close to 1 during depolarization (Figure 5.25). α_n and β_n are constants of the transition from closed to open and from open to closed channels, respectively. Having experimentally fixed certain parameters, such as temperature and the effect of sodium substitution by choline, α_n and β_n are described by the exponential equations:

$$\alpha_n = 0.01 \cdot \frac{V_m + 10}{\exp\frac{V_m + 10}{10} - 1};\ \beta_n = 0.125 \cdot \exp\frac{V_m}{80}, \tag{5.14}$$

and are voltage-dependent, as the transitions occur due to conformational change following the variation of the electric field by membrane depolarization. They assume a value close to 0 for negative potentials and close to 1 during depolarization (Figure 5.25).

The final form of Equation 5.11 therefore becomes:

$$I_K = \bar{G}_K \cdot n^4 \cdot \left(V_m - E_K\right) \tag{5.15}$$

While the conductance for potassium G_K increases in the presence of membrane depolarization and decreases when the membrane becomes hyperpolarized, the conductance for sodium G_{Na} shows a rapid increase that lasts for about 0.5 ms and is followed by a reduction until it is completely cancelled, even if depolarization persists (Figure 5.23). This behavior can be described by two variables that obey two first-order differential equations. The variable m controls the activation process that produces the transition from **closed to open channel** and vice versa; the variable h controls the conformational change of a domain, a different

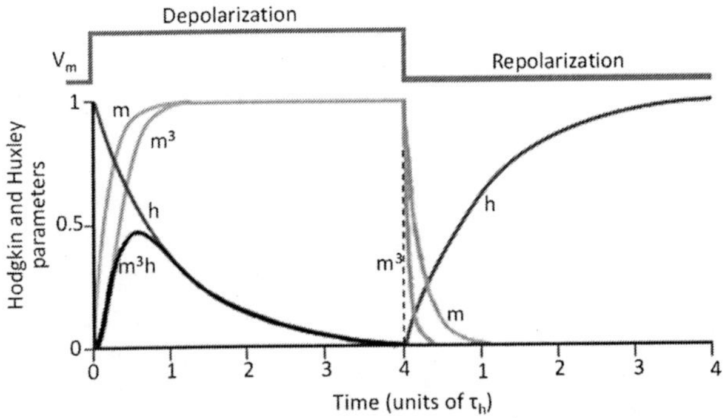

Figure 5.26 Hypothetical time course, according to the Hodgkin and Huxley model, of m, m^3, h, and m^3h in the presence of depolarization followed by repolarization. τ_h is the time constant of the h curve.

change from the previous one, that determines the transition from **non-inactivated to inactivated channel** and vice versa, as demonstrated by the conductance trend for time-dependent sodium (Figure 5.23). Therefore $G_{Na(Vm,t)}$ of Equation 5.11 becomes:

$$G_{Na(V_m,t)} = \bar{G}_{Na} \cdot m^3 \cdot h \qquad (5.16)$$

which is graphically represented by the curve m^3h of Figure 5.26, whose temporal trend is perfectly superimposable with that represented by the continuous lines of Figure 5.24A, which in turn are perfectly compatible with the experimental points. The constant G_{Na}, proportional to the third power of m, is the maximum conductance when all of the sodium channels available in the membrane are open. m and h are percentages, dimensionless, vary from 0 to 1, represent the fraction of channels actually open and inactivated, respectively (Figure 2.26). They can be calculated, if their initial value is known, with the differential equations:

$$\frac{dm}{dt} = \alpha_m \cdot (1 - m) - \beta_m \cdot m \qquad (5.17)$$

and

$$\frac{dh}{dt} = \alpha_h \cdot (1 - h) - \beta_h \cdot h \quad (5.18)$$

respectively, where m is the fraction of open channels, 1 - m is the fraction of closed channels, h is the fraction of non-inactivated channels and 1 - h is the fraction of inactivated channels (Figure 5.24A). m is time-dependent, indicating at each time interval dt the fraction of open channels; it determines the increase in G_{Na}. h is time-dependent, indicating at each time interval dt the fraction of inactivated channels and the decrease in G_{Na}. At negative potentials, m is close to zero, as the conductance for sodium is low, and h is close to 1, as the channels are non-inactivated (Figure 5.26). α_m and β_m are the transition constants of the channels from closed to open; α_h and β_h are the transition constants of the channels from non-inactivated to inactivated and vice versa.

Once experimentally set, some parameters such as temperature and the values of α_m and β_m, when the potential is at rest, are described by the exponential equations

$$\alpha_m = 0.1 \cdot \frac{V_m + 25}{\exp\frac{V_m + 25}{10} - 1}; \ \beta_m = 4 \cdot \exp\frac{V_m}{18} \quad (5.19)$$

and α_h and β_h by the exponential equations

$$\alpha_h = 0.07 \cdot \exp\frac{V_m}{20}; \ \beta_h = \frac{1}{\exp\frac{V_m + 30}{10} + 1}. \quad (5.20)$$

α_m, β_m, α_h and β_h are voltage-dependent, as transitions occur due to simultaneous and **instantaneous** conformational changes of membrane domains as a result of change in the electric field due to membrane depolarization.

The final form of Equation 5.10 therefore becomes:

$$I_{Na} = \bar{G}Na \cdot \left(m^3 h\right) \cdot \left(V - E_{Na}\right) \qquad (5.21)$$

Finally, taking into account Equations 5.15 and 5.21 and that by Ohm's law, we have:

$$I_L = G_L \cdot \left(V - E_L\right) \qquad (5.22)$$

where G_L is the constant leakage conductance (Paragraph 5.8), Equation 5.7 becomes:

$$I_{TOT} = C_m \frac{dV_m}{dt} + \bar{G}_K n^4 \left(V_m - E_K\right) + \bar{G}_{Na} m^3 h \left(V_m - E_{Na}\right) + \bar{G}_L \left(V_m - E_L\right) \qquad (5.23)$$

The total membrane current is a function of time, since m (Equation 5.17), h (Equation 5.18), n (Equation 5.13) and the capacitive current (Equation S2. 2) are time-dependent, although the latter is negligible in the propagation of the action potential in an axon, and it is a function of voltage, since α_m and β_m (equation 5.19), α_h and β_h (equation 5.20), α_n and β_n (equation 5.14) are voltage-dependent.

Now, consider the following elements: 1) Ohm's law, which links current to voltage and resistance, 2) the specific membrane resistance expressed as resistance per unit length of the fiber, 3) the fiber radius and the estimated conduction velocity in an axon, combined in the constant K. With these three elements in mind, and with a series of mathematical steps and considerations that can be explored further (Hodgkin and Huxley, 1952b), Equation 5.23 gives the overall equation for the Hodgkin and Huxley model:

$$\frac{d^2V_m}{dt^2} = K\left\{\frac{dV_m}{dt} + \frac{1}{C_m}\left[\bar{G}_K n^4 \left(V_m - E_K\right) + \bar{G}_{Na} m^3 h \left(V_m - E_{Na}\right) + \bar{G}_L \left(V_m - E_L\right)\right]\right\} \qquad (5.24)$$

The propagated action potential can be reconstructed by numerically integrating Equations 5.13, 5.17, 5.18 and 5.24. We start from an estimated membrane potential value at time t + dt, for an initial membrane depolarization of 0.1 mV with respect to $V_{m(t=0)}$ and with a dt of 0.01 ms. With this potential value, we determine the voltage-dependent parameters α_n, β_n, α_m, β_m, α_h and β_h with Equations 5. 14, 5.19 and 5.20, taking into account a Q_{10} of 3, and then the time-dependent parameters n, m and h by integrating Equations 5.13, 5.17 respectively. We then calculate the values G_K and G_{Na} from Equations 5.12 and 5.16, respectively, and dV_m/dt from Equation 5.24. If the value of the potential is consistent with that measured experimentally, the mathematical procedure is repeated with a new $V_{m(t+dt)}$, otherwise the calculations are repeated by appropriately modifying the parameters. Figure 5.27 shows the result of this complex mathematical procedure, assuming a stimulus depolarization of 15 mV and integration intervals of 0.01 ms in the initial faster part of the action potential, 0.02 ms in the intermediate phase and 1 ms for the final slow repolarization.

Figure 5.27 allows the following observations to be made:

a) the trend of V_m is perfectly superimposable with the experimental one of Figure 5.15;

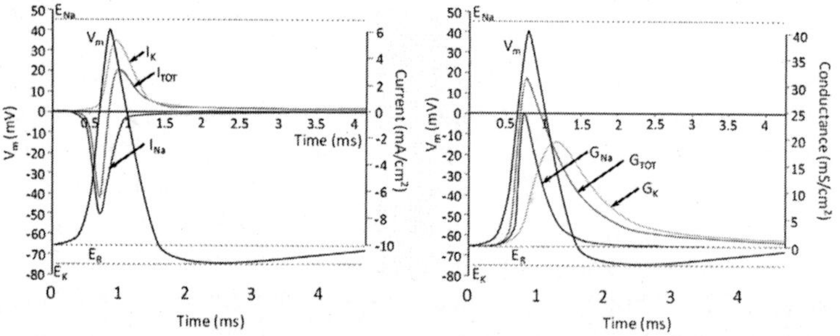

Figure 5.27 Reconstruction according to the Hodgkin and Huxley model of the propagated action potential calculated in a node of Ranvier of a squid axon. In A, the total current (I_{TOT}), sodium current (I_{Na}) and potassium current (I_K) are shown. In B, the total conductance (G_{TOT}), conductance for sodium (G_{Na}) and conductance for potassium (G_K) are shown. V_m = membrane potential, V_R = resting potential, E_{Na} = equilibrium potential for sodium, E_K = equilibrium potential for potassium. The phenomenon of after hyperpolarization is due to the clear predominance of the potassium current in that phase of the propagated action potential.

b) the curves G_{Na} and G_K are compatible with the experimental ones of Figure 5.23, taking into account that the latter were obtained with a long and constant depolarization in voltage-clamp with respect to V_m of Figure 5.27, which varies continuously in time;
c) the action potential remains between the equilibrium potential of potassium E_K and sodium E_{Na};
d) G_{Na} reaches the maximum value when the action potential has the greatest variation in time (dV/dt) during the fast depolarization (flexus). G_{Na} decreases due to time- and voltage-dependent early inactivation;
e) after the peak, which is reached about halfway through repolarization of the action potential, the potassium conductance deactivates. The outward potassium ion flux decreases (I_K in Figure 5.27) due to the progressive reduction in the driving force as the membrane potential approaches the equilibrium potential for potassium.

5.11 The inactivation of the sodium current

Sodium current inactivation can be characterized by a voltage-clamp experiment, such as that illustrated in Figure 5.28, using the vaseline-gap method (Data Sheet 5.4) in a node of Ranvier of a myelinated frog axon. A fixed depolarization at -15 mV (V_{TEST}) is preceded by a conditioning pre-pulse at various potentials between -96 and -50 mV lasting at least 50 ms (V_{PRE}), which allows inactivation to reach a time-independent steady state.

5.11.1 The h∞ curve

The sodium currents recorded at the various potentials (Figure 5.28 A) show that the peak current decreases as the preconditioning depolarization increases. The ratio between the current recorded at -15 mV (V_{TEST} of Figure 5.28 A) and the currents recorded at the various preconditioning potentials (V_{PRE} of Figure 5. 28 A), gives the steady-state inactivation curve, or h_∞ curve, for the sodium current as a function of voltage. In agreement with the h_∞ curve at membrane potentials below -90 mV, the non-inactivated channels comprise about 100% of the population and the inactivated channels are about 0%. Moreover, at the

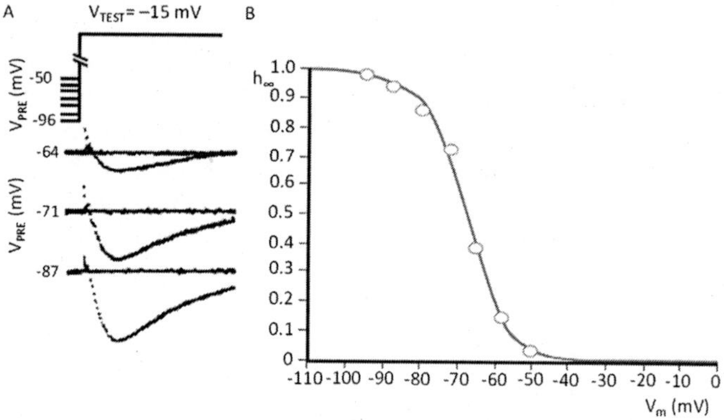

Figure 5.28 A) The peak evoked sodium current at 15 mV (V_{TEST}) in a myelinated frog axon node after conditioning pre-pulses at various potentials (V_{PRE}) decreases as depolarization increases. B) By taking the ratio of the maximum peak sodium current to that at the various V_{PRE} potentials, the h∞ curve is obtained. The arrow indicates the resting potential (V_{rip}) of the cell analyzed.

resting potential of -75 mV, the non-inactivated channels are about 70% and the inactivated channels are about 30%. Finally, at potentials around 0 mV, the non-inactivated channels are about 0% and the inactivated channels are about 100% (Figure 5. 28B). Therefore, with a hyperpolarization at -95 mV, there is a complete removal of inactivation.

The parameter h∞ , which is time-independent, gives the percentage of inactivation of the sodium channels equal to that defined by h of Equation 5.24. In contrast, the parameter h is time-dependent, but only at time t = 0 of the action potential. The cell has been "preconditioned" for a long time to the negative resting potential close to -80 mV, therefore in an experimental situation similar to that of Figure 5.28. As time passes, the membrane potential increases steadily. Rapidly, the number of channels activated according to parameter m increases; the number of inactivated channels increases with slower kinetics according to parameter h (Equation 5.18 and Figure 5.26) and a sufficiently high amount of sodium current can cross the membrane and depolarize the cell.

In general, in excitable cells, particularly in axons and conduction processes with peripheral targets, the time-dependent behavior of sodium current inactivation is not significant to understand the action potential generated by sodium and potassium currents. In fact, as mentioned above h∞ is studied by preconditioning cells with long depolarizations at various potentials chosen by the investigator, whereas depolarization under physiological conditions is given by the continuous variation of the action potential.

5.11.2 Recovery from inactivation

It is useful, from a physiological point of view, to evaluate the time course of recovery from sodium channel inactivation. The experiment has been done examining voltage, using the vaseline-gap configuration (Data Sheet 5.4). Sodium and potassium currents are activated in the squid axon by a stimulus from the holding potential of -75 mV to a test potential of -15 mV (pulse 1 of Figure 5.29). Maintaining the cell depolarized for few milliseconds completely inactivates the inward sodium current. A subsequent depolarization identical to the first, after 0.5 ms (pulse 2 of Figure 5.29), evokes only the outward potassium current (track 1 of Figure 5.29). With a progressive increase in the time between the two impulses, there is a progressive increase of the peak of the sodium current (track 2 and 3 of Figure 5.29) until it becomes identical to that of the current induced by impulse 1 (track 4 of Figure 5.29). From the ratio of the peaks of the sodium current produced by pulse 2 and pulse 1 at the various time intervals between the two pulses (I_{NaREL} of Figure 5.29), we obtain the exponential time trend of the recovery from the inactivation of the sodium current (Figure 5.29B) according to the equation:

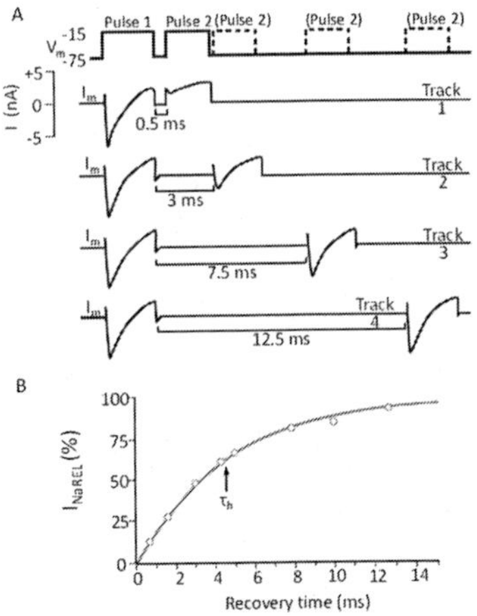

Figure 5.29 With a two-pulse experiment operating in voltage-clamp in the vaseline-gap configuration (A), it is possible to construct the recovery curve from sodium current inactivation (B).

$$I_{NaREL} = 1 - \exp\left(\frac{t}{\tau_h}\right) \quad (5.25)$$

where I_{REL} is the ratio between the sodium current activated by pulse 2 and pulse 1 (Figure 5.29A) and τ_h is the time constant of the inactivation recovery. In the experiment of Figure 5.29, τ_h = 4.6 ms, which represents the time necessary for 66.67% of the sodium channels to recover from inactivation, a percentage that could be sufficient for the cell to generate a new action potential; 12 ms are necessary for complete recovery (Figure 5.29B).

5.11.3 The physiology of inactivation

The phenomenon of the inactivation of the sodium current is an important factor that allows cell excitability to function correctly for various reasons. Without inactivation of the sodium current, after an initial fast depolarization, the membrane potential would stabilize at an intermediate value between the equilibrium potentials of sodium and potassium (Paragraphs 5.3). Repolarization towards the resting potential would be impossible, and the potential energy given by the electrochemical gradients that provide the driving force for the sodium and potassium ions would be compromised. Furthermore, the inactivation of the sodium current lasts as long as the membrane potential is depolarized. This allows potassium to be the only active membrane current after the peak of the action potential. In this way, the outward current is free to repolarize the membrane towards the cell resting potential in order to prepare the cell for a new action potential. In some excitable cell types (Figure 5.27), hyperpolarization reaches the potassium equilibrium potential, a phenomenon called "after hyperpolarization". This lasts until the voltage-dependent removal of the inactivation of the sodium current, the restore of the small depolarizing current that slightly balances the potassium current and move the membrane potential back to around its resting value.

Finally, an excitable cell, and in particular an axon, must be able to produce action potentials that are all the same in form, duration and amplitude, in order to send frequency-coded messages to target cells, such as muscle cells, while a single action potential carries no useful information. In the language used by the nervous system, the upper limit at which messages can be sent, i.e., the maximum frequency that a sequence of action potentials can reach (firing, Paragraphs 6.4), is determined by the charge of the membrane capacitors and the duration of the action potential and thus also by the inactivation of the sodium current (Paragraphs 6.4).

Finally, inactivation of the sodium current causes the conduction of the action potential along the axons to be unidirectional, as will be seen below (Paragraphs 6.2.4).

5.12 From membrane current to channel population dynamics

Hodgkin and Huxley had already foreseen that the study of excitability mechanisms had to be done at the molecular level (Paragraphs 5.10). However, it was Edwin Neher and Bert Sakmann (1976) who, taking advantage of technological advances in electronics, micromechanics, microscopy and cell biology techniques, were able to isolate cells from various tissues and study them individually, developing the patch-clamp technique rather than using the voltage-clamp (Paragraphs 5.5). Their first results from molecular dynamics studies revealed that the single-channel protein, with a conformational change, allows or denies ions to pass through a biological membrane, studies for which they received the Nobel Prize for Medicine in 1991. Later, the patch-clamp technique was perfected in order to be able to study even intact tissues or intracellular membranes, such as the endoplasmic reticulum and the nuclear envelope, leaving the mechanical, chemical and physical interactions within and between cells untouched. In addition to the above-mentioned functional aspects, the introduction of molecular genetics and molecular biology techniques in the study of excitability phenomena has led to significant progress on structure. These techniques made it possible to identify and sequence the many membrane proteins that form ion channels, to reconstruct their secondary, tertiary and quaternary structure, and to discover, through targeted mutations of individual amino acids, the sites of ion permeation, regulatory sites and residues that can be activated by phosphorylation.

Spontaneous mutations of individual amino acids that are responsible for the malfunctioning of particular ion channels, leading to various diseases, have also been identified, including serious conditions such as cystic fibrosis, for which a chloride channel is responsible. The combined studies of structure and function have provided, and are still providing, very useful data for the synthesis of increasingly targeted drugs and anaesthetics with fewer and fewer side effects. A typical example is a substance such as lidocaine, a transient blocker of voltage- and time-dependent sodium channels, which is widely used in dentistry as a local anaesthetic.

In order to understand the relationship between membrane channels and macroscopic currents such as those of sodium or potassium (Figure 5.17), which are a few hundred nanoamperes (10^{-9} amperes) large, we can hypothesize that these currents are due to the sum of many individual microcurrents, each in the order of a few tens of picoamperes (10^{-12} amperes), carried by individual channels. This first hypothesis, which we could call **analogue**, would imply that each channel should respond to a stimulus with a microscopic fraction of the macroscopic current. It should therefore possess a molecular mechanism of opening and closing, or inactivation, that is voltage- and time-dependent. In practice, each individual element should have a structure that allows it to open and close the channel in a discretionary manner. In this way, the set of openings of individual channels responding to a stimulus could generate a macro-current with the kinetics described by the model of Hodgkin and Huxley (Equations 5.15 and 5.21). This would mean, at the molecular level, the production by the organism of very elaborate proteins, with a considerable expenditure of energy and a higher probability of error.

The second hypothesis, which we could call **digital**, assumes that the selective channels for a particular ionic species are all the same and can only assume two conformations: open channel or ON or closed channel or OFF. The individual protein, in this case, could always oscillate randomly between the two conformations in dynamic equilibrium. Voltage and time would act as modulators of the probability of the channel spending more time in one position than the other. In this hypothesis, each channel would contribute a "quantum" of the macroscopic current.

Patch-clamp experiments have shown with direct recordings that individual ion channels behave according to a digital logic, supporting the second hypothesis. Each ion channel at a certain temperature oscillates randomly between two states. Changes in voltage and time act as modulators of these random transitions. For voltage- and time-dependent ion channels, the probability of a protein assuming a certain conformation depends on the membrane potential and the time spent at that potential value. At a resting potential of -70 mV, for example, the probability of opening the TTX-sensitive sodium and delay rectifier potassium channels is generally very low, whereas the probability of opening the inward rectifier potassium channel is high, so the resting potential is stabilized at a value close to the equilibrium potential of potassium.

The voltage dependence and the time dependence of the channel, which in the model of Hodgkin and Huxley are described by the parameters α and β (Equations 5.14, 5.19 and 5.20), in reality, consist of a modulation of the probability that the channel is in a certain conformation.

In order to verify the above concepts, it is possible to construct a computer simulation that takes into consideration a theoretical model of a membrane, in which a generic ion channel is inserted and given the possibility of oscillating randomly between two states: closed and open. The probability that the ion channel assumes an open or closed position is voltage- and time-dependent. In addition, the ion channel is attributed the characteristic of responding with a certain delay to a sudden change in potential (Figure 5.30).

The simulation starts from a membrane potential of -100 mV where, due to the characteristics conferred on the hypothetical ion channel, the probability of passing from the closed to the open state is very low. If ten depolarizations are applied to the cells, each bringing the membrane potential to -40 mV, the probability of the transition from the closed to open state will increase, while still maintaining the property of randomness. Thus, we obtain some traces in which there are long openings (traces 2, 3, 4, 7, 8 of Figure 5.30) and others in which there is no change (traces 1, 5, 6, 9, 10 of Figure 5.30). Ultimately, the channel switches from the closed to the open conformation and vice versa in a random manner, both in terms of the moment of opening or closing and the duration. Each channel protein has its own specific property of increasing or decreasing the probability of one of the conformations depending on the level of depolarization and the time elapsed since the change in potential.

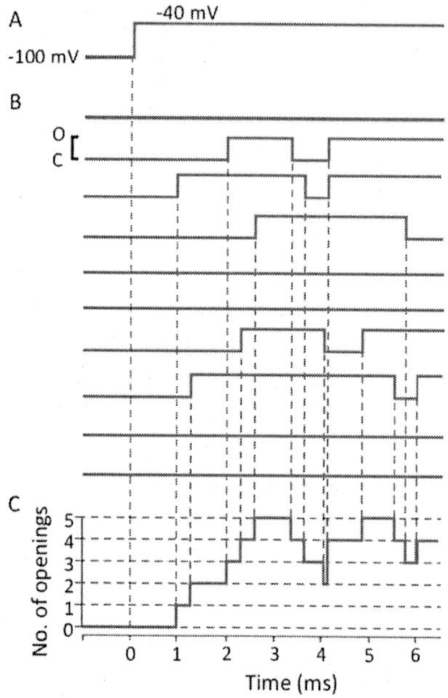

Figure 5.30 A simulation of the depolarization of a cell with ten stimuli (A) evokes single-channel currents by random conformational change and transition from the closed state (indicated by C) to the open state (indicated by O) and vice versa at different times (B). C) Summation of the ten currents above.

The macro-current, resulting from the summation of the single traces (Figure 5.30C), despite being the result of a few simulations, already shows defined kinetics: the current flow has a delayed activation, a subsequent increase in time and constant maintenance of the current throughout the duration of the stimulus. For the summation to no longer show a trace of the single events, in practice to pass from a digital to an analogue situation, it is sufficient to increase the number of active channels. Figure 5.31 presents the same simulation as Figure 5.30, in which the elements contributing to the summation increase from 5 to 10, to 25, to 100 and finally to 1000, with a progressive decrease, until extinction, of the noise accompanying the currents' trend.

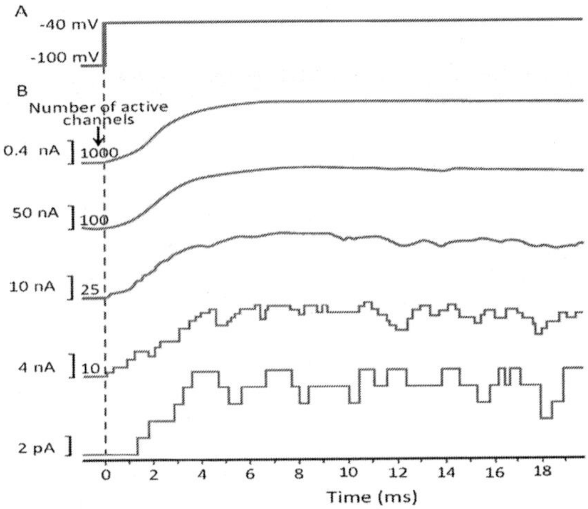

Figure 5.31 In a membrane model with a variable number of equally voltage- and time-dependent channels, depolarization (A) produces a family of currents whose amplitude increases as the number of activated channels increases (B). nA = nanoampere; pA = picoampere.

The voltage dependence of the membrane currents, seen as modulation of the probability that the channels are in a certain conformation, is described in the model of Hodgkin and Huxley by the parameters α and β in Equations 5.13, 5.17 and 5.18. The modulation by voltage and time of the probability of individual channel protein conformations is specific to the protein residues involved in the conformational changes and are therefore **intrinsic characteristics** of individual proteins. The voltage-dependent and time-dependent potassium and sodium channels that are involved in generating the action potentials of excitable cells have dynamics similar to those described above, with the addition, for the sodium channel, of the inactivated channel conformation, in addition to the closed and open channel conformations. The link between voltage dependence, time dependence and probability, which determines the behavior of the channels, makes it necessary to update the parameters of the Hodgkin and Huxley model, which describes the kinetics of membrane currents during an action potential (Equation 5.24).

5.13 New equations for voltage and time dependent currents

Following the experiments using the patch-clamp technique by Neher and Sakmann (1976), which characterized the voltage dependence and time dependence of the channels involved in the action potential (Paragraphs 5.10), it is necessary to update Equation 5.15 for potassium currents and Equation 5.21 for sodium currents.

5.13.1 Sodium channels

To study the behaviour of the sodium current, in particular the fast TTX-sensitive behavior, it is possible to extrapolate the whole cell kinetics from single-channel patch-clamp recordings, with appropriate mathematical models (Data Sheet 5.4) In the cell-attached configuration (Figure 5.32 A),

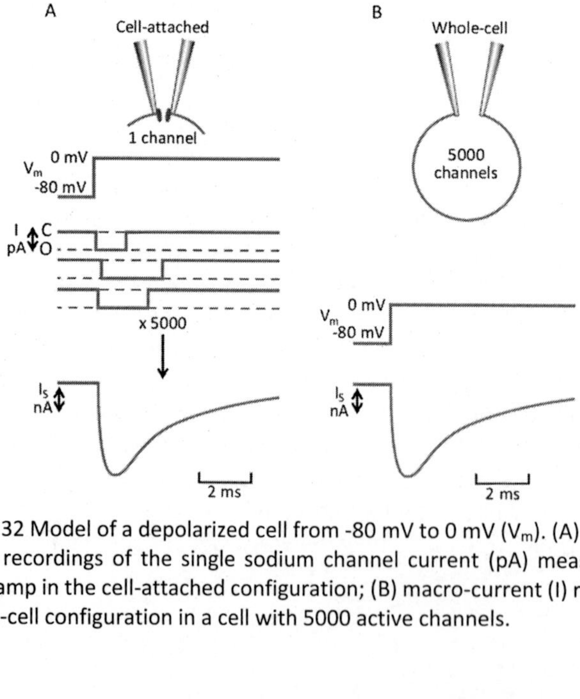

Figure 5.32 Model of a depolarized cell from -80 mV to 0 mV (V_m). (A) Sum (I_S) of 5000 recordings of the single sodium channel current (pA) measured in patch-clamp in the cell-attached configuration; (B) macro-current (I) recorded in whole-cell configuration in a cell with 5000 active channels.

and of all the channels present on the plasma membrane in the whole-cell configuration (Figure 5.32 B), in the presence of a depolarization from the resting potential of -80 to 0 mV (V_m of Figure 5.32).

From the analysis of the single-channel currents, the probability of the channel to be in the open (O) configuration increases due to the voltage dependence. The opening of the channel is more likely during the first 2-2.5 ms of the depolarization due to the time dependence of the channel. In addition, the algebraic sum of the current traces evoked by 5000 successive identical depolarizations yields the graph of a macro-current (I_S of Figure 5.32A) of nanoampere amplitude, which is activated in less than 1 ms after the voltage- and time-dependent increase in the probability of switching from the closed to the open channel conformation. Subsequently, the inward current is inactivated, despite that depolarization is maintained at a potential that promotes an increase in the probability of opening, by a mechanism that is primarily time-dependent. The mechanism implies that for channels in an open configuration, the probability of switching from the open to inactive configuration increases with time. In parallel, a recording of a single current induced by a single depolarizing voltage step in a hypothetical cell that contains 5000 channels shows a macro-current equal to the sum of those of a single channel (I of Figure 5.32 B). In practice, the summation of 5000 currents recorded in sequence in cell-attached configuration is essentially equal to the simultaneous recording of 5000 channels in whole-cell. Even in a real cell-attached experiment, in which the current of a single sodium channel of an excitable cell is recorded in the presence of a depolarization from -80 mV to -40 mV, the channel openings are all concentrated in the first 3 ms (I_m of Figure 5.33A). The average of the currents (Figure 5.33B) has a time course of activation and inactivation comparable to that of the cell model in Figure 5.32. Keep in mind that in an experiment of this kind, it is not possible to establish the exact number of channels present in the patch without recordings that last tens of minutes and special statistical models.

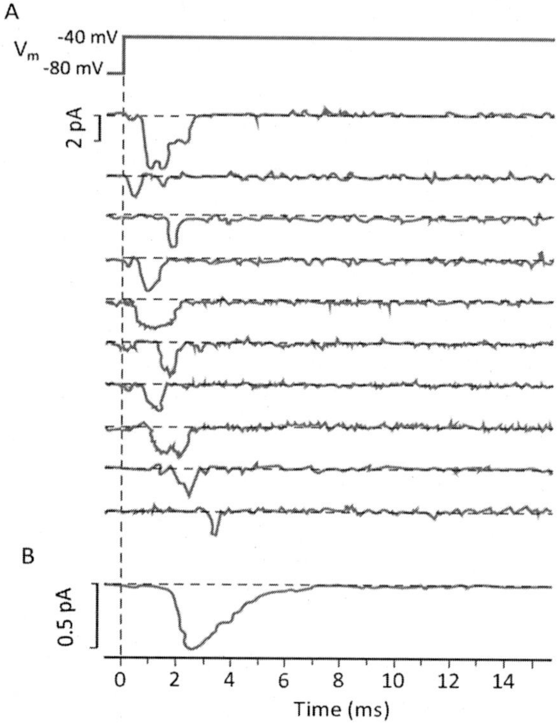

Figure 5.33 Multiple single-channel sodium current recordings triggered by a depolarization from -80 mV to -40 mV (A) and the average of all recorded currents (B).

However, observing the first trace at the top, we can affirm that in this case, there are at least 2 channels, since the current has an amplitude double that of the others and goes to zero with two successive jumps of I_m.

Therefore, the conformations of both closed and inactivated channels give the same result, i.e., they prevent the passage of current. The substantial difference, from the molecular point of view, is that the charged amino acid residues that prevent the channel from conducting ions by changing conformation are different in the two cases (Data Sheet 4.4). From a kinetic point of view, and again depending on voltage and time, a channel can switch from closed to open configuration and vice versa. Open or closed channels can switch to the inactive configuration mainly in a time-dependent manner. Conversely, an inactivated channel must close to return to the open configuration. This conformational change is ensured by repolarization during the action potential, when the inactivated channel is 'forced' to the closed channel configuration in a voltage-dependent manner, ready to reopen upon subsequent depolarization.

5.13.2 Potassium channels

An experiment similar to that in Figure 5.32 can be set up to study the potassium current as the delayed rectifier that repolarizes the action potential. A depolarization from -100 mV to 0 mV (Vm of Figure 5.34) is

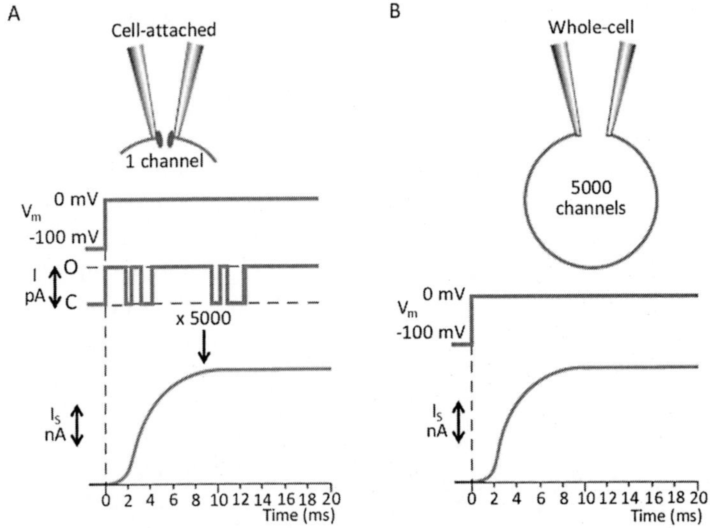

Figure 5.34 Model of a cell depolarized from -100 mV to 0 mV (V_m). A) Sum (I_S) of 5000 recordings of a single potassium channel current (pA) measured in patch-clamp in the cell-attached configuration; (B) macro-current (I) recorded in whole-cell configuration in a cell with 5000 identical active channels.

applied to a generic cell model with a potassium channel using the patch-clamp technique in the cell-attached configuration (Figure 5.34 A), which increases the probability that the channel switches from the closed (C) to the open (O) state. Unlike the sodium channel, the openings and closures follow each other for the duration of depolarization. If we then depolarize the cell 5000 times and take the algebraic sum of the recorded currents, we obtain a macro-current (I_S of Figure 5.34 A) that is identical in amplitude and time course to that recorded in the whole-cell with the same depolarizing step (V_m) in a cell that contains 5000 identical potassium channels (I of Figure 5.32 B) The activation of potassium channels is voltage- and time-dependent and passes a constant current throughout the duration of depolarization. The behavior of potassium channels is confirmed if a depolarization of -100 mV to +50 mV in the cell-attached configuration is applied to an excitable cell. In the single-channel current recordings of Figure 5.35A, it is observed that at the beginning of the depolarization and throughout its duration, the probability for the channel to switch from the closed to the open state becomes very high. In fact, the potassium channel remains in the open conformation for long periods, compared to the sodium channel (Figure 5.33), returning to low values of probability of being in the open state when the potential hyperpolarizes. Furthermore, the average current of all recordings (Figure 5.35B) shows a voltage- and time-dependent activation, given by the intrinsic biophysical characteristics of

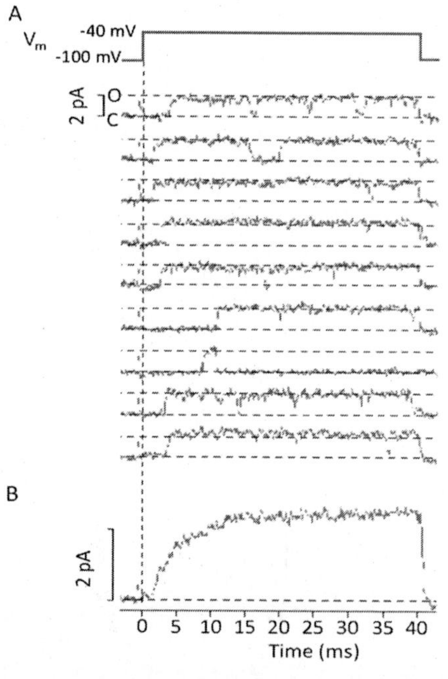

Figure 5.35 Multiple recordings of a single potassium channel current (I_m) triggered by a depolarization from -100 mV to +50 mV (A) and average of all recorded currents (B).

the channel-forming protein, i.e., that it is instantaneous with respect to the onset of depolarization, but the relative current effectively becomes large, with a delay of about 2.5 ms. The average current also shows deactivation due to extrinsic factors such as repolarization towards the potassium equilibrium potential, with restoration of the low probability of opening.

5.13.3 The new kinetic parameters

The many thousands of ion channels embedded in the plasma membrane of an excitable cell form various populations, distinct according to the ion species for which they have the greatest specificity. These distinct populations of channels have components with identical biophysical and functional properties. They switch from a closed to an open state or vice versa, in examples such as potassium channels and sodium channels. Sodium channels can also switch from the open or closed state to the inactivated state. State transitions are always instantaneous and due to conformational changes in parts of the protein.

All known channels have, for each value of membrane potential and for each time, a precise probability of being in one state rather than another. In addition, the magnitude and time course of macro-currents depend on the set of single-channel currents and thus on the number and state of the channels of each species (open, closed or inactivated).

This new probabilistic model for the ionic current makes it essential to revise the model of Hodgkin and Huxley and to redefine the ionic permeability parameters of the two main membrane currents, sodium and potassium, that are involved in the genesis of the action potential. Thus Equation 5.21 for the sodium current I_{Na} becomes:

$$I_{Na} = i_{Na} \cdot N_{Na} \cdot p_{Na(V,t)} \qquad (5.26)$$

where i_{Na} is the single channel current defined by the equation, which reflects the generalized Ohm's law (Equation 5.2), N_{Na} is the total number of sodium channels present in the membrane and $p_{Na(V,t)}$ is the probability for each channel to be in the closed, open or inactivated state. This parameter is voltage- and time-dependent and replaces the parameter

m^3h in Equation 5.21. The driving force is contained in i_{Na}

$$i_{Na} = \gamma_{Na} \cdot (V - E_{Na}) \qquad (5.27)$$

where γ_{Na} is the single-channel conductance and $V_m - E_{Na}$ is the electromotive force for the sodium ion.

The single-channel current i_{Na} can be measured experimentally (Figure 5.33) and the total number of N_{Na} channels can be estimated by comparing the single-channel current i_{Na} with the total current I_{Na} measured in whole-cell configuration. Then, from Equation 5.26, one can calculate $p_{Na(V,\,t)}$, the probability of the individual sodium channel to be in a definite state.

The same transformation can be applied for Equation 5.15 concerning the potassium current, I_K, which becomes:

$$I_K = i_K \cdot N_K \cdot p_{K(V,t)} \qquad (5.28)$$

where i_K is the single-channel current, N_K is the total number of potassium channels present in the membrane and $p_{K(V,\,t)}$ is the probability for each K channel to be in the closed or open state. This parameter is voltage- and time-dependent and replaces parameter n^4 of Equation 5.15. i_K is defined by the equation:

$$i_K = \gamma_K \cdot (V - E_K) \qquad (5.29)$$

where γ_K is the single-channel conductance and $V_m - E_K$ is the electromotive force for the potassium ion. The single channel current i_K can be measured experimentally (Figure 5.35) and the total number of N_K channels can be estimated by comparing the single-channel current i_K with the total current I_K measured in whole-cell configuration. Then, from Equation 5.28, one can calculate $p_{K(V,\,t)}$, the probability of the individual potassium channel to be in a definite state.

The probabilities of the individual sodium channel and the individual potassium channel, $p_{Na(V,\,t)}$ and $p_{K(V,\,t)}$, respectively, being in a definite state determine the biophysical characteristics of the sodium and potassium currents during the action potential.

5.14 Shaping the action potential

Equations 5.26 and 5.27 for sodium and 5.28 and 5.29 for potassium provide a new model that can accurately describing the voltage- and time-dependent behavior of the main currents involved in the action potential in a squid axon. Using the patch-clamp technique in a squid axon

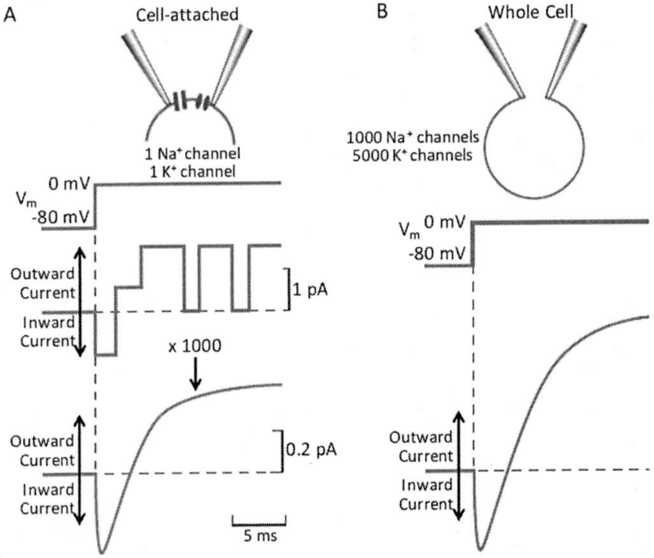

Figure 5.36 Model of a cell depolarized from -80 mV to 0 mV (V_m). (A) Sum (I_S) of 1000 single sodium and potassium channel current recordings (pA) measured in patch-clamp in the cell-attached configuration; (B) macro-current (I) recorded in whole-cell configuration in a cell with 1000 identical sodium channels and 5000 identical potassium channels active.

membrane model, it is possible to observe a temporal relationship between the two main ionic currents responsible for the action potential. In cell-attached configuration, applying 1000 depolarizations from -80 mV to 0 mV to one Na and one K channel is qualitatively similar to applying a single voltage step to a whole-cell model (Figure 5.36).

Ultimately, we characterized a macroscopic phenomenon, the action potential, at a cellular level (Figure 5.1), with a sequence of mechanisms highlighted at a molecular level. The task was to move from the mathematical description of the membrane current kinetics, calculated from voltage-clamp experiments and integrated into the Hodgkin and Huxley model, to a description based on single-channel data and analyzed in accordance with population dynamics, in which ion channels of the same species increase or decrease their probability of being closed, opened and in some cases inactivated, as a function of voltage and time (Equations 5.26 and 5.28).

Finally, it is possible to correlate the biophysical properties of only two ion channels, the TTX-sensitive sodium channel and the delay rectifier potassium channel, with the time course of the action potential in a giant squid axon.

5.14.1 Sodium and potassium channels during the action potential

The time-dependent variations of the membrane potential that characterize the squid axon action potential derive from the particular voltage- and time-dependent properties of the sodium and potassium channels. These channels continuously and randomly change conformation from the open to the closed state and vice versa and from the non-inactivated to the inactivated state and vice versa (Paragraphs 5.10), with a typical voltage- and time-dependent transition probability $p_{Na(V,t)}$ for the sodium channel (Equation 5.26) and $p_{K(V,t)}$ for the potassium channel (Equation 5.28). The different phases of the action potential described in Figure 5.37A can find their molecular version in the behavior of different populations of ion channels. The action potential of the squid axon (Figure 5.1) can develop using only two types of ion conductance, the TTX-sensitive sodium and the delay rectifier potassium, and this allows a very simplified description from a molecular point of view. Figure 5.37B shows the single-channel sequence of events for sodium and potassium. The different configurations of open (O), closed (C) and inactivated (I) are integrated into the probability parameter p, which is voltage- and time-dependent, of Equations 5.26 and 5.28. The equations show the probability that an individual channel is in one of three configurations for the sodium channel (O, C and I) and one of two configurations for the potassium channel (O and C).

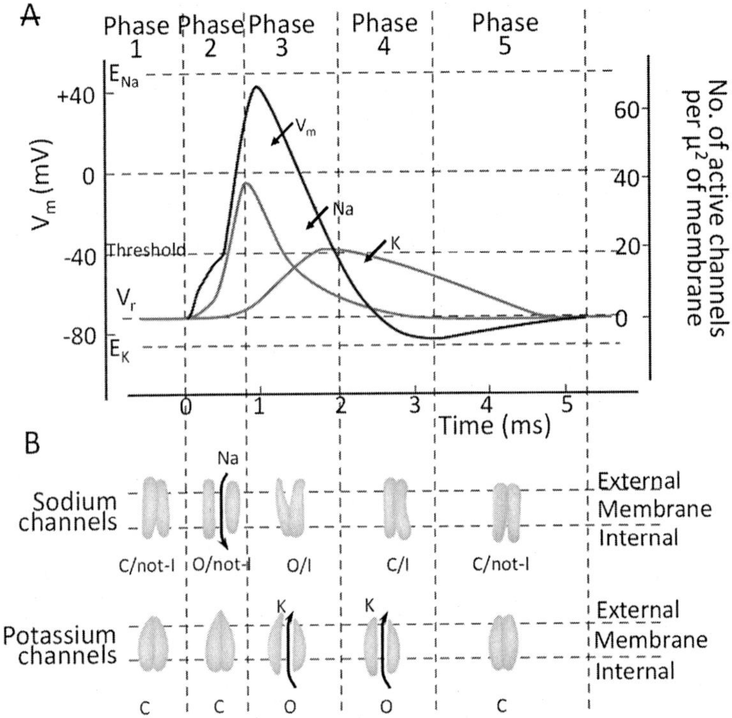

Figure 5.37 (A) Schematic of the relationship between the squid axon action potential (V_m) and density of active sodium (Na^+) and potassium (K^+) channels. Time- and voltage-dependent cycling of sodium and potassium channels (B) in relation to the time course of the action potential; C = closed; O = open; non-I = non-inactivated; I = inactivated.

The model proposed by Qin et al (2000), which uses the Markov property (Andrej Andreevič Markov, 1856-1922) for the variation of the probability of conformational change, can be described, with due simplification, by the sequence of states in which the sodium channel and the potassium channel can be found according to the diagram in Figure 5.38.

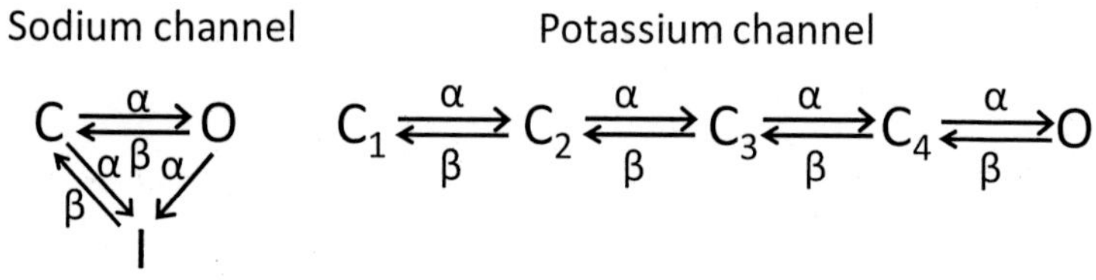

Figure 5.38 Sequence of states and relative probabilities in which the sodium channel and potassium channel can be found by conformational change. α and β are the probabilities that the channel will have a conformational change at a particular voltage and time.

The probability is a parameter that can be obtained directly from single-channel analysis in patch-clamp experiments (Data Sheet 5.5). As far as the TTX-sensitive sodium channel is concerned, the model predicts a C-O-I cycle. This implies that a channel can go from closed to open state and vice versa as well as from inactivated to closed and vice versa. However, the transition from open to inactivated is one-way. In fact, the open inactivated channel must first close and become non-inactivated before it can become open again.

For the delay rectifier potassium channel, the sequence of events is different. The model considers four closed states that the protein must activate before it can finally open. Since here too, α and β are voltage- and time- dependent probabilities, reaching the open channel configuration involves a delay. The channel will close when the repolarization, induced by the same channel, reaches negative values such that the most probable configurations will be the four closures, of which the most probable will be, at each closure, determined in a decreasing way by the level of hyperpolarization.

5.14.2 The physiology of inactivation

The inactivation of sodium channels and the subsequent recovery from inactivation (Paragraphs 5.11) are two of the most important factors for excitability to function properly in various cellular regions. Inactivation is mostly a time-dependent phenomenon but also has an important voltage-dependent component. Sodium channel inactivation starts immediately after the open channel probability reaches its maximum dV/dt during the fast depolarization. In neurons, inactivation has three particular functions. First, it allows membrane repolarization by the potassium channels after the upstroke of the action potential. Second, it is responsible for the unidirectional progression of the excitation wave in the axon (Paragraphs 6.2.4.2). Third, inactivation regulates the production of repetitive action potentials. Mutations that compromise sodium current inactivation are often fatal.

In neurons, the single action potential carries no useful information. The communication code used by the nervous system is based on the production of different sequences of action potentials at different frequencies (firing, Paragraphs 6.4). In the language used by the nervous system, the upper limit at which messages can be sent, and thus the maximum frequency it can be reach, is determined by the charge of the membrane capacitors and the duration of the action potential, and thus also by the inactivation of the sodium current (Paragraphs 6.4).

5.14.3 Particular properties of action potentials

The squid axon generates action potentials that are always identical to the those described due to the presence of sodium and potassium channels in the membrane, with the biophysical properties remaining constant as they are genetically determined. However, numerous excitable cells are known to produce action potentials with properties and characteristics that are very different from those of the axon, because they have, in the course of evolution, selected different sodium and potassium membrane channels or other specific channels, for example, for calcium or chloride. Figure 5.39 compares the propagated action potential of the squid axon with that of the striated muscle cell, the cardiac ventricle, the cardiac sinus node and the neurons of the inferior olive of the central nervous system.

In **skeletal muscle**, the action potential is produced, as in the squid axon, by the sodium current I_{Na} and the potassium current I_{KDR} with biophysical properties very similar to those described. However, the repolarization of 10-15 mV after the peak is generated by an outward chloride current carried by chloride ions entering the cell and completed, up to the resting potential, by the potassium current.

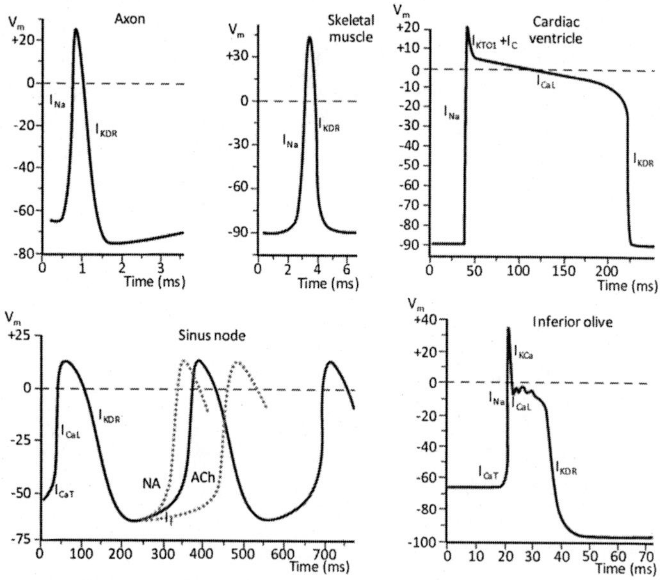

Figure 5.39 Action potentials in skeletal muscle, cardiac ventricle, sinus node and inferior olive compared with that propagated in the squid axon. The various I-currents responsible for potential changes and the effects of norepinephrine (NA) and acetylcholine (ACh) are shown.

In the **cardiac ventricle**, depolarization is originated by the inward sodium current I_{Na} and hyperpolarization by the outward potassium current I_{KDR}, which is however compensated, for about 200 ms, by an inward depolarizing calcium current of the long-lasting type, I_{CaL}. I_{CaL} maintains the potential at depolarized values around zero, the so-called plateau. It performs the dual function of allowing complete contraction of the whole ventricle and of delaying repolarization by prolonging the occurrence of the next action potential (Data Sheet 7.3), during which the muscle cells, not being excitable, are not at risk of ventricular extrasystoles due to action potentials that are too close in time. This current also introduces calcium ions into the cell, which are essential for activating the mechanism of muscular contraction (Paragraphs 7.4).

In the **sinoatrial node** (Data Sheet 7.3), which is a small group of cells that give the heart its autorhythmicity, the peak of the action potential is due to a transient calcium current or I_{CaT} followed by a long-lasting calcium current or I_{CaL}. Repolarization is due to the outgoing potassium current I_{KDR}, followed by a mixed current of incoming sodium and outgoing potassium, the I_f (Brown and DiFrancesco, 1980), which activates in hyperpolarization and depolarizes the potential to the threshold value for the I_{CaT} current. Acetylcholine (ACh of Figure 5.39), released by the vagus nerve, increases the time needed to reach the action potential threshold and therefore slows down the heart rate, defined as a negative chronotropic effect. At the same time, adrenaline and noradrenaline (NA of Figure 5.39), released by the cardiac sympathetic nerve, have the opposite effect by decreasing the time needed to reach the action potential threshold and increasing the heart rate with a positive chronotropic effect.

In neurons of the **inferior olive** (Llinás and Yarom, 1981) of the central nervous system, the initial depolarization up to the threshold of the action potential is generated by the calcium current of the transient type I_{CaT}, while the potential is brought to the peak by the sodium current I_{Na}, then by the potassium current of the type I_{KDR}, which begins to hyperpolarize the membrane. The long-lasting calcium current or I_{CaL} is activated, which keeps the cell depolarized for 10-15 ms; the hyperpolarization due to the I_{KDR} potassium current is then completed.

Bibliography

Adrian RH, Chandler WK, Hodgkin AL. Voltage clamp experiments in striated muscle fibres. J. Physiol. (1970) 208: 607-644

Brown HF and DiFrancesco D. Voltage-clamp investigations of membrane currents underlying pace-maker activity in rabbit sino-atrial node. J. Physiol. (1980) 308: 331-351.

Chen W and Lee RC. An Improved Double Vaseline Gap Voltage Clamp to Study Electroporated Skeletal Muscle Fibers. Biophys. J. (1994) 66: 700-709.

Cole KS and Curtis HJ. Electric impedance of the squid giant axon during activity. *J. Gen. Physiol. (1939) 22 (5): 649-670.*

Cole KS and Moore JW. Ionic current measurements in the squid giant axon membrane. J. Gen. Physiol. (1960) 44:123-167.

Hille B. The selective inhibition of delayed potassium currents in nerve by tetraethylammonium ion. J. Gen. Physiol. (1967) 50: 1287-1302.

Hodgkin AL and Huxley AF. Resting and action potential in single nerve fibres. J. Physiol. (1945) 104: 176-195

Hodgkin AL and Huxley AF. Currents carried by sodium and potassium ions through the membrane of the giant axon of *Loligo*. J. Physiol. (1952a) 116, 449-472.

Hodgkin AL and Huxley AF. A quantitative description of membrane current and its application to conduction and excitation in nerve. J. Physiol. (1952b) 117, 500-544.

Hodgkin AL, Huxley AF, Katz B. Measurement of current-voltage relations in the membrane of the giant axon of *Loligo*. J. Physiol. (1952) 116: 424-448.

Hodgkin AL and Katz A.L. The effect of sodium ions on the electrical activity of the giant axon of the squid. J. Physiol. (1949) 108: 37-77.

Llinás R and Yarom Y. Electrophysiology of mammalian inferior olivary neurones in vitro. Different types of voltage-dependent ionic conductances. J. Physiol. (1981) 315: 549-567.

Neher E and Sakmann B. Single-channel currents recorded from membrane of denervated frog muscle fibres. Nature. (1976) 260(5554): 799-802.

Qin F, Auerbach A, Sachs F. A direct optimization approach to hidden Markov modeling for single channel kinetics. Biophys. J. (2000), 79 (4): 1915-1927.

Takata M, Moore JW, Kao CY, Fuhrman FA. Increase in Lobster giant axon by tarichatoxin (tetrodotoxin). J. Gen. Phisiol. (1966) 49: 977-988.

Data Sheet 5.1

The impedance in a giant squid axon

The measuring cell used by Cole and Curtis (1939) to measure changes in the resistance of the squid axon membrane (Figure 5.13) is shown in Figure S5.1. The recording chamber, made of transparent plastic material, is placed in a metal box. A constant flow of seawater keeps the temperature at the appropriate value. The chamber has a groove along its whole length and is completed by a series of platinum electrodes. The electrode distribution is as follows: a pair of stimulus electrodes (a) to trigger a change in the membrane potential of the axon and evoke the action potential, a pair of membrane resistance measurement electrodes (b and b') to allow the measurement of the unknown resistance of the axon (R_X) when the cell is inserted into the Wheatstone bridge (Figure 5. 13) and the potential measurement electrodes (c_1 and c_2) to measure the potential variations on the axon surface with respect to the reference c' electrodes, choosing at each measurement the one that is in a non-active zone of the axon. With the cell, it is also possible to measure the conduction velocity by determining the delay of the peak of the action potential measured between the electrode c_1 with respect to the reference electrode c' and between the electrode c_2 also with respect to c'.

Once the axon has been positioned, the cell is covered with a coverslip to minimize evaporation during recordings when the flow of seawater is suspended. The water inside the groove is removed to increase the resistance between the various electrodes as much as possible and to be able to measure significant potential differences.

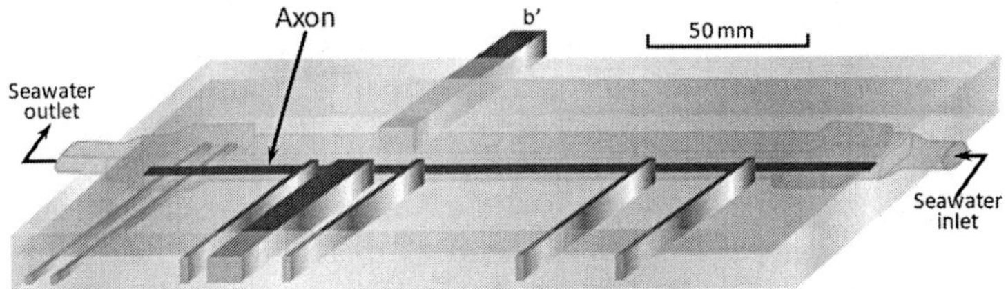

Figure S5.1 The giant squid axon (in black, indicated with an arrow) is arranged in the central recording chamber and perfused with seawater (black arrows). The horizontal bars represent the electrodes: stimulus electrodes (a), impedance measurement (b and b') and action potential recording (c1, c2 and c').

Data Sheet 5.2

The oscilloscope

An indispensable instrument for observing the rapid electrical phenomena affecting excitable cells is the oscilloscope (Figure S5.2). It is ideal for measuring potential differences in biological systems as it is characterized by a very high input resistance ("impedance"), which does not alter even the small currents affecting the systems under examination,

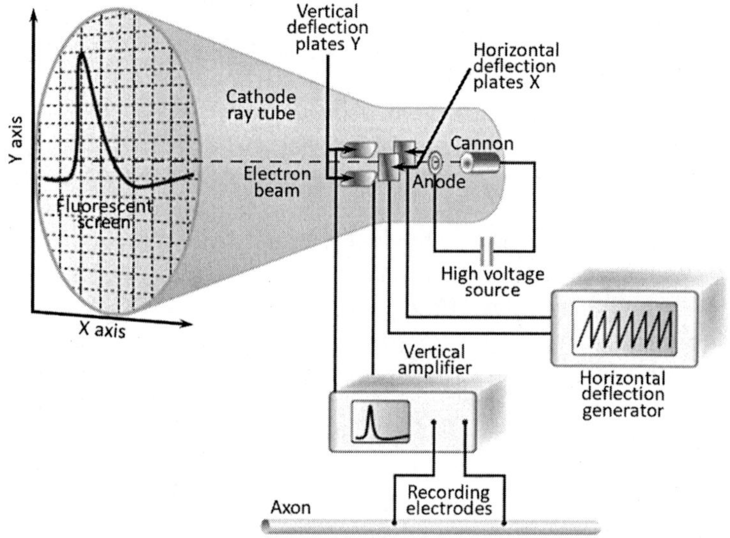

Figure S5.2 Schematic representation of an oscilloscope in a potential measurement apparatus.

and by the rapidity of the response because, as we will see later, it works with beams of electrons moving in a high vacuum.

The oscilloscope consists of: 1) a glass tube roughly in the shape of a truncated cone, in which a high vacuum is created, 2) the cannon, which in fact acts as a positively charged cathode, consisting of a resistor which, upon heating, emits electrons, 3) the anode, negatively charged, which focuses the electron beam and directs it, 4) the vertical deflection plates, which deflect the electron beam along the y-axis and 5) the horizontal deflection plates, which deflect the electron beam along the x-axis. The electron beam collapses against the fluorescent screen (6), which emits light when the negative particles produced by the cannon collide with it. The screen plane has a calibrated grid so that times can be measured on the x-axis and potential differences can be measured on the y-axis.

The resistance of the cannon is heated to 800-1000 °C by a current of 10-15 amperes generated by a potential difference of 90-100 volts, while the potential difference between the cathode of the cannon and the anode, which compacts and accelerates the electron beam, is $10\text{-}12 \cdot 10^3$ volts and is generated by a current of 0.1-0.2 amperes. The combination of the displacement of the electron beam produced by the vertical deflection plates and that produced by the horizontal deflection plates causes the luminous point to be in a certain position in the plane of the screen at each instant.

When the instrument is switched on, a bright spot is formed in the center of the fluorescent screen. The horizontal deflection generator sends, through the horizontal deflection plates (Figure S5.2), a cyclic saw tooth signal (Figure S5.3).

The values of the potential that the saw tooth takes on in time are such that when the potential value is -1, the bright spot is at the extreme left of the screen. As the potential rises to the value of +1, the spot sweeps to the right at a speed that depends on the slope of the ramp in phase 1 (Figure S5.3) and constitutes the reference time for the potentials to be measured. Then, the bright spot returns instantaneously to the left when the potential returns to the value of -1 (phase 2) and begins a new sweep from left to right (phase 3). Phase 2 of the cycle is so fast that it is not actually visible.

On the other hand, the signals coming from the biological preparation are sent to the vertical deflection plates (Figure S5.2) and are picked up at the output of the differential amplifier for the measurement of membrane potentials. The amplitude of the membrane potential varies over time (Figure 5.39). Therefore, the point on the screen of the oscilloscope will undergo a vertical deflection, the greater the amplitude of the signal of the biological preparation. At the same time, it will move horizontally because of the ramp that marks the reference time (phase 1 of Figure S5.3).

It should be kept in mind that the reference time is chosen by the operator on the basis of the time course of the electrical phenomena under examination by modifying the slope of the sawtooth. However, for the potential differences to be measured, the operator simply chooses the signal amplification factor so as to make the most of the height of the oscilloscope screen. It should also be borne in mind that the combination of the value of the two potentials that cause the light spot to deflect will draw the time course of the biological phenomenon on the fluorescent screen.

Data Sheet 5.3

Measuring the membrane potential in a giant squid axon

The first measurements of the membrane potential and the action potential in a nerve cell were carried out by Hodgkin and Huxley (1945) on the giant squid (*Loligo forbesi*) axon, which allows non-damaging

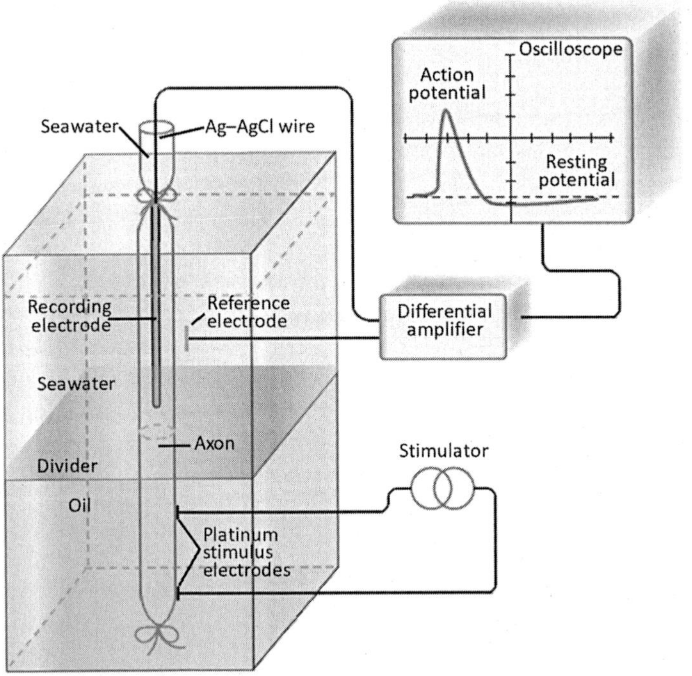

Figure S5.4 Schematic of the experimental apparatus for the measurement of membrane potential in the giant squid axon.

manipulations as its diameter can be around 1.0 mm.

The recording electrode made from a glass capillary with a diameter of 1-2 mm, of which a part with a diameter of 50-100 μm is forged by hot stretching, is inserted into the axon (Figure S5.4). A suitable ligature is then made so that the axon adheres perfectly to the capillary and another ligature is made at the other end so that the internal solution does not escape.

The axon prepared in this way is placed in a tray divided by a sealed partition into two compartments. In the compartment where the end of the bound axon is located, and to which oil is added, two platinum electrodes connected to a suitable stimulator are placed. In the other compartment, seawater and a reference electrode, consisting of a silver wire (Ag) coated with silver chloride (AgCl), are placed (Data Sheet 3.4). The reference electrode, with the recording electrode, is connected to an amplifier and the oscilloscope. The recording electrode is filled with seawater and is also connected to the measuring system with a silver electrode covered with silver chloride. The potential detected by the recording electrode and the constant-zero potential detected by the reference electrode are sent to a differential amplifier. The output potential of the differential amplifier is the potential difference between the inside and outside of the axon and is sent to the oscilloscope for measurement. In this way, the resting potential of the axon and, after appropriate stimulation, the action potential can be recorded.

It is important to bear in mind that with an experimental preparation such as that in Figure S5.4, it is possible, given the large size of the axon, to replace the intracellular solution with a solution known to the experimenter. It is also possible to immerse the axon in a solution other than seawater. This allows specific blockers to be used or the concentrations of the ions involved in the phenomena under study to be changed.

Data Sheet 5.4

The voltage-clamp

The problem posed by the Hodgkin and Huxley equations for sodium and potassium currents (Paragraphs 5.10) is that although one can measure the potential with an experiment such as that in Figure S3.2 and can calculate, at least in a giant axon, the equilibrium potentials E_K and E_{Na} using known external and internal solutions (Data Sheet 3. 2), it is not possible to calculate the current. This is because every time the excitable cell is depolarized, the action potential described by the time-dependent currents of Equation 5.4 is established.

This equation cannot be solved because it has three variables and during the action potential, both the potential V and the current I and the membrane conductance G change in time. The solution proposed by Hodgkin, Huxley and Katz (1952) was to make the potential $V_{(t)}$ constant and known in order to measure the current $I_{(t)}$ and to be able to calculate from Equation 5.4 the variation of the membrane conductance as a function of time $G_{(t)}$.

The voltage-clamp technique therefore makes it possible to measure the currents flowing through the membrane over time at a constant voltage known to the experimenter, using a feedback circuit capable of supplying, by means of a feedback mechanism, a current of the appropriate value to keep the imposed potential constant, even in the presence of an active response of the cell. The feedback circuit uses differential amplifiers (AD), special integrated circuits that provide an output potential value given by the difference, instant by instant, between the two potentials present at the two inputs. They may be multiplied by an amplification factor established by the operator, without subtracting significant amounts of current from the circuit in which they are inserted.

If, for example, the membrane potential of a neuron is changed from -70 mV to -50 mV, an inward current is generated by activation of the sodium channels (Paragraphs 5.7), which produces a further depolarization. The differential amplifier compares, instant by instant, the potential set by the experimenter and the membrane potential actually recorded in the cell by means of a potential electrode. If the two potentials are different, the feedback circuit generates a current of appropriate direction, which is injected into the cell by another electrode and which is able to cancel the sodium current that would tend to depolarize the cell further. If we measure this current, we do not have the membrane current directly, but the current that the feedback circuit provides to cancel the current generated by the cell and which would produce the potential variations, thus maintaining the voltage-clamp condition.

Squid axon. In the squid axon, placed in a suitable experimental dish, the electrode for measuring the potential difference and the current electrode are inserted, both coupled to an external reference electrode (Figure S5.5). Given the size of the axon, the electrodes are in fact wires with a diameter of 0.1 mm with practically no resistance.

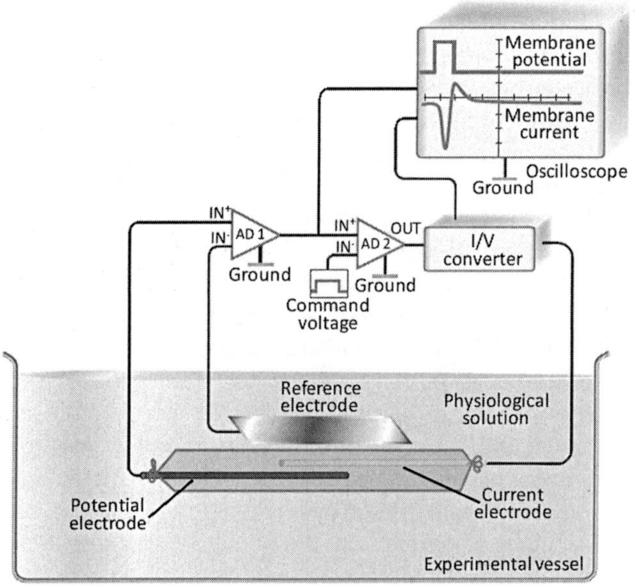

Figure S5.5 Voltage-clamp in the squid axon. AD 1 and AD 2 are two differential amplifiers.

To the positive input IN^+ of the differential amplifier, AD 1 is sent the potential detected by the potential microelectrode, while to the negative input IN^-, the potential of the reference electrode is sent. The output OUT of AD 1, the difference between the two inputs, is sent on one side to an oscilloscope for control and eventual measurement, and on the other side, to one of the two inputs of the differential amplifier AD 2. The other input of AD 2 has the control signal set by the experimenter. If the potential of the cell is different from the command potential, for example due to currents caused by the activation of voltage-dependent channels, at the output OUT of AD 2, there will be a non-zero potential which will create a potential difference between the interior of the cell and the reference electrode. A current of appropriate direction will be generated, which will tend to cancel the difference. This current will be equal and of opposite sign with respect to the membrane currents that have been activated and have induced the potential variations recorded in the cell. The I/V converter, inserted between the output of the differential amplifier AD 2 and the cell and without interfering with the current flowing through it, generates a potential proportional to the current. Therefore, the measurement of this potential on the oscilloscope is in fact equivalent to the measurement of the membrane currents, taking due account of the

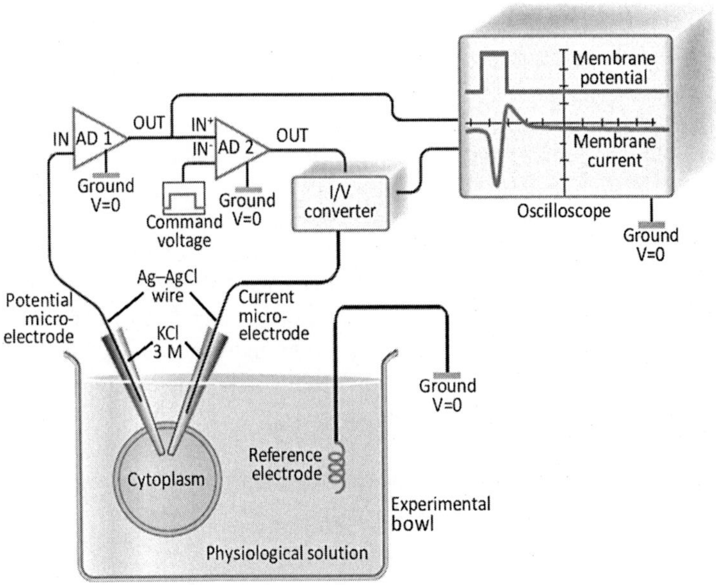

Figure S5.6 Schematic of the experimental apparatus for the two-electrode voltage-clamp. AD 1 and AD 2 are two differential amplifiers.

signs. At this point, from Equations 5.5 and 5.6, we know, instant by instant, the membrane potential V_m as imposed by the voltage-clamp, the membrane currents because they are measured and the respective equilibrium potentials because the solutions used are known. We can therefore calculate the conductances as a function of voltage and time.

Oocyte. The two-electrode voltage-clamp technique (Adrian, Chandler and Hodgkin, 1970) should be used to study a large cell, such as a frog oocyte or a muscle cell. Of the two microelectrodes, one is used to measure the potential, with a circuit similar to that used for the axon, and the other is used to inject the current necessary to maintain the membrane potential at the required value (Figure S5.6). The problem created by the use of microelectrodes is their high resistance due to their small diameter, on the order of less than 0.1 μm, which can reach up to 50 MΩ. While this resistance does not create problems for the measurement of the potential, it greatly limits the amount of current that can be injected to maintain the potential at the desired value.

Vaseline-gap. Voltage-clamping using the vaseline-gap technique (Chen and Lee, 1994) is indispensable with long cells, such as muscle fibers, in order to have a better space clamp, i.e., a constant and homogeneous potential along the whole length of the fiber. The cell under examination is placed in the experimental dish (Figure S5.7) in which three environments are created, electrically and hydraulically isolated by two vaseline septa.

The cell is positioned in such a way that the two cut ends are in sectors A and C, into which a solution identical to that of the cytoplasm is introduced, while the part positioned in sector B is bathed by a solution identical to that of the extracellular solution. With this arrangement, it is as if the electrodes in A and B were positioned, from an electrical point of view, inside the cell. As in the case of other voltage-clamp configurations, the differential amplifier AD 2 circulates a current from sectors A and C through the part of the membrane in sector B until the potential electrodes of sectors A and C measure, by means of the differential amplifier AD 1, a potential that is identical to the command potential.

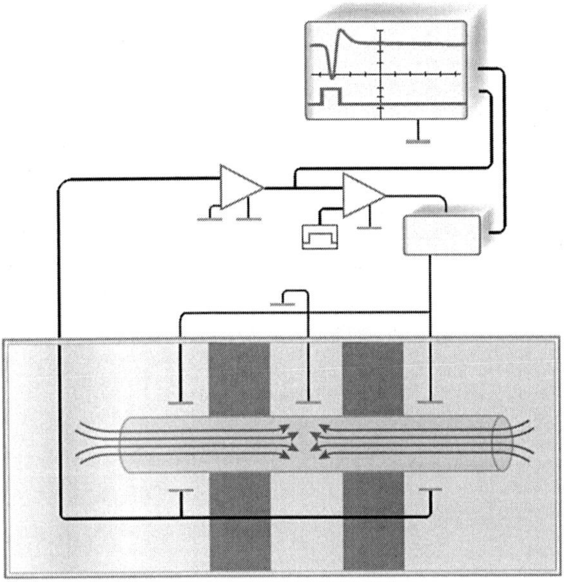

Figure S5.7 Vaseline-gap technique. AD 1 and AD 2 are two differential amplifiers.

Data Sheet 5.5

The patch-clamp

To carry out patch-clamp experiments, micropipettes are used, forged from glass capillaries with a diameter of 1.2 mm, using a procedure that is

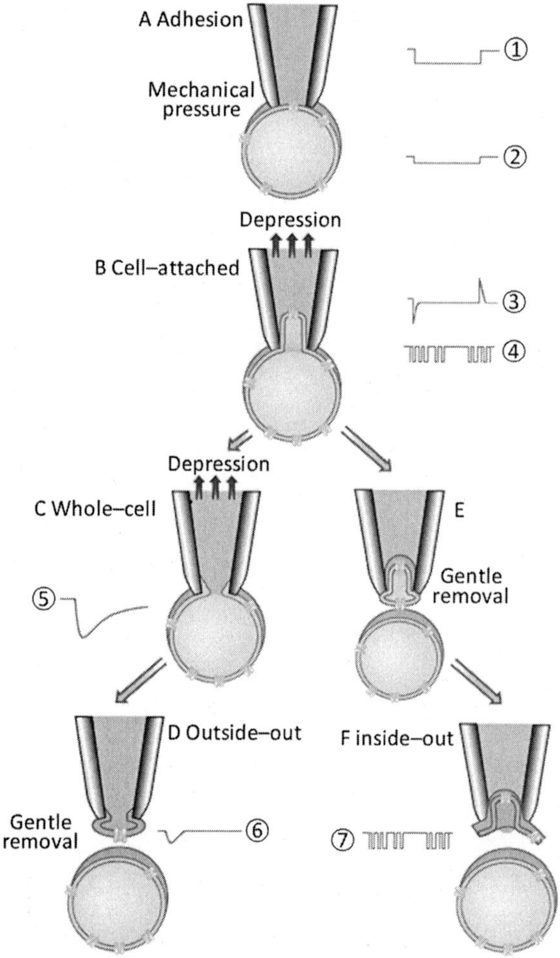

Figure S5.8 The various patch-clamp configurations.

very similar to that used for microelectrodes (Data Sheet 3.4). These micropipettes have a tip with a diameter of 0.5-1 μm and a resistance of 3-5 MΩ (MΩ = 10^6 Ω). Using a micromanipulator, an instrument that allows micrometric movements, the micropipette is immersed in the perfusion solution and the measured potential is reset to zero to compensate for any unwanted potentials. It is then switched to voltage-clamp, and while sending a step of hyperpolarizing potential (1 of Figure S5.8), the micropipette is placed on the surface of the cell and lightly pressed until the amplitude of the step decreases (2 of Figure S5.8A), which indicates the increase in resistance due to adhesion to the plasma membrane.

Cell-attached. This configuration allows functional recording of the individual membrane channel proteins that randomly switch from the open to the closed state and vice versa. In order to establish this experimental condition, gentle pressure must be applied to 'seal' the micropipette to the cell and establish the cell-attached configuration (Figure S5.8B), which can be monitored by the appearance of capacitive peaks (3 of Figure S5.8B). During mechanical pressure of the electrode tip against the cell membrane, the resistance of the micropipette increases from 3-5 MΩ to 10-20 MΩ and even to 20 GΩ (GΩ= 10^9 Ω) when pressure is applied. The micropipette, which electrically connects the outside of the cell to the control and measurement circuit, must be filled with a suitable solution outside the cell, and allows the recording of the single-channel currents (4 of Figure S5.8B) of one or several channels present in the patch of membrane that is electrically isolated by the micropipette.

Whole-cell. This configuration is useful to observe the macroscopic current produced by all of the channels present on the plasma membrane (5 of Figure S5.8C). To achieve the whole-cell configuration, greater pressure must be applied to tear the part of the membrane that faces the inside of the micropipette, thus putting in electrical and chemical communication the internal solution of the micropipette with the cytoplasm (Figure S5.8C), with reciprocal exchange of solutes. For this type of experiment, the micropipette must be filled with a solution with a composition compatible with the internal solution of the cell. Alternatively, a different solution can be used so that the diffusion of the solution in the cytoplasm, which occurs after the whole-cell configuration has been achieved, makes it possible to set the internal solution of the cell modified by the dialisys of the micropipette solution inside the cell.

Outside-out. With this configuration, it is possible to study the action of hydrophilic compounds that could act directly on individual ion channels that expose their extracellular components to the external (and therefore modifiable) solution. From the whole-cell configuration, the gentle removal of the micropipette causes the detachment of a small membrane vesicle and leads to the outside-out configuration (Figure S5.8D). The inner face of the membrane is in contact, as in the whole-cell configuration, with the micropipette solution. Small membrane currents are measured from the few remaining channels in the vesicle attached to the micropipette (6 of Figure S5.8D).

Inside-out. With this configuration, it is possible to study the action of cytoplasmic second messengers that act directly by activating, modulating or blocking the action of individual membrane ion channels. Starting from the cell-attached configuration (Figure S5.8B) and with the gentle removal of the micropipette, a small membrane vesicle is detached (Figure S5.8E). Once exposed to the air for a few seconds, the outermost membrane of the vesicle ruptures, leaving only the inner half of the vesicle which, once re-immersed in the solution, will juxtapose the inner part of the membrane and the outer solution. Thus, the inside-out configuration is achieved (Figure S5.8F). In this configuration, single-channel currents are measured (7 of Figure S5.8F) and the intracellular modulations of single ion channels can be studied. The control circuit of the patch-clamp consists of: 1) a high-input resistance differential amplifier with a negative input IN^- to which the micropipette is connected and a positive input IN^+ to which the control potential is sent; 2) a feedback resistor, which must be at least 1 GΩ and can range up to 100 GΩ, because it must abundantly prevail over the resistances of the micropipette and the cell. The feedback resistor is used to increase the amplifier gain; it connects the negative input with the output OUT (Figure S5.9).

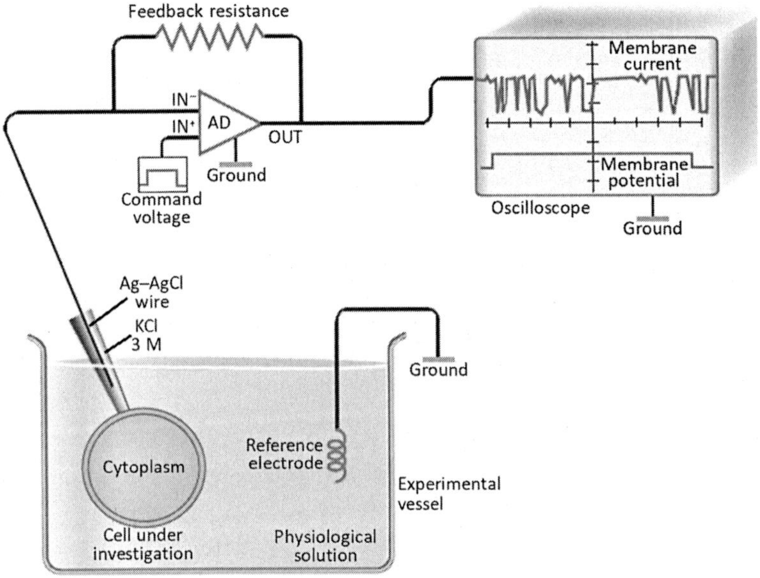

Figure S5.9 Schematic of an experimental apparatus for patch-clamp. AD is a differential amplifier. The micropipette should be filled with an external or intracellular solution for the studied cell depending on the experimental setup used (Figure S5.8).

The micropipette measures the potential of the cell because it is assumed that the potential measured at the IN^- input is equal to that present at the tip of the micropipette, since the currents involved are negligible due to the large resistances in the circuit. At the instant when the potential read by the micropipette becomes different from the command potential, at the output OUT there is a non-zero potential, which generates a current through the feedback resistor to the micropipette, to the cell under study and to the reference electrode to earth. The current modifies the potential until it returns to a value identical to the command potential, thus opposing the current, for example, due to the activation of channels, which has disturbed the membrane potential. The potential present in OUT is therefore proportional, according to Ohm's law, to the current that flows through the circuit to the cell that cancels the endogenous current that has changed the membrane potential; it can be measured with an oscilloscope or other suitable measuring system.

CHAPTER 6

Communicators 2: communication between cells

6.1 Communication mechanisms: electrical communication
6.2 Communication mechanisms: chemical communication
6.3 The chemical synapse
6.4 Neuronal firing as a language in the nervous system
6.5 Autorhythmicity and pacemaker activity

The development of multicellular biological organisms has most likely taken place in two fundamental and sequential phases that have unfolded over millions of years. First, an early phase, when from unicellular organisms, there were the first aggregations of identical cells, with very little functional variability among them. Second, an evolutionarily more advanced phase, in which organisms assumed larger sizes and a higher degree of complexity, with the formation of groups of cells with different characteristics and specializations to perform different functions within the organism.

It is almost certain, and there is a large body of evidence, that one of the fundamental steps that enabled organisms to grow in size, and consequently in functional complexity, was that multicellular organisms developed communication systems. At the beginning, these systems were between neighbor cells and later, between different, far apart cells, which enabled communication and coordination between different regions of the organism.

It is possible to hypothesize how evolution has operated from a functional point of view, using a logical sequence of possible events that could have led to the development of communication systems between cells. These, in turn, have allowed the development of increasingly complex living organisms with greater adaptive capacity. The observation is interesting, for example, that biological entities during evolution continuously grew in size up to the point where evolution favored, instead of dimensions, the functional complexity of successful organisms, which effectively drastically reduced their size. There are certainly multiple reasons for this last consequence, with communication one of the most important. A clear example is the paradox of the great reptiles. The **brontosaurus** (Figure 6.1) is an upper Jurassic herbivorous dinosaur about 24 m long, 8 m high and weighing 35 tonnes. An animal with such an enormous body mass could continue to graze the grass without realizing that a **ceratosaurus**, an 8 m long and 2.5 m high Jurassic carnivorous predator, is eating its tail. This is possible because the brontosaurus has a slow communication system that causes the tail's danger signal to be perceived with a considerable delay, rendering the animal's reaction ineffective. It can reasonably be said that the efficiency of communication between different areas of the organism was certainly one of the decisive factors that led, over the course of evolution, to the supremacy of mammals over large reptiles.

Figure 6.1 Reconstruction of a brontosaurus, an herbivorous dinosaur, being attacked by two ceratosaurs, carnivorous predators. Both lived in the Upper Jurassic.

Communication between cells, the ability to process stimuli coming from the surrounding environment and the ability to control all of the functions of the biological organism have all been fundamental steps during evolution.

6.1 Communication modes: electrical communication

The primary event that enables cells to generate a message is the single action potential (Chapter 5). From a functional point of view, the single action potential is not significant, but it is certainly the trigger for a more or less complex cascade of electrical, chemical or mechanical events which produce action. In order to perform a biological function, the single action potential must be modulated in duration. The cardiac ventricle, for example, elongates for the duration of the depolarization (Paragraph 5.14.3), since the amplitude cannot be modified (Figure 5.27).

The solution to the communication problem presupposes a certain degree of specialization between the two cells involved. The generating cell must have the capacity to generate the message and make it easily available to the receiving cell, which may be very close to the generating cell – sometimes even attached – or very distant.

It is essential for the functionality of an organism that the primary event, the action potential produced by one cell, is transferred to another cell or group of cells. Over the course of evolution, various mechanisms have been selected to perform this function.

The most direct and immediate transfer of information between a generating cell and one or more receptor cells, which can also become generating cells themselves, is certainly when they are in physical contact, so there is the possibility of establishing a direct connection. This mode of transmission, a clear example of the close relationship between structure and function, is present in many epithelia, in cardiac muscle, in the vessels and in the digestive tract of mammals; it is also found, among molluscs, in the mantle of squid and in the common octopus.

6.1.1 Gap junctions

From an evolutionary point of view, the oldest specialized structures that have guaranteed the transmission of information in living organisms are the gap junctions (Figure 6.2). Gap junctions can be described as cytoplasmic bridges with low electrical resistance that allow the exchange of ions and small molecules between cells.

From a structural point of view, gap junctions (Data Sheet 6.1) make the tissue mechanically resistant, particularly in epithelia where, by

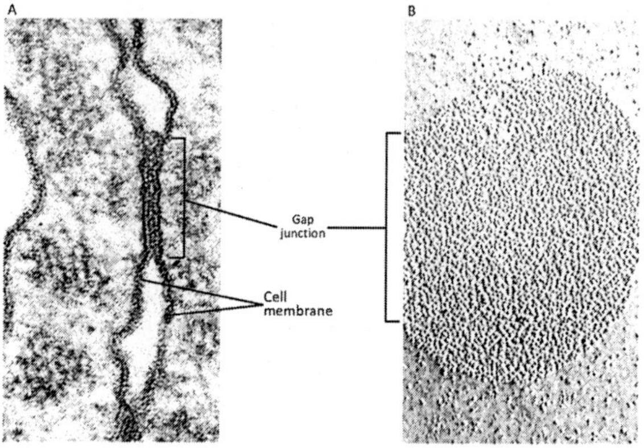

Figure 6.2 Electron microscopy images of a gap junction in fibroblasts in culture in a thin section (A) and with the freeze-fracture technique (B).

reducing the space between cells, they prevent solutes from permeating extracellularly, except for water in the intestinal epithelium (Paragraph 4.2.5).

From a functional point of view, gap junctions allow a homogeneous distribution of solutes in all of the cells that make up an epithelium. In excitable cells like cardiac muscle cells or smooth muscle cells in blood vessels, they allow electrical communication between adjacent cells, as they are low-resistance pathways.

Gap junctions can also perform an important defence function in a tissue. Indeed, if any traumatic situation irreparably damages the plasma membrane of a cell, there is an uncontrolled increase in the concentration of intracellular calcium, which can rise from 10^{-7} M to a concentration of 1.8 mM (Table 3.2), incompatible with proper cellular function.

The increase in calcium concentration by diffusion across cytoplasmic bridges would affect all cells in contact with the damaged cell. This can be prevented by gap junctions, which can block the diffusion of these divalent ions into neighbor intact cells, though a calcium-dependent conformational change.

6.1.2 Communication between cells with comparable properties

How the electrical message spreads between neighbouring cells joined by cytoplasmic bridges is through the use of electrotonic currents

(Paragraph 6.2.4.1). These are the ionic currents that flow between two points at different electrical potential located in the same conductive environment (Paragraph 6.2.4.1). Figure 6.3 shows a schematic model of a tissue formed by cells joined by the low-resistance cytoplasmic bridges formed by gap junctions. Assume that an appropriate electrical stimulus applied to cell 1 raises its membrane potential from the resting value of -70 mV to ≅-50 mV. The potential difference between ≅+30 mV of cell 1 (peak of the action potential) and -70 mV of neighboring cells induces electrotonic currents through the gap junctions that charge the membrane capacitors of cells 2 and 5. An action potential is generated in each cell if the electrotonic currents are sufficiently intense. In addition, the charging of the membrane capacitors (tab. 2.1) must be sufficiently fast to depolarize the membranes of cells 2 and 5 to the threshold value before the sodium channels are inactivated (Paragraph 5.11). In turn, cells 2 and 5 depolarize cells 3, 6 and 9 with electrotonic currents, which in turn generate action potentials. In a relatively short time, all of the cells in the tissue joined by gap junctions are the site of an action potential.

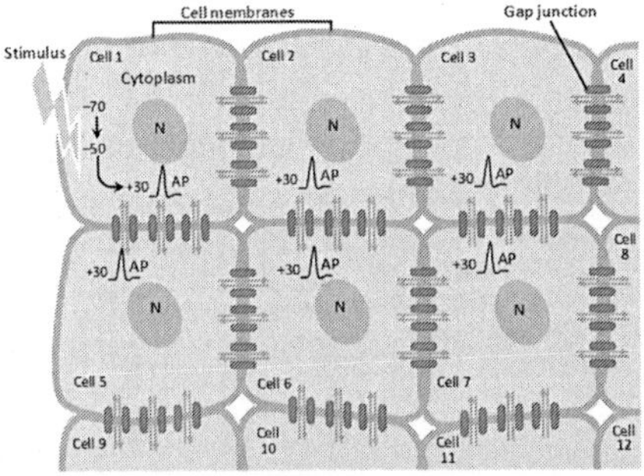

Figure 6.3 Schematic model of the communication between cells through gap junctions. The values shown are potential differences from the external value of zero (ground) and are expressed in millivolts. The black arrows represent time sequences. The double grey arrows represent the ionic fluxes through the gap junctions. AP symbols represent action potentials. N indicates the cell nucleus.

Electrical communication via gap junctions can only take place between cells of comparable size, effectively preventing one cell from transmitting the excitatory event to a larger cell. Consider the schematic model of two cells joined by gap junctions in Figure 6.4. Cell 1 produces an action potential in response to a suitable stimulus and an electrotonic current is

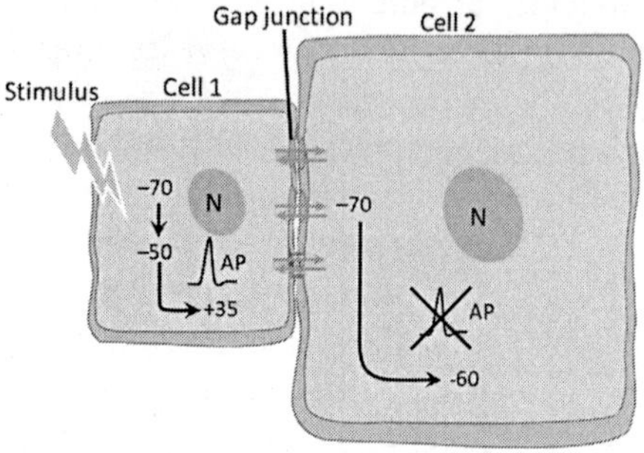

Figure 6.4 Schematic model of communication through gap junctions between two cells of different sizes. The values shown are potential differences from the external value of zero and are expressed in millivolts. The thin black arrows represent time sequences. The double grey arrows represent the ionic fluxes through the gap junctions. AP symbols indicate action potentials. N indicates the cell nucleus.

generated, which charges the membrane capacitor and depolarizes cell 2. Cell 2 is much larger than cell 1. The membrane capacitance is directly proportional to the surface area of the cell; the time to charge the capacitance depends on the time constant τ, which is directly proportional to the membrane capacitance. In this case, the electrotonic current flowing from cell 1 and 2 will be adsorbed by the membrane capacitor of cell 2 without producing a sufficient depolarization to trigger an action potential. This is for two reasons. Charge distribution on a larger surface produces a smaller change in membrane voltage. Furthermore, the charge of the capacitor takes time. The capacitor time constant is proportional to the membrane extension (Equation 2.1). A slower time constant could induce the inactivation of the sodium channels, impeding the formation of an action potential.

The communication mode through gap junctions in a tissue such as the one schematized in Figure 6.3 has a series of extremely interesting characteristics from a functional point of view:

1) for short distances, the transmission of depolarization by electrotonic currents is practically immediate, at a speed just below that of light;
2) the action potentials are generated successively from the cell that has received an adequate stimulus (cell 1 of Figure 6.3) to all of the others, until in an extremely short time, an entire population of cells that form the tissue is involved;
3) the stimulus follows a preferential direction, longitudinally and radially, guaranteed by the proximity of the cells of the same tissue joined by

the low-resistance cytoplasmic bridges formed by gap junctions;

4) no cell can generate a second action potential induced by the electrotonic currents, for example from cell 3 to cell 2 of Figure 6.3, because the inactivation of the sodium channels renders it refractory (Paragraph 5.14.2).

Two important physiological functions that require synchronization of the electrical signal use the communication mode of gap junctions: the contraction of the atria and ventricles during the cardiac cycle (tab. 7.3) and the peristalsis of the digestive tract, i.e., the mechanism that allows the contents of the digestive tract to pass continuously through the small and large intestines, proceeding from the duodenum to the rectum and anus (Data Sheet 6.2).

The importance of direct cell-to-cell communication is evident in the two examples just discussed, which relate to the vital mechanisms of food intake and adsorption through the digestion track and the distribution of nutrients through the circulatory system. The myocytes of the longitudinal and circular layers of the smooth muscle of the digestive tract (Paragraph 7.3) as well as the cardiac myocytes (Paragraph 7.4) while remaining, histologically, single contractile cells, are functionally connected by gap junctions that ensure their electrical continuity and coordinate functionality. However, there are also some drawbacks to these systems. Communication via cytoplasmic contacts can occur in either direction. For example, malfunctions in sodium current inactivation can lead to several negative consequences if the primary stimulus is generated in a tissue or organ while it is in a pathological state. In such abnormal situations, it is possible to develop ventricular extrasystoles, which cause the contraction of the ventricles following a stimulus not produced by the sinoatrial node and before the ventricles have been completely filled (Paragraph 7.4.1). Moreover, the phenomenon of vomiting, which does not always have negative connotations, is caused by an antidromic peristaltic wave that tends to carry the contents of the digestive tract towards the oral cavity.

Some organisms, such as the squid and the common octopus (Figure 6.5), have selected, during the course of evolution, particular ways of exploiting the typical characteristics of purely electrical signal propagation, involving several organs or even the whole organism. The squid has most of its muscular system organized by contacts between electrical cells that

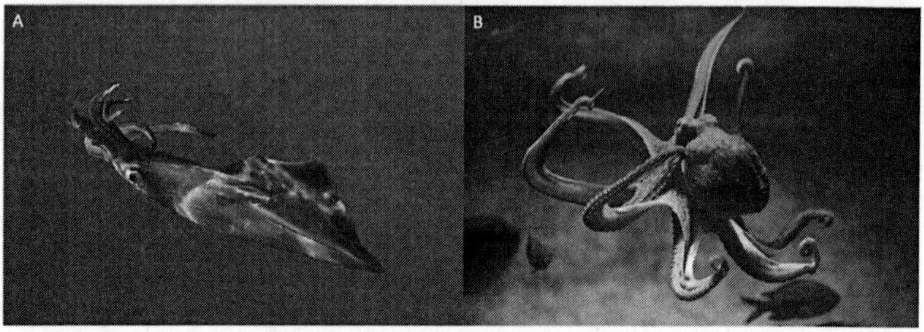

Figure 6.5 A) A squid (*Loligo forbesi*); B) common octopus (*Octopus vulgaris*).

are capable of making the mantle contract upon stimuli conducted by a giant axon at high conduction velocity (Paragraph 6.2.4.2). Water is violently expelled through the siphon and the animal can move and escape from danger. With a contractile and nerve-conduction apparatus comparable to that of the squid, the common octopus forcefully expels water from the siphon to move and expels ink to escape from danger.

Over the course of evolution, gap junctions and chemical mediator communication between cells have been selected for at very different times, in step with the evolution of multicellular organisms, in order to improve functional control by the organism. The most advanced communication systems did not replace the previous ones, but were added to them, and all have retained their irreplaceable specificity in their various functions.

This has made it possible to retain the most primordial communication systems for the vital functions of the organism, reserving increasingly complex and specialized tasks for the more evolved systems.

6.2 Communication mechanisms: chemical communication

In order to ensure the correct evolution towards the acquisition of more complex and organized functionalities and towards the formation and development of multicellular organisms composed of tissues of different embryonic origins and cell sizes, it is necessary to overcome the functional limitations of communication with electrical synapses. Electrical communication is only possible between cells in direct contact, between cells of comparable size, in simple organisms and in simple physiological processes that do not require a fast time frame (a few milliseconds), too large a range of action or complex modulation and control. Moreover, in cell-to-cell communication based on electrical contacts, the limitations are relative not only to the individual tissue, but to the possible specialization of an increasingly large and complex organism, where the speed of transmission of the primary signal becomes crucial. In fact, while for short distances and in cells of the same tissue, electrotonic currents are the fastest means of transferring information, as the size of the organism increases, this mode of transmission becomes incompatible with fast communication (6.2.4.1). Therefore, there was a need to develop more complex communication systems that could modulate the response to a stimulus, communicate between cells of different tissues, increase the distance to be covered and at the same time reduce the signal transmission time. Lastly, regarding the most advanced specialization, two different cells are put into direct communication, where a first cell 'commands' a second cell to perform a specific task, with the ability to modulate the instructions even over long distances.

6.2.1 Communication between groups of cells via a mediator

A second way of passing information between two cells is where following various stimuli, a cell, in its own right as a stimulator, produces and releases a generally small quantity of molecules that perform the function of a chemical mediator (Figure 6.6). The molecules reach other cells, which can be defined as receptors or, more generally, target cells.

Over the course of evolution, multiple possibilities of specialization and modulation have been selected, which are present in different ways and in variable numbers in the various structures:

1) stimulator cells can produce and release different chemical mediators in different concentrations only as a response to a

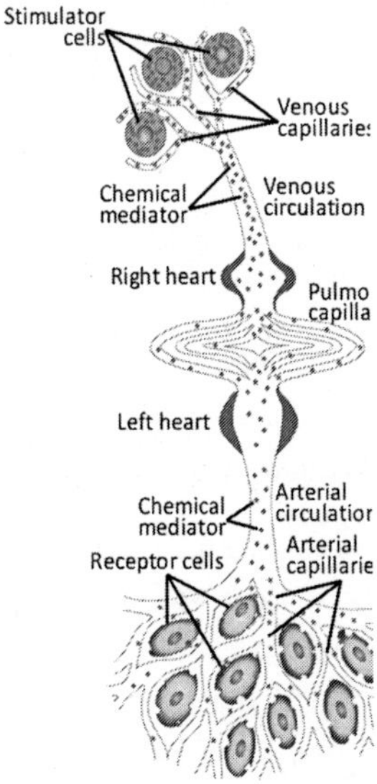

Figure 6.6 Schematic representation chemical mediators communication modes.

chemical, physical or electrical trigger;
2) the chemical mediators may be target cells transported by the circulatory system or passing directly from tissue to tissue;
3) the distance between the stimulating cells and the receptor cells may vary, affecting the reaction time of the latter;
4) if the chemical mediator is hydrophobic, it may penetrate directly through the plasma membrane and interact directly with cytoplasmic and nuclear molecules and organelles;
5) if the chemical mediator is hydrophilic in nature, it must necessarily interact with specific membrane receptors that may be directly coupled to ionic conductances and activate ionic currents, or it may interact with membrane or cytoplasmic proteins that act as intracellular second messengers;
6) target cells can interact in different ways with chemical mediators, for example by modulating the receptor molecules on their plasma membrane.

In an information transfer system such as the one just described, the chemical mediator is produced with relatively long times, varying from a few hundred milliseconds to several seconds or even minutes, by different stimulator cells (3 in the schematic in Figure 6.6), and interacts with different cells (9 in the schematic in Figure 6.6). Control of the system by the organism must take into account the overall activity of the stimulating cells, the efficiency and speed of transfer of the circulatory system, the degree of reactivity of the target cells and their sensitivity to the chemical mediator. The chemical mediator acts as a hormone because it carries a chemical message and is an agonist to a receptor with which it can interact.

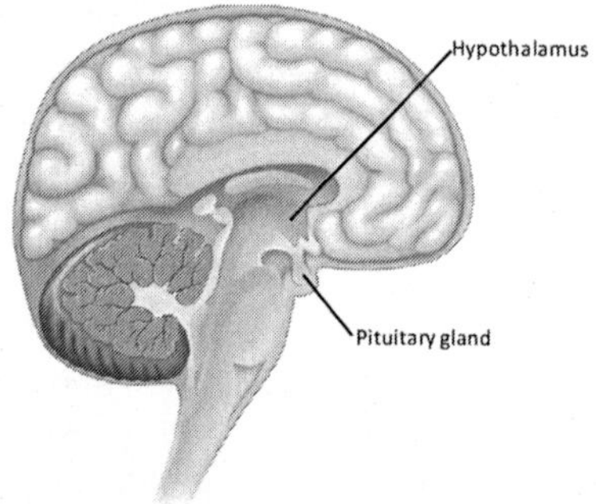

Figure 6.7 Schematic representation of the human encephalon. The hypothalamus and pituitary gland are highlighted.

One of the mammalian systems that uses the structure described in Figure 6.6 to regulate the pressure and volume of circulating fluids is the kidney (Paragraph 4.3.7). A reduction in blood volume recorded by the atrial baroreceptors, which are sensitive to pressure variation, or a reduction in pressure detected by the baroreceptors of the aortic arch and carotid sinus is the physical trigger that leads to the regulation of the two functional parameters, plasma volume and blood pressure. The trigger stimulates the cells of the supraoptic and paraventricular nuclei of the hypothalamus (Figure 6.7) – the stimulator cells – to produce the vasopressor precursor or ADH (anti-diuretic hormone). The chemical mediator ADH is delivered into the bloodstream through the neurohypophysis. It undergoes a series of transformations, reaches the receptor cells located in the distal part of the renal tubules and collector ducts and binds to specific membrane proteins, causing renal water reabsorption and consequently causing blood volume and pressure to increase.

6.2.2 The hormonal system

It is reasonable to assume that during the course of evolution, the first chemical communications took place between cells of different natures, but which were present in the same aqueous environment. In this case, the chemical mediator was released into the external medium by one or more cells and by diffusion reached the target cells, which must express the specific receptors. Subsequently, with the development of organisms in which it was possible to recognize an internal environment separate from the external environment, the need arose to diffuse the mediators to the various parts of the organism. This task was performed by the circulatory system, which, in addition to its primary function of distributing nutrients and oxygen, has been adapted as a means of transport that is capable of reaching all regions of the organism.

The development of a system capable of carrying the chemical mediator from the stimulator cells to the receptor cells, in which even a single chemical mediator is able to stimulate cells that are physically distant by synchronizing their action, has been of extreme importance for the development and improvement of the body's overall functions.

This complex and widespread system of chemical communication, generally referred to as the hormonal system, has developed and specialized to the point where it oversees a large number of vital functions, which essentially consist of coordinating the time and space to complete the project written in the genome of each biological organism. Many examples confirm this important function: the harmonious development of the organism, the balanced growth of the different components of an individual, the completion and optimal function of organs and systems even with very different development times, the control of the concentration of electrolytes in the extracellular fluids, and the balance between acid and base in tissues.

Despite the efficiency of its action, the precision of the mechanism of action of the chemical mediator on the receptor, and the speed of execution, which can even be in the order of milliseconds, the primordial form of this mode of communication lacks a centre of coordination and the possibility of signal processing or modulation, including of the quantity of hormone produced. However, the greatest impediment for the hormonal system to be responsible for communication within an organism is that it cannot be specific.

The circulatory system reaches every part of the organism. To be able to interact with only one cell or a group of cells would require a chemical mediator-receptor relationship, specific to each function. This would entail the production of a large number of agonists and an equally large number of receptors, all at an enormous expenditure of energy. The consequence for the evolving organism would be to have only a limited number of possible functions and the difficulty or impossibility of evolving functions of increasing complexity and adaptation to the environment, which by nature is constantly changing.

In addition to the problem of the relationship between the mode of stimulus and the complexity of the function, there is also the problem of the speed with which the message is carried. While the response time following the interaction between chemical mediator and receptor can be in the order of milliseconds, as the size and complexity of the organism increases, the limiting factor becomes the time it takes the circulatory system to distribute the agonists to the most distant regions, times that can be in the order of seconds or even minutes. In this case, the activation of a series of actions necessary for predation, escape from immediate danger, a response to a harmful stimulus, or a voluntary action by a small number of cells would be completely inadequate.

6.2.3 The nervous system

In parallel with the chemical communication system based on the generic and hard-to-control relationship between agonists and receptors, widely implemented in the organism, a system has been selected and developed over the course of evolution that favours the direct relationship between only two cells. Electrical or chemical information, both before and after the interaction between chemical mediator and receptor, follows pre-established favored pathways.

Cellular extensions have developed from the stimulating cell, which selectively and directly contact the receptor cell: these extensions are able to carry the message over distances of tens of centimetres or more in time frames in the order of milliseconds. In this way, the nervous system was constructed, first consisting of just a few cells and then becoming increasingly complex, with a command centre that coordinates, integrates and stimulates functions in all regions of the body.

The chemical synapse, i.e., the direct relationship between two cells, allows a virtually infinite number of functions using a few types of agonists and a few types of receptors, with the ability to select new and more efficient circuits rather than inventing new agonists and new receptors.

Communication using chemical synapses has some overlapping properties with the hormonal system. As in the hormonal system, communication using chemical synapses involves the release and diffusion of a chemical mediator. The interaction between the chemical mediator and the receptor on the target cell activates the receptor, which may act directly as an ion channel, or activates a transmembrane protein, which in turn activates a cascade of intracellular second messengers (Paragraph 3.3).

Unlike the hormonal system, for the chemical synapse, it is the structure that fully determines how it functions. In fact, the direct relationship between two cells is realized in the narrow intersynaptic space of 10-20 nm. The space is confined to a few square microns between the presynaptic membrane of the stimulating cell that releases the chemical mediator and the postsynaptic membrane of the receptor cell. This arrangement of the two cells is of fundamental importance in ensuring that the diffusion of the chemical mediator occurs without dispersion and with timescales of the same order of magnitude as both the release of the chemical mediator and its interaction with the receptor. The direct relationship between two cells also guarantees a precision in message delivery that would be impossible with the chemical communication of the hormonal system.

As chemical synapses evolved, it became possible to build neuronal circuits to process signals from the external environment – the sensory system (Chapter 8) – and to generate a response with characteristics that take into account numerous variables within the organism related to the complexity of the circuit. All of this leads to the generation of a rapid reaction, in the order of milliseconds, as a synthesis of all of the variables that are integrated in the neuronal circuits.

To simplify, once the rapid response process has been triggered, for example, the reaction to an immediate danger or painful stimulus, the electrical message is transported from the stimulating cell to the receptor cell by the cellular extensions that form the nervous system. Here, the chemical mediator, released from the end of the stimulator cell, activates receptors that in turn initiate muscle contraction and the body's escape reaction.

In order to understand how information is circulated in the nervous system, it is necessary to describe the process of message transfer, which consists of several steps, each with its own particularities and different stages of modulation.

It is then necessary to focus on the language that the nervous system uses to modulate different information, both incoming (the stimuli coming from the external environment) and outgoing (the stimuli produced within the organism).

The single-neuron action potential, like the one occurring in skeletal muscle, cardiac cells (Paragraph 5.14.3) or that linked to the phenomenon of fertilization (Paragraph 3.3), cannot be modulated in amplitude. The ability to vary the duration of a single action potential depends on the activation of specific currents. In ventricular cells and oocytes, the maintained calcium current is capable of creating a plateau at depolarized potentials that can last from a few seconds to several minutes (Figure 5.38). In the case of cells that modulate the duration of action potentials, the signal is directly linked to a function. The single excitable wave in cardiac cells has a duration of several milliseconds to stimulate and maintain the contraction of the ventricles for the necessary time. In oocytes, after the acrosome reaction due to the interaction of the spermatozoon with the oocyte membrane, the excitation wave due to the calcium current lasts several minutes and participates in the process that prevents polyspermy.

In the nervous system, the single action potential has no meaning. The membrane potential of the nerve cell is maintained in a condition of unstable dynamic equilibrium, which produces individual spontaneous action potentials even when the cell is at rest. The nervous system transmits messages and modulates them based on the frequencies at which action potentials are produced. Bursts of action potentials, known as firing (Paragraph 6.4), encode messages through their frequency, which can be finely modulated.

6.2.4 The propagation of the action potential in an axon

The excitable cells of an organism are responsible for producing changes in membrane potential in response to a stimulus within or, more often, external to the cell, e.g., from a peripheral receptor or coming from another neuron. The response mechanism to a stimulus takes advantage of the potential energy stored in the electrochemical ion gradients, the biophysical characteristics of the plasma membrane and those of membrane ion channels.

The responses of the excitable cell are of two types: passive and active. Passive responses, such as graded potentials whose amplitude is proportional to the intensity of the stimulus, can be hyperpolarizing or subthreshold depolarizing. They can be modified by other graded potentials that affect the same cell and decrease in amplitude as a function of the distance travelled. Active responses, such as action potentials, respond to the stimulus in an all-or-none mode, i.e., they arise irrespective of the amplitude of the stimulus, provided that the stimulus reaches the threshold. Neuronal action potentials have peculiar characteristics, such as self-generation and maintaining constant amplitude and duration regardless of the distance covered by the electrical signal propagating along the axon.

6.2.4.1 Graded potential

The axon of a nerve cell is electrically formed by a conductor (the cytoplasmic electrolytic solution) surrounded by an insulator (the phospholipid membrane; Paragraph 2.2), all of which is immersed in the conductor constituted by the external solution.

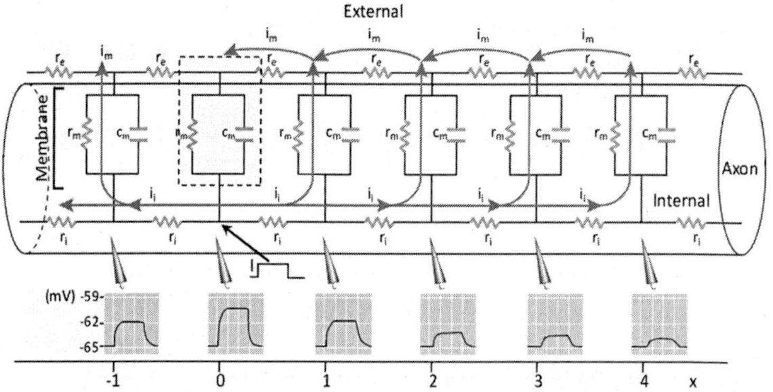

Figure 6.8 A current I injected into an axon at the point x = 0 generates a graded passive response, which propagates through the internal resistances r_i, decreasing in amplitude due to current leakage (i_m) through the capacitances c_m and the membrane resistances r_m. The grey area shows an RC circuit consisting of a resistor (r_m) and a capacitor (c_m) arranged in parallel.

The axon can be represented hypothetically by a sequence of identical parallel circuits, called local circuits (Figure 6.8). Each portion "dx" can be schematized by an RC circuit formed by the membrane resistor r_m and the membrane capacitor c_m placed in parallel. Each RC circuit is connected to the neighboring circuit inside the axon by the cytoplasm resistance r_i and outside by the external solution resistance r_e. rm represents the membrane ion permeability resistance and cm, the phospholipid bilayer capacitance.

If a hyperpolarizing or depolarizing subthreshold current (I of Figure 6.8) is injected at the point x=0, the axon responds with a potential variation (tab 2.1). Thus, a potential difference is established between point x = 0 and points x = 1 and x = -1, which are still at resting potential. For the moment, we consider only the direction x = 1, neglecting the phenomena occurring in the opposite direction, which are identical.

Because there is a potential difference between point 0 and 1, an electrotonic current is generated towards the area x = 1. Here, the current follows two paths: the internal resistance r_i, with a non-zero value, and the membrane RC circuit, i.e., the membrane resistance r_m, with a non-zero value, and the membrane capacitance c_m. At time t = 0, the membrane capacitance has zero resistance because it is unloaded (Data Sheet 2.1). In agreement with the principle that the current goes where it encounters less resistance, therefore, at t=0, charge is predominantly flowing towards the membrane capacitor. As time goes by, the capacitor resistance increases. So, the current through r_m constantly increases up to the point at which c_m resistance become infinite. There will be a partial depolarization at x = 1, but with a smaller amplitude. Part of the electrotonic current follows the cytoplasmic direction through r_i. The circuit loops towards x = 0 through the external resistance r_e. This "local circuit" current loop is repeated between x = 1 and x = 2, and so on until the end of the axon. In each infinitesimal stretch dx, therefore, there will be part of the current i_m that crosses the membrane and part, i_i, that follows the cytoplasmic path. The propagation is characterized by a progressive decrease in the amplitude of i_i and consequently in the membrane depolarization.

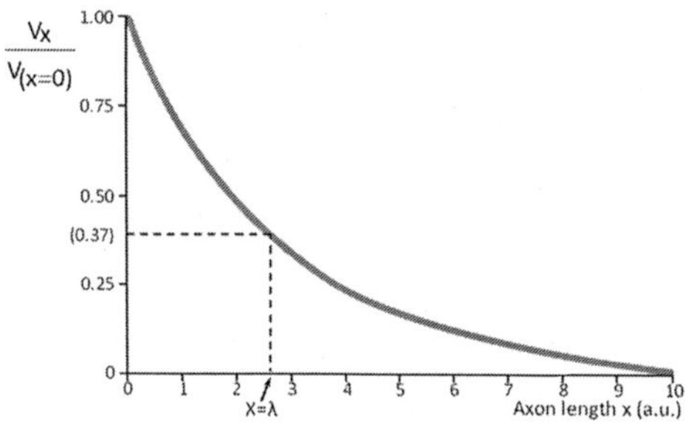

Figure 6.9 The electrotonic potential decays along the axon with a negative exponential trend and is characterized by the space constant λ. λ is the distance from the current injection point $V_{(x=0)}$ and V_x is where the membrane potential decreases by 37% of the initial value.

The electrotonic potential shows a negative exponential relationship along x with respect to the injection point x = 0 (Figure 6.9), described by the relation:

$$V_x = V_{(x=0)} \cdot \exp\left(-\frac{x}{\lambda}\right) \tag{6.1}$$

The relation derives from the cable theory proposed at the beginning of the 1900s to describe the behavior of submarine telegraph cables; it is independent from the potential difference value and from the time. V_x represents the membrane potential variation at distance x along the axon. λ is a parameter, dimensionally a length, defined as the distance from $V_{(x=0)}$ and the point along the axon in which there is a 37% decay of the $V_{(x=0)}$ according to the equation

$$V_{x=\lambda} = V_{(x=0)} \cdot \exp(-1) = V_{(x=0)} \cdot 0.37 \tag{6.2}$$

λ is described by the relation $\lambda = \sqrt{\frac{r_m}{r_i}}$ (6.3)

where r_m and r_i are, respectively, the membrane and cytoplasmic resistance per unit axon length. From Equation 6.3, it can be deduced that λ increases if r_m increases, if, for example, the membrane of the axon decreases its permeability. λ can also increase if r_i decreases, for example due to an increase in the axon diameter.

A different parameter is the time constant ?, dimensionally a time. The membrane time constant is the time for the potential to fall from resting to 63% of its final value in the charging curve during the application of a small current pulse. The membrane time constant is given by the relation

$$\tau_m = r_m \cdot c_m \tag{6.4}$$

where r_m and r_i are, respectively, the membrane and cytoplasmic resistance. Because excitable cells must have a fast time constant, ideally, the membrane resistance must be very high. In neurons, membrane resistance should be high and the surface should be extended. Membrane resistance ranges from 50 MΩ and 2 GΩ. The cell membrane area is in the order of thousands of square microns.

In the nerve cell, λ generally has a value of a few tens of micrometers and [illegible] is in the order of a few microseconds. A graded potential, which propagates electrotonically at a speed close to that of light, is functional only for small distances, such as those between the dendrites and the cell body of the nerve cells, those between the receptive zone and the synapse zone of an afferent neuron in the photoreceptors (Paragraph 8.6), and those between the cell body and the insertion cone of the axon. In the latter case, electrotonic currents are essential, as they enable signal modulation by temporal and spatial integration between synapses or other signals originating at different distances and at different times in the nerve cell, as well as between signals of opposite polarity (Paragraph 6.3.2).

6.2.4.2 The 'propagation' of action potentials

The electrotonic potential (EP of Figure 6.10), is formed at the level of

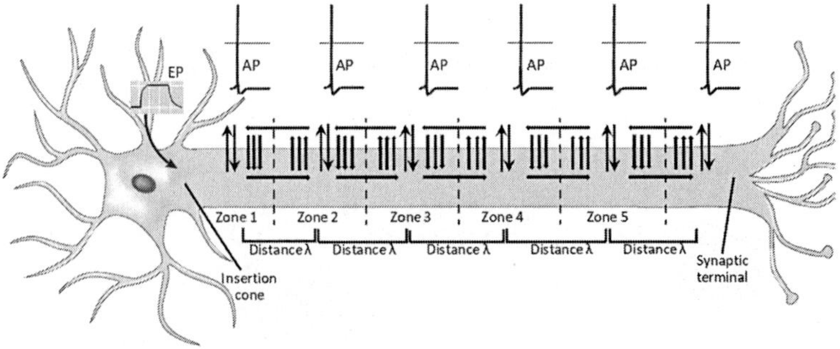

Figure 6.10 Schematic diagram of an unmyelinated axon in which an electrotonic potential (EP) above threshold at the level of the insertion cone activates an action potential (AP). Alternating local electrotonic currents (thick black arrows) between contiguous zones and generation of action potentials in each zone, the depolarization reaches the synaptic terminal. The double black arrows represent the inward sodium and outward potassium currents.

the insertion cone of the axon, whose membrane is rich in voltage- and time-dependent sodium and potassium channels. The electrotonic potential is able to trigger the launch of an action potential when the depolarization is above the threshold value (AP of Figure 6.10).

The action potential creates a potential difference between zone 1 and zone 2, which is at resting potential. The electrotonic current flowing through the internal resistance r_i depolarizes zone 2 by charging the membrane capacitor (Paragraph 6.2.4). The current closes the circuit by flowing through the external resistor r_e (grey arrows in Figure 6.10) to zone 1. The action potential of zone 1 raises the membrane potential from the resting value of ≅ -70 mV to a peak of ≅ +30 mV. 100 mV of potential difference between zone 1 and zone 2 induces an electrotonic current, which in the local circuit depolarizes the membrane up to the point of generating a new action potential in zone 2. It should be noted that an electrotonic current, not shown in the Figure, is also generated between zone 2 and zone 1. However, with the latter, it is not possible to generate a new action potential, since zone 1 is refractory due to the inactivation of the sodium channels (Paragraph 5.11). This is the reason why nerve conduction can only occur in one direction, from the insertion cone to the synaptic terminal. At the synaptic terminal, there is the formation of an action potential that results from the continuous alternation of the electrotonic currents from one zone to another and of the action potential in the various zones, identical to that of zone 1. In fact, the same channels with constant biophysical characteristics are present all along the axon.

While the electrotonic current propagates at a speed close to that of light, the onset of each action potential requires about 0.1-0.3 ms (Figure 5.27) and delays the onset of the electrotonic current that must depolarize another zone of the axon above threshold. Thus, the action potential moves along the axon as an "excitation wave" as every dx stimulates the launch of an action potential. The entire process slows down the progression of the depolarization that tries to reach the end of the axon. Since the majority of this time is devoted to reproducing the action potential, the logical consequence is to reduce the number of action potentials. This can be done by increasing λ. There are two possible ways to do this: decreasing the internal resistance r_i or increasing the membrane resistance r_m. Decreasing r_i would require increasing the diameter of the fiber, but the space occupied by the nerve fibers would lead to an unfavorable structure-function relationship. The giant squid axon, which has a diameter of 0.8-1.0 mm and conducts electrical impulses at a maximum speed of 20 m/s, is still too slow in its excitation conduction to meet the needs of more complex and evolved organisms.

The solution selected by evolution has been to increase precisely the resistance of the membrane r_m, not by varying its biophysical characteristics, but by surrounding some areas of the axon with sleeves of

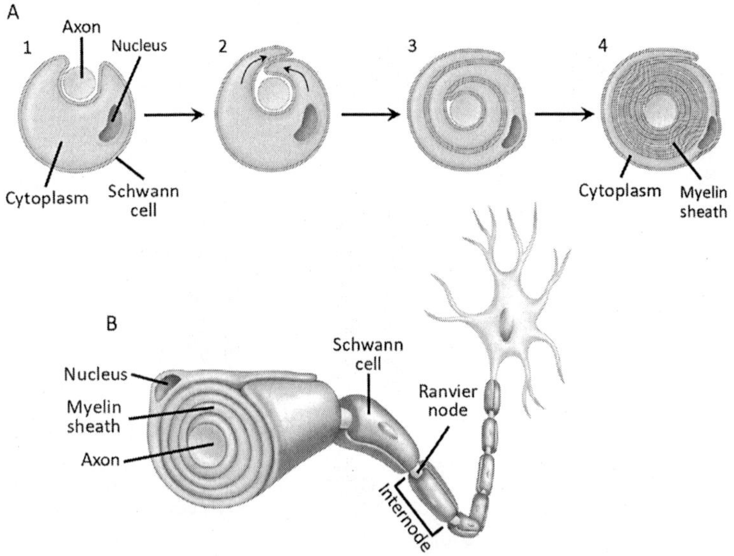

Figure 6.11 A) A Schwann cell (1) produces large amounts of membrane, (2) depositing it in layers on the axon (3 and 4) to form myelin sheath sleeves. B) Nodes of Ranvier, points of interruption of the myelin sheath, are rich in sodium and potassium channels.

myelin sheath, made of Schwann cells or oligodendrocytes (Figure 6.11 A). The sleeves form the internodes and consist of several dozen layers of membrane that is poor or lacking in ion channels, interspersed with nodes of Ranvier, bare areas rich in sodium and potassium channels. This forms a myelinated axon (Figure 6.11 B).

6.2.4.3 Jumping 'propagation' of action potentials

In a myelinated axon, the space constant λ increases considerably compared to that of unmyelinated fibers, from a few tens of micrometers to 1.5-2.0 mm. In myelinated axons, the internodal distance practically coincides with these values.

The mechanism of conduction of the neural message in these axons is very similar to that already described for those that are unmyelinated (Figure 6.9). The electrotonic potential EP, generated at the insertion cone level, must be adequately intense to promote a depolarization above the threshold to trigger an action potential (AP of Figure 6.12). The electrotonic currents in myelinated axons flow between one node and the next (circuit of grey arrows of Figure 6.12). The action potential formed in each node maintains the depolarization to the maximum level.

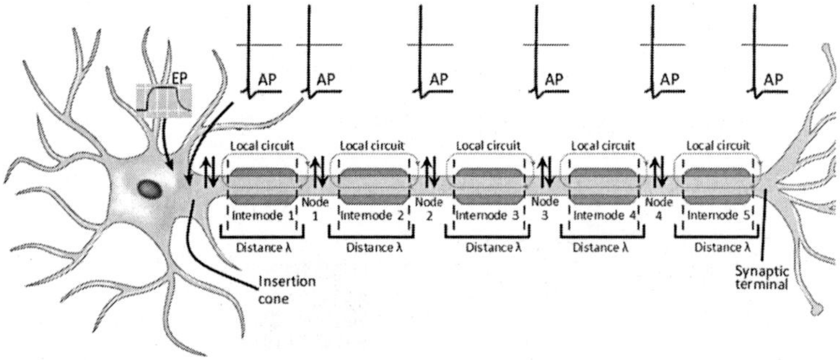

Figure 6.12 Schematic diagram of a myelinated axon in which an electrotonic potential (EP) above threshold at the insertion cone activates an action potential, (AP) which, with alternating local electrotonic currents (grey arrows) in the internodes and action potentials in the nodes of Ranvier, is generated at the synaptic terminal. The double black arrows represent the inward sodium and outward potassium currents.

This mode of propagation of the action potential, which is characterized by a conduction velocity that in some cases reaches 200 m/s, is generally referred to as jumping conduction, as the action potential appears to jump from one node to another. In fact, it is the electrotonic currents that move along the axon and only in the direction from the insertion cone to the synaptic terminal. Even the jumping conduction is unidirectional. At the point of action potential formation in each node, the electrotonic currents flow in both directions. However, in the previous node, the action potential is not yet over and the sodium channels are still in the inactivated mode (Paragraph 5.11).

Nerve signal conduction would be instantaneous if it were due exclusively to the electrotonic current and if the membrane were a perfect insulator. Action potentials, as mentioned, are indispensable for re-establishing the potential difference that generates the electrotonic current each time, even if they slow down conduction. Myelination enables electrical signals to be conducted at high speed. Logistically, to carry the signal to the end of the axon, an unmyelinated axon 100 mm long with a space constant of 0.15 mm would have to generate about 660 action potentials, each of which would introduce a delay of about 0.4 ms, and would therefore take 162 ms. A myelinated axon, also 100 mm long and with a space constant of 2 mm, must generate about 50 action potentials and would take 20 ms to bring the signal to the end of the axon.

6.2.5 Cell-to-cell communication via chemical mediator

The information transfer system selected in the course of evolution, which has made it possible to overcome most of the problems inherent in the selective control of complex organisms, is schematized in Figure 6.13. Its properties represent a further evolution with respect to those described in paragraph 6.2.1:

1) a single stimulator cell continuously produces chemical mediator molecules (neurotransmitter) of one type which it accumulates in intracellular vesicles. Vesicles are accumulated in the proximity of the axon ends (presynaptic membrane) ready to be released into the small space separating it from the receptor cell;
2) the neurotransmitter is produced with relatively long timescales, which may vary from a few hundred milliseconds to several seconds or even minutes;
3) the neurotransmitter is released from the presynaptic membrane only in response to an exclusively electrical trigger that does not stimulate production, but rather the release from intracellular vesicles following an increase in intracellular calcium concentration;
4) the neurotransmitter is released in excess of the amount needed to ensure its effectiveness;
5) the neurotransmitter is released within a few hundred microseconds and diffuses towards the membrane of the receptor cell (postsynaptic membrane) within a few tens of milliseconds;

6) the neurotransmitter is hydrophilic and it must interact with specific membrane receptors;
7) the receptors can be directly coupled to ionic conductances and activate ionic currents (ionotropic) or they can interact with membrane or cytoplasmic proteins that act as intracellular second messengers (metabotropic);
8) target cells can interact in different ways with neurotransmitters, for example by modulating the number of receptor molecules on the postsynaptic membrane.

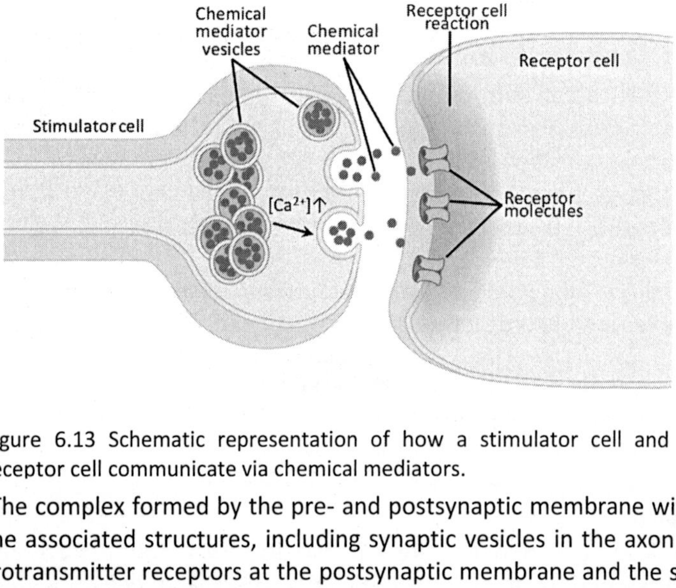

Figure 6.13 Schematic representation of how a stimulator cell and a receptor cell communicate via chemical mediators.

The complex formed by the pre- and postsynaptic membrane with all of the associated structures, including synaptic vesicles in the axon end, neurotransmitter receptors at the postsynaptic membrane and the space between the two cells (Figure 6.13) is called, from the Greek term σύναψις that means connection, a chemical synapse. It is the most common mode of communication between cells.

6.3 The chemical synapse

Over the course of evolution, there has been a shift from communication being based essentially on the electrical properties of cells (Paragraph 6.1) to that based essentially on chemical communication

(Paragraph 6.2). Each of the two types of communication is characterized by specific functions that have been conserved in evolution and are used in the performance of tasks that are essential to life. The fusion of these two different modes of information transfer is in the chemical synapse. The chemical synapse, through a specific cell-to-cell relationship, is able to operate a mode of control that is absent in less evolved organisms. In practical terms, this translates into the possibility of precise stimulation resulting in highly specific and localized actions.

Chemical synapses add or refine two other factors, which are partially or totally absent in the more primitive modes of electrical and chemical communication: the ability of the organism to respond to external stimuli, and to execute a voluntary action.

Both electrical and chemical synaptic communication are capable of responding to stimuli from the external environment, such as noxious stimuli, osmotic changes, temperature variations or even circadian rhythms. But what chemical synapses introduce is the possibility of responding to an external stimulus more quickly, in a very precise way, and, more importantly, being able to modulate the response by making it more appropriate to the type and intensity of the stimulus received. This property, completely absent in electrical and chemical communication, is that of 'voluntariness'. Voluntariness at the cellular level means the ability to produce a targeted response autonomously, either after having received and processed information from the external environment or simply so that the organism itself can benefit. These behaviours are present in organisms with a primordial nervous system, made up of clusters of a few excitable cells, but with limited efficiency. In more advanced organisms, where a distinction is made between control cells, command cells and executive cells, voluntariness becomes increasingly important, both in controlling the organism and in coordinating multiple actions. However, the morpho-functional pattern of chemical synapses between neurons and between neurons and non-neuronal cells remains identical: a stimulating presynaptic cell that releases the chemical mediator, an intersynaptic space, and a postsynaptic cell that is stimulated by the presence of receptors for the chemical mediator (Paragraph 6.2.1). All of this occurs in the space of a few square microns. We can recognize synapses coupled to sensory neurons that collect information from the outside, synapses between neurons responsible for processing, modulation and control, and finally synapses between neurons and other cell types that perform actions directed towards the surrounding environment. While those of the first two types are essentially synapses that modulate the transmission of information, the latter are synapses that, depending on the command they produce, promote actions through different cells that may be muscular, glandular or other.

6.3.1 Release of neurotransmitter from presynaptic membranes

The mode of neurotransmitter synthesis and release involves such a sequence of steps and large number of proteins with enzymatic functions that it represents the first level of modulation in the general function of communication in chemical synapses.

Two types of neurotransmitters can be distinguished: large-molecule neurotransmitters and small-molecule neurotransmitters.

Large-molecule neurotransmitters require the production of a precursor propeptide that is synthesized in the rough endoplasmic reticulum of the cell body of the presynaptic neuron, transferred to the Golgi apparatus (1 of Figure 6.14), where it is activated by appropriate post-translational modifications, and selected, with respect to the other types of molecules present, for incorporation into secretory granules that are released by budding. Secretory granules are transported by axonal

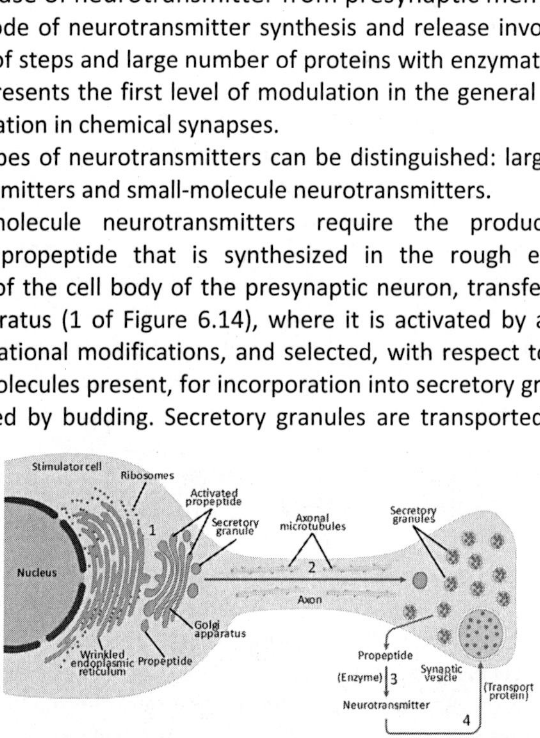

Figure 6.14 Schematic representation of the most significant steps, highlighted by numbers, in the process of synthesis, transformation and storage in synaptic vesicles of a large-molecule neurotransmitter in the presynaptic neuron. The grey arrow indicates transport by molecular motors.

microtubules at a rate of 3-4.5 mm/s along the axon to the synaptic terminal (2 of Figure 6.14) where they are stored. Here, precursor propeptides are converted by specific enzymes into neurotransmitter molecules (3 of Figure 6.14) and stored in the terminal synaptic vesicles by appropriate transport proteins (4 of Figure 6.14). Conversely, small-molecule neurotransmitters, such as acetylcholine, are synthesized directly in the synaptic terminal using enzymes produced by the cell body, transported along the axon by microtubules and stored in small vesicles between 40 and 60 nm in diameter.

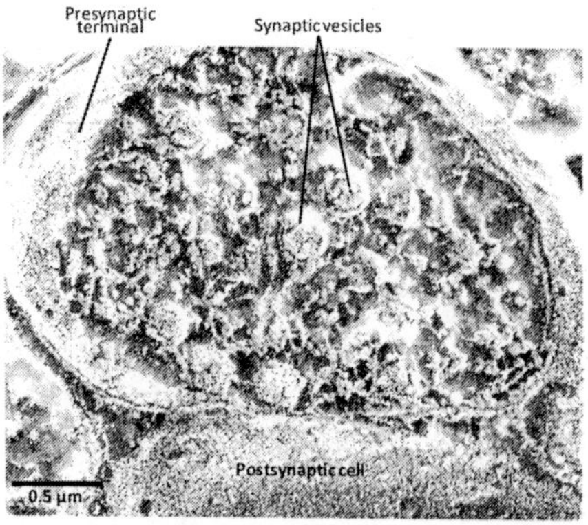

Figure 6.15 Electron microscope image of a fragment of a chemical synapse.

The presynaptic neuron must be in a position to respond appropriately to any demand for neurotransmitter release. Therefore, the synaptic terminal must continuously produce neurotransmitter vesicles in order to have a sufficiently large number accumulated, as shown by electron microscopy images (Figure 6.15). A synaptic terminal with a

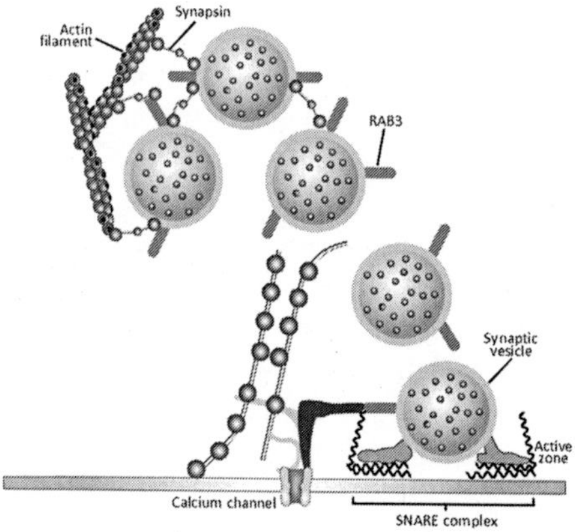

Figure 6.16 Schematic detail of the active zone of neurotransmitter release, the binding between actin filaments and synaptic vesicles, and some proteins involved in the mechanism of synaptic release. RAB3 is a protein complex that regulates Ca^{2+}-dependent release.

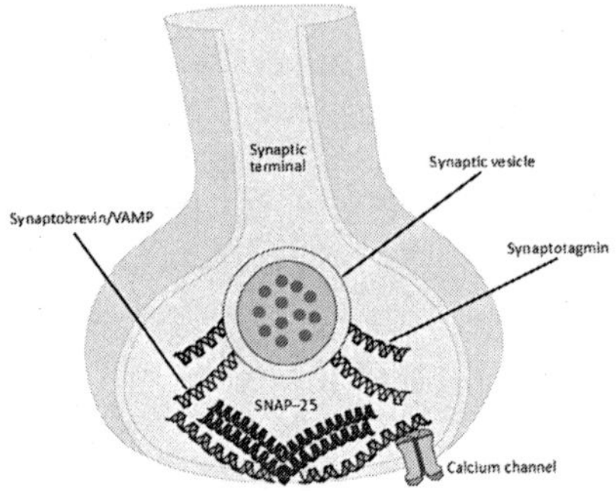

Figure 6.17 Some protein molecules of the synaptic vesicle and the intracellular face of the synaptic terminal involved in the mechanism of neurotransmitter release are highlighted.

volume of 0.1 to 0.3 μm^3 can contain on average 200 to 500 vesicles that are held in place by the actin microfilaments of the intracellular matrix (Figure 6.16). In this area of the presynaptic membrane, voltage-dependent calcium channels are very abundant.

The main proteins involved in the release of synaptic vesicles onto the presynaptic membrane (Figure 6.17) are: synaptobrevin/VAMP, synaptotagmin which functions as a sensor for calcium on the vesicle membrane, syntaxin-1 and SNAP-25. When the presynaptic neuron is at rest and with a low calcium concentration, neurotransmitter release does occur, but this release is of negligible magnitude and with no temporal control other than randomness, so it generally produces no appreciable effect. It is only in the presence of an adequate stimulus that the release of the neurotransmitter becomes consistent. The depolarization produced by the succession of action potentials activates voltage-dependent calcium

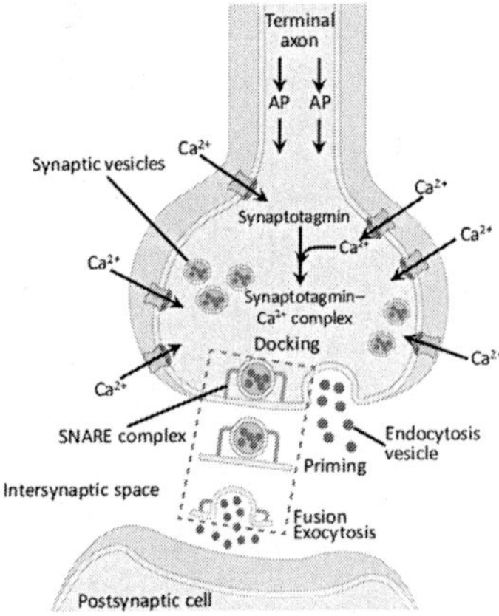

Figure 6.18 Schematic details of the series of events from the action potential (AP) to the release of the neurotransmitter at a chemical synapse. In the diagram, the intersynaptic space appears dilated due to the explosion outside the active zone (area delimited by the hatching), which is in fact a large area close to the plasma membrane of the presynaptic neuron.

channels that are diffusely present throughout the membrane of the synaptic terminal, but particularly in the active zone. The calcium ion enters the synaptic terminal (Figure 6.18) where it initially frees the vesicles from the bond with the actin network that holds them in place. It then forms the Ca^{2+}-synaptotagmin protein complex that promotes a cascade of events:

1) it induces interaction between syntaxin-1, SNAP-25 and synaptobrevin, the first two on the plasma membrane and the third on the vesicle membrane;
2) it promotes the formation of another protein complex called SNARE that induces vesicle docking, i.e., its initial attachment to the plasma membrane;
3) it induces priming of the vesicle, making it competent for fusion to the plasma membrane.

At this point, fusion and exocytosis occur, with release of the neurotransmitter into the intersynaptic space. To prevent the plasma membrane from expanding uncontrollably, due to the incorporation of synaptic vesicles after exocytosis, the process of endocytosis is initiated and new vesicles are formed to increase the pool available for subsequent release cycles.

When depolarization of the action potential ceases, calcium ions are rapidly restored to their resting intracellular concentrations, 10^{-8} M, in three different ways, with each method in different proportions depending on the tissue. A large proportion of calcium is rapidly sequestered by binding to specific endogenous buffer proteins. A further proportion can be removed by certain cytoplasmic organelles, including the mitochondria, by active transport, and the endoplasmic reticulum by a Ca^{2+}-dependent ATPase called SERCA (Sarcoendoplasmatic Reticulum Calcium ATPase). Finally, calcium close to the plasma membrane is carried outside the neuron by a sodium/calcium exchanger and a Ca^{2+}-dependent ATPase called PMCA (Plasma Membrane Calcium ATPase).

Generally, the release of neurotransmitter is massive in the intersynaptic space, to ensure an effect. However, for control of nerve transmission to depend exclusively on the state of excitation of the presynaptic neuron and not on the amount of neurotransmitter released, it is necessary to have a mechanism of continuous subtraction of the excess mediator. There are essentially three ways in which the neurotransmitter is removed: by diffusion to the sides of the synaptic cleft, by enzymatic degradation, and by reuptake by the presynaptic terminal or by glial cells in the central nervous system.

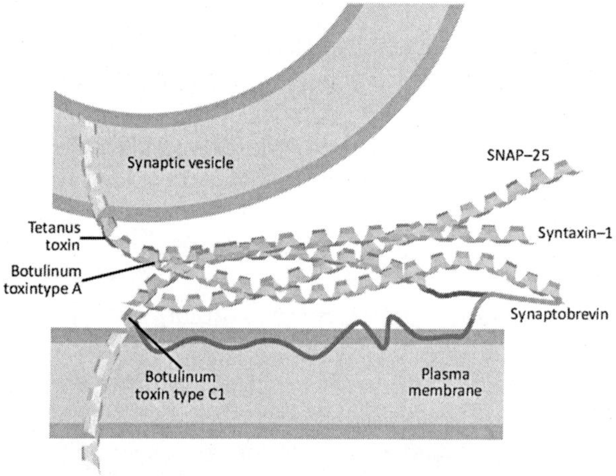

Figure 6.19 Schematic representation of botulinum toxin and tetanus toxin binding sites at chemical synapses. SNAP-25, syntaxin-1 and synaptobrevin form theSNARE protein complex, which is essential for synaptic release.

The complete neurotransmitter release cycle of each vesicle lasts about 60 s in total, almost all of which is required for translocation within the presynaptic axon and recovery by endocytosis, with 10 to 20 ms for docking and priming, and less than 1 ms for fusion. At the same time, hundreds of vesicles enter the cycle, which are carried to the numerous active zones of the terminal and which, by exocytosis, release the large quantity of neurotransmitter required due to the stimulus characteristics of the presynaptic neuron.

Considerable light was shed on all of the components of the mechanism of release of the chemical mediator by the presynaptic membrane by the study of botulinum toxin and tetanus toxin (Figure 6.19). Botulinum toxin, particularly type A and type C1, two of the seven known varieties, is a neurotoxic protein with protease action produced by the bacterium *Clostridium botulinum.* It is considered to be the most toxic substance identified to date. Botulinum toxin type A acts on the SNAP-25 protein of the SNARE complex, while type C1 acts on syntaxin. Both prevent vesicles from fusing with the presynaptic membrane, consequently blocking the release of the neurotransmitter acetylcholine and causing paralysis of the affected muscle.

Tetanus toxin, a neurotoxin produced by the bacillus *Clostridium tetani*, is carried by the bloodstream from the point of infection to the peripheral nerve fibers, where it is stored in vesicles by endocytosis and then returned to the central nervous system. It interacts with synaptobrevin of the SNARE complex, preventing vesicle fusion and the secretion of the neurotransmitters GABA (Data Sheet 6.4) and glycine, which modulate or inhibit muscle contraction. It causes continuous and painful muscle spasms affecting all the muscles of the body, starting with the face and limbs, and can cause death by spasm of the respiratory muscles.

6.3.2 Neurotransmitter receptors in the nervous system and muscular system

Over the course of evolution, the postsynaptic cell has had to develop receptors for chemical mediators and the complex molecular mechanisms for their operation. The vast majority of the molecules that act as chemical mediators are hydrophilic, and therefore unable to enter the cell directly. The cell must therefore construct membrane receptors to which the chemical mediator is able to bind and that activate cytoplasmic pathways for signal transduction. In communication between cells by means of chemical synapses, the relationship between the chemical mediator and the receptor is another very important aspect of the mechanism by which not only selectivity but also the mode of passage of information is decided. Agonists may be simple molecules or more or less complex peptides. A single chemical mediator can act on several receptors, acting as a stimulant or inhibitor depending on the type of receptor, the type of cell and its functional state. Receptors, in turn, are almost always proteins located on cell membranes. The numerous possible combinations of different agonists and receptors give rise to virtually limitless possibilities for the execution and modulation of communication processes.

Receptors can be divided into ionotropic and metabotropic. **Ionotropic receptors** are both receptors and ion channels. Once the neurotransmitter is bound, the probability increases that the receptor will change conformation and become an open ion channel that can generate depolarization or hyperpolarization, depending on the permeating ion and the conditions in the cell. **Metabotropic receptors**, after activation when the neurotransmitter is bound, cause the formation of second messengers with a cascade of intracellular reactions leading to the activation of endocellular proteins. In some cases, activation of metabotropic receptors can mediate the function of ion channels by promoting an increase in the probability of opening or closing promoted by second messengers.

The functional difference between ionotropic and metabotropic receptors, in addition to the different modes of activation and timing, concerns the physiological effect on the target cell.

Stimulation of ionotropic receptors usually results in a local depolarization of the plasma membrane, which in some cases results in the generation of an action potential.

In the case of metabotropic receptors, however, synaptic stimulation has an effect that is amplified during signal transduction. In fact, a single receptor can activate different types of proteins associated with it, each of which in turn activates different cytoplasmic proteins, with an amplifying effect on the signal throughout the cell.

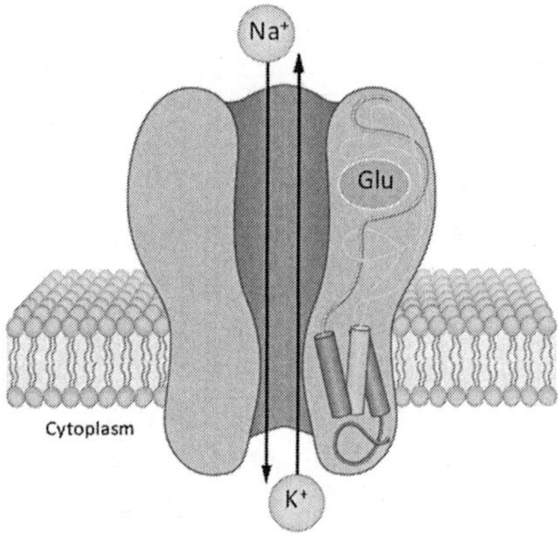

Figure 6.20. AMPA receptor. Glu = L-glutamate.

6.3.2.1 Glutamate receptors

Glutamate receptors can be metabotropic and ionotropic. **Ionotropic receptors** are also known as ROCs (Receptor Operated Channels) and are classified, on the basis of the synthetic chemical mediator that activates them, as postsynaptic AMPA and kainate receptors, also known as non-NMDA receptors, or as NMDA receptors.

AMPA receptors are activated by the synthetic compound α-amino-3-hydroxy-5-methyl-4-isoxazolpropionic acid, while kainate receptors are activated by the synthetic molecule kainic acid, found in its natural state in the alga *Digenea simplex*. These two receptors are important for rapid synaptic transmission in the central nervous system. Both are excitatory, because the conformational change activates sodium and potassium channels with single channel permeability of approximately 20 pS (Figure 6.20). A very rapid, but weak EPSP (Paragraph 6.3.3) is produced, with the peak close to 0 mV.

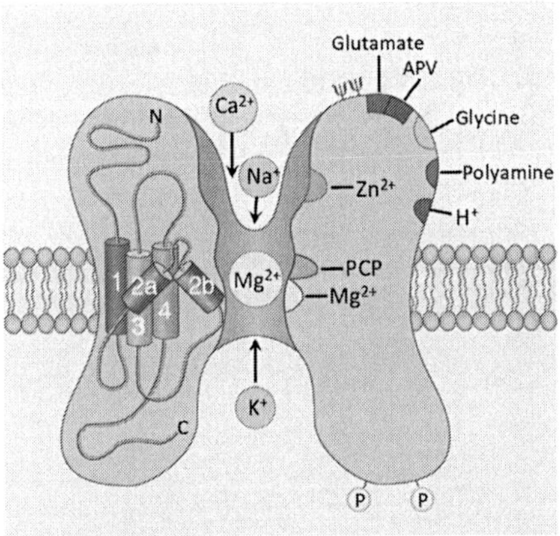

Figure 6.21 NMDA receptor. APV = 2-amino-5-phosphovalarate; PCP = phencyclidine; P = phosphorylation sites.

The postsynaptic **NMDA receptors**, so called because the synthetic agonist is N-methyl-d-aspartate, are excitatory and have APV (2-amino-5-phosphonovalerate) as a highly specific blocker. They are highly complex integral membrane proteins. In addition to the binding site for glutamate, polyamines and hydrogen ions, postsynaptic NMDA receptors have binding sites for numerous modulators including zinc, drugs such as PCP (phencyclidine), also known as angel dust, and especially magnesium (Figure 6.21), which is always bound to the receptor at rest and tends to block it.

Glutamate binding makes the receptor a channel for sodium, potassium and calcium ions, provided the postsynaptic cell is depolarized, so as to eliminate a magnesium blockade. This activates a single channel conductance of 50 pS and generates an incoming depolarizing sodium and calcium ion current and an outgoing hyperpolarizing potassium ion current, both of which are regulated by their respective equilibrium potentials.

When glutamate binds to a metabotropic receptor, a sequence of cytoplasmic second messengers is activated (Figure 6.22). These, in turn, activate various functions that involve different cytoplasmic reactions, up to and including stimulating protein synthesis, with fast responses in the order of milliseconds or very few seconds, or with slow responses that can occur after several seconds or several minutes.

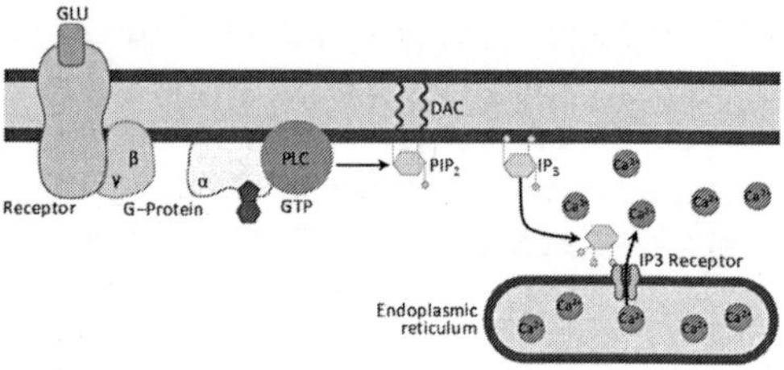

Figure 6.22 Schematic model of the mode of operation of a metabotropic glutamate receptor. Glu = glutamate; α, β and γ = G-protein subunits; PLC = phospholipase C; DAC = diacylglycerol; GTP = guanidine-triphosphate; PIP2 = phosphor-inositolo-diphosphate; IP3 = inositolo-triphosphate; Ca^{2+} = calcium ions.

Among the fast responses that modulate cell function, a very important mechanism is that involving phosphor-lipase C. When the receptor binds glutamate, it activates a G-protein associated with it, which stimulates phospholipase C, which in turn, by acting on the membrane lipid complex formed by diacylglycerol and phosphoinositol diphosphate, converts the latter into inositol triphosphate. The diacyl-glycerol remains in the membrane while the inositol-triphosphate is released and rapidly diffuses into the cytoplasm, binds to specific receptors located on the membrane of the endoplasmic reticulum and promotes the release of the second messenger calcium from the reticulum reserves.

If cytoplasmic calcium reaches a sufficiently high concentration, the process of modulating cellular activity can continue by following two cascades of intracellular events involving its regulatory and catalytic phosphokinase C sites. In the first pathway, calcium binds to the regulatory site, which changes conformation and binds to the diacylglycerol remaining in the membrane. The catalytic site of phosphokinase C, exposed by the conformational change induced by calcium binding, then phosphorylates a substrate protein with an ATP molecule, which is then capable of eliciting a cellular response.

In the second mode of modulation of the cellular response, calcium is selectively bound by calmodulin, acting as a second messenger, which binds to the regulatory site of phosphokinase C. This binding makes accessible a site that catalyzes the phosphorylation of a substrate protein, which can enact an appropriate cellular response.

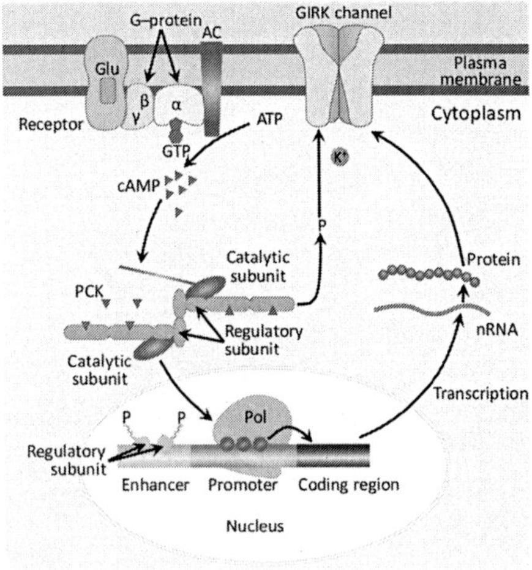

Figure 6.23 Cascades of slow modulation events induced by the binding of a glutamate (Glu) molecule to its membrane receptor. α, β and γ = G-protein subunits; AC = adenylate cyclase; PKC = phosphokinase C; Pol = RNA polymerase.

When modulation of cellular activity requires the synthesis of particular proteins, such as those that will form membrane channels, metabotropic glutamate receptors activate a cascade of intracellular events that constitute a slow response pathway (Figure 6.23):

1) the glutamate-activated receptor activates its associated G-protein, which in turn activates adenylate cyclase, which catalyzes the transformation of ATP into cAMP;
2) at the same time, a phosphate radical is transferred to GDP, associated with the G-protein, which is transformed into GTP;
3) cAMP, in turn, activates the cAMP-dependent phosphokinase C, which penetrates the nucleus, with its catalytic subunit activating, by phosphorylation, the regulatory subunit of the enhancer, and transcription is initiated;
4) the transcribed mRNA exits the nucleus and the protein will be synthesized to create an integral membrane protein, e.g., a potassium channel.

6.3.2.2 GABA receptors

The most common neurotransmitter in the central nervous system and one that can activate receptors that inhibit neuronal excitability is the amino acid γ-aminobutyric acid (GABA). GABA is synthesized by decarboxylation of glutamic acid and is finally degraded to succinic acid, an intermediate in the Krebs cycle, according to the following reactions:

GABA is able to activate two types of receptors: the ionotropic GABAA receptor and the metabotropic GABAB receptor.

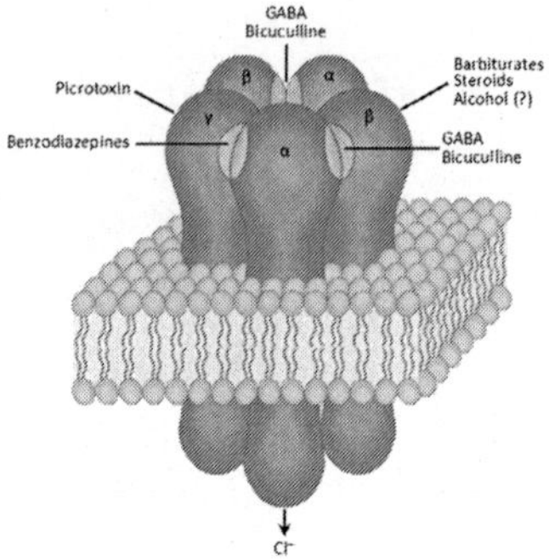

Figure 6.24 Model of the GABAA receptor showing the two α subunits, the two β subunits and the γ subunit, the specific sites for GABA, for barbiturates, for steroids and probably for alcohol, for picrotoxin and for benzodiazepines.

GABA bound to its postsynaptic **GABAA receptor** causes it to become, by conformational change, a chloride channel (Figure 6.24). In the adult, an inhibitory incoming chloride current is generated that hyperpolarizes the membrane and moves the potential away from the excitability threshold, which is generally between -60 and -40 mV, towards the chloride equilibrium potential at -85 to -90 mV. In embryonic neurons, however, a Na^+/Cl^-dependent ATPase, called KCC1, keeps the cytoplasmic chloride concentration high and shifts the equilibrium potential towards more positive potentials. In embryonic neurons, therefore, when chloride channels open, a depolarizing outgoing current is generated, with excitatory effects on the postsynaptic cell. This different behaviour between adult and embryonic neurons may suggest that the forming nervous system needs all synapses to be excitatory during development. The switching of the adult neuron's GABAA receptors to inhibitory function may therefore be considered a key indicator of the development of the central nervous system, so much so that its malfunction appears to underlie several pathological states, such as various forms of autism.

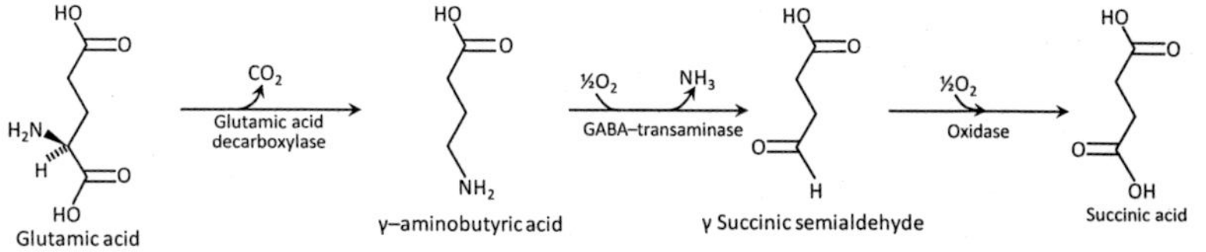

Several classes of compounds are known to interact with GABAA receptors (Figure 6.24). In particular, benzodiazepines, barbiturates and alcohol are able to modulate the response of the receptor to GABA by increasing the chloride current and making the inhibition of neuronal excitability stronger.

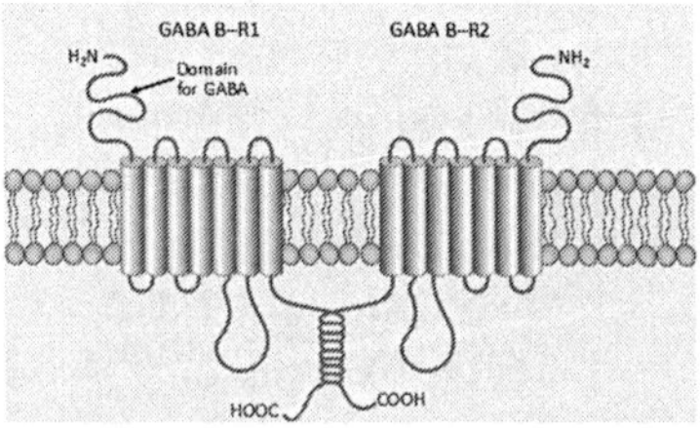

Figure 6.25 Structure of the GABAB receptor.

The GABAB receptor can be present on both the postsynaptic and presynaptic membranes. In this case, it functions as an auto receptor, with the mediator produced by the cell itself.

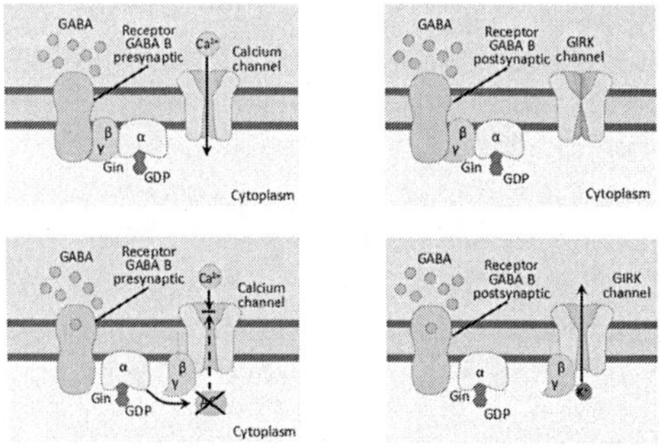

Figure 6.26 Mode of operation of presynaptic GABAB receptor (A and B) and postsynaptic GABAB receptor (C and D). Gin = inhibitory G-protein; α, β, γ = inhibitory G-protein subunits; GDP = guanosyn-diphosphate; GTP = guanosyntriphosphate; AC = adenylate cyclase.

The **GABAB receptor** is a heterodimer formed by the R1 and R2 subunits, which interact with each other at their respective C-terminals. The N-terminal domain of the R1 subunit has the GABA binding site, while the C-terminal domain of the R2 subunit interacts with the inhibitory G-protein (Figure 6.25).

Binding of a GABA molecule to the receptor activates an inhibitory G-protein, formed by the three α, β and γ subunits, anchored to the inner face of the plasma membrane near the receptor itself.

The G-protein splits into the α-subunit, which catalyzes the phosphorylation of GDP into GTP, and the βγ-complex (Figures 6.26A and 6.26C).

In the case of the presynaptic receptor, the α-subunit of the inhibitory G-protein inhibits adenylate cyclase, which normally blocks calcium entry into the cell and has an inhibitory effect on neurotransmitter release. Inhibition of adenylate cyclase acts as an intracellular second messenger and initiates an enzymatic cascade that leads to, among other things, the activation of calcium channels (Figure 6.26B). If, on the other hand, the GABAB receptor is postsynaptic, the βγ complex interacts with the potassium channel of the inward-rectifier GIRK type. Interacting with this channel has an inhibitory action on cell excitability due to the activation of an outgoing current of potassium, which tends to hyperpolarize the membrane potential (Figure 6.26D).

6.3.2.3 Acetylcholine receptors

Acetylcholine receptors, or cholinergic receptors, are found at many synapses in the central nervous system, where they can be either presynaptic or postsynaptic. In the neuromuscular junction, on the other hand, they are found only on the membranes of muscle cells.

Depending on the mechanism of action and the response following interaction with the neurotransmitter, two types of acetylcholine receptor can be recognized: ionotropic and metabotropic.

The **ionotropic acetylcholine receptor**, also called the **nicotinic receptor** after the selective antagonist nicotine, is an integral membrane protein of 290 kDa. It is a receptor consisting of five subunits. In the nervous system, the five subunits are either all the same α7-type subunit or the receptor is composed of two α4-type subunits and three β2-type subunits. In the neuromuscular synapse, probably more evolved than synapses within the nervous system, the receptor is formed by two α-type subunits, one β-type subunit (different from the neuronal α and β subunits composition), one γ-type subunit and one δ-type subunit (Figure 6.27A). Each subunit consists of four transmembrane domains with the N-terminal and C-terminal on the extracellular side (Figure 6.27B).

When two acetylcholine molecules bind to the specific binding sites present in the two α-subunits (Figure 6.27C), the receptor undergoes a conformational change to become a cation channel. In the presence of a resting potential of the postsynaptic cell that is known to be lower than -60 mV, an incoming flow of sodium ions is generated in the skeletal muscle and an incoming flow of calcium ions occurs in the nervous system. Subsequently, an outflow of potassium ions is generated in both (Figure 6.27D). In the nervous system, depolarization has the function of increasing cytoplasmic calcium; in skeletal muscle, the result of depolarization is the formation of an excitatory postsynaptic potential (Paragraph 6.3.3) and a muscle cell action potential (Paragraph 6.3.4).

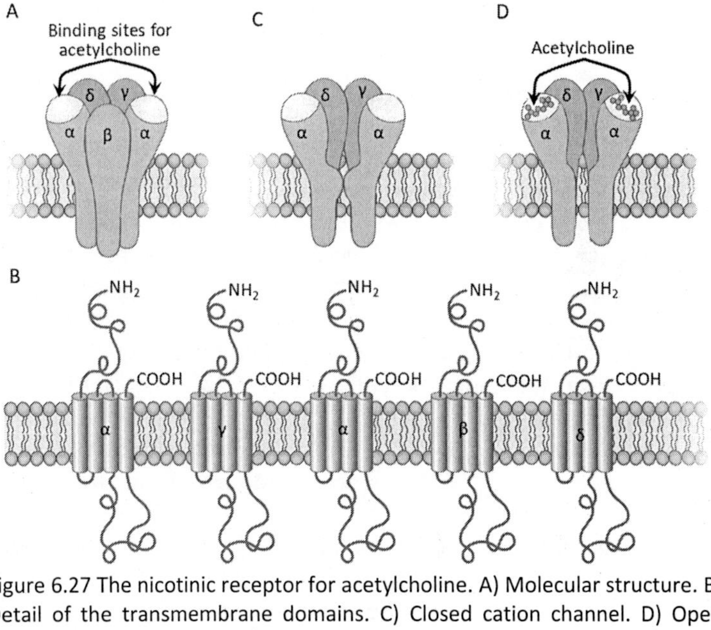

Figure 6.27 The nicotinic receptor for acetylcholine. A) Molecular structure. B) Detail of the transmembrane domains. C) Closed cation channel. D) Open cation channel. α, β, γ, δ = subunits forming the receptor. In C and D, the β subunits are not shown for claritv.

The **metabotropic acetylcholine receptor**, also called the **muscarinic receptor** after the selective antagonist muscarine, is present in the nervous system in the M1, M4 and M5 isoforms, in cardiac muscle in the M2 isoform and in smooth muscle in the M3 isoform.

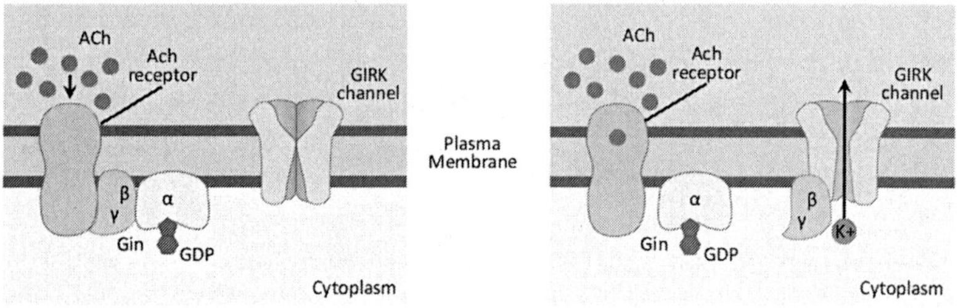

Figure 6.28 The action of the muscarinic receptor for acetylcholine (ACh). Gin = inhibitory G-protein; α, β, γ = inhibitory G-protein subunits; GDP = guanosyphosphate; GTP = guanosyntriphosphate.

Binding of an acetylcholine molecule causes the receptor to interact with an inhibitory G-protein, formed by the three subunits α, β and γ and present on the inner face of the membrane (Figure 6.28A). The enzymatic activity of the α-subunit catalyzes the transformation of a GDP molecule, bound to the α-subunit, into GTP. The G-protein splits into the α-subunit and the βγ-complex. The latter interacts with the GIRK potassium channel, an inward-rectifier type channel, but which is activated by the G-protein. There is a conformational change that increases the probability of opening the channel and activating an outgoing current of potassium that tends to hyperpolarize the membrane potential (Figure 6.28B), reducing the degree of cellular excitability. In sinus node pacemaker cells, hyperpolarization causes a lengthening of the duration of the slow diastolic depolarization given by the I_f current and thus a bradycardic effect, i.e., a decrease in heart rate (Paragraph 5.14.3).

In the nervous system, the M1 and M3 isoforms of the muscarinic receptor may induce excitation by depolarization of the membrane potential through inhibition of potassium conductance. Inhibition of potassium conductance is due to an increase in cyclic AMP through activation of adenylate cyclase by the α-subunit of G-protein, which converts ATP into the second messenger cyclic AMP. The potassium channel is inhibited by phosphorylation catalyzed by protein kinase A, which in turn is activated by cyclic AMP.

6.3.3 Presynaptic and postsynaptic currents and potentials

In the nervous system, excitation of the chemical synapse is determined by the arrival at the terminal axon of the action potential generated by the neuron and conducted along the axon. The depolarization produced by the action potential activates the voltage-dependent calcium channels; calcium ions enter the nervous system and initiate and control the neurotransmitter release process (Paragraph 6.3.1).

The evolutionary advantage that led organisms to select communication via chemical synapses over communication via electrical synapses, and to some extent communication via the hormonal system, probably lies in the numerous ways that the responses of the postsynaptic cell can be modulated. From a biochemical point of view, the nature of the neurotransmitter and the ability to release it in saturating quantities, thanks to optimal synthesis and storage mechanisms, are certainly of considerable importance.

Even more decisive are the properties that exploit the electrical nature of certain events in synaptic transmission, in particular the action potential of the neuron and the response of the postsynaptic cell.

The chemical synapse is capable of transforming high frequencies of action potentials, the code by which the nervous system communicates, into proportional amounts of released transmitter. If the frequency is high enough that the time interval between one action potential and the next is shorter than the time it takes for the entering calcium for depolarization to be removed, subsequent depolarizations cause the intracellular calcium concentration to increase more and more, producing the release of additional neurotransmitter. Calcium, in turn, binds to a greater number of receptors in the postsynaptic cell. Thus, the higher the frequency of the action potentials, the greater the response generated by the postsynaptic cell, compared to that produced by a single action potential.

As in almost all cells of an organism, in the external solution of the postsynaptic cell, the predominant ions are the sodium cation and the chloride anion, at a concentration of 145.0 mM and 115.0 mM, respectively, while in the cytoplasm there is a high concentration of potassium ions, 140.0 mM (Table 3.2). The neurotransmitter released into the intersynaptic space from the presynaptic terminal (Figure 6.29A) can produce two opposing effects, depending on the properties of the postsynaptic cell receptor and the ions involved in the communication process: the excitatory postsynaptic potential and the inhibitory postsynaptic potential.

Excitatory postsynaptic potential (EPSP). A first type of receptor, which changes conformation by binding to the neurotransmitter, has the property of becoming a cation channel, such as the nicotinic ionotropic receptor for acetylcholine (Paragraph 6.3.2.3) or the ionotropic glutamate receptors of the AMPA type or kainate receptors (Paragraph 6.3.2.1). Through these channels, there is a flow of sodium ions inwards and a flow of potassium ions outwards from the cell (Figure 6.29B), with a practically equal permeability coefficient, therefore with flows of potentially the same magnitude.

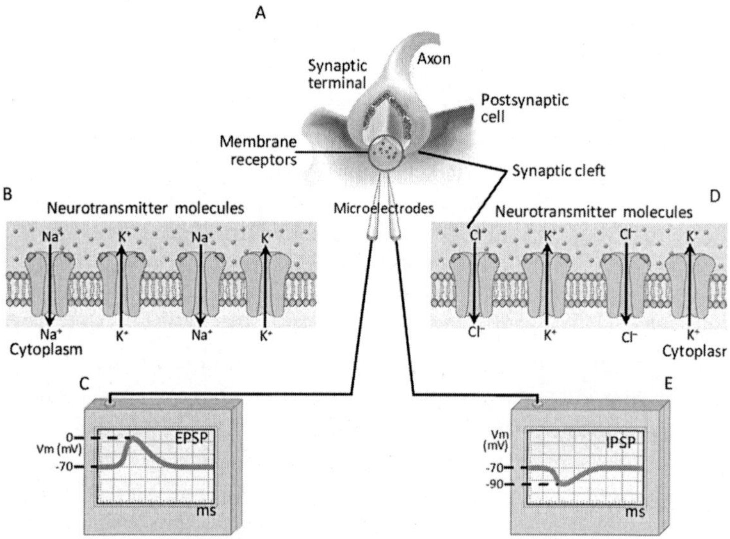

Figure 6.29 In a chemical synapse, several receptors for the neurotransmitter released from the synaptic terminal (A) activate, directly or indirectly, specific membrane ion channels (B and D) and induce opposite potential changes (C and E).

As the input of sodium depolarizes, tending to bring the membrane potential towards its equilibrium potential, the electrochemical gradient for sodium decreases and that for potassium increases; in parallel, the hyperpolarizing outgoing current of potassium increases and the depolarizing current of sodium decreases. The depolarization continues until the membrane potential reaches a value close to the average of the two equilibrium potentials (-20 mV in Figure 6.29C). Then, there is a return to the resting value of the potential due to the cessation of the release and complete removal of the neurotransmitter.

This variation of potential is obviously the excitatory electric message of the postsynaptic cell. It is therefore called the Excitatory Postsynaptic Potential (EPSP) and propagates electrotonically (paragraph 6.2.4.1).

Inhibitory postsynaptic potential (IPSP). A second type of receptor, changing conformation due to neurotransmitter binding, has the property of becoming an ion channel through which there is either an inward flow of chloride ions (e.g., the ionotropic receptor for γ-aminobutyric acid of the GABAA type (Paragraph 6. 3.2.2)), or an outward flow of potassium (e.g., non-NMDA-type glutamate ionotropic receptors; Figure 6.29D (Paragraph 6.3.2.1)). Both chloride and potassium fluxes result in an inhibitory hyperpolarization, termed Inhibitory Postsynaptic Potential (IPSP, -90 mV in Figure 6.29E), which propagates electrotonically in the postsynaptic cell. In the case of IPSP, too, there is a return to the resting value of the potential by cessation of the release and complete removal of the neurotransmitter.

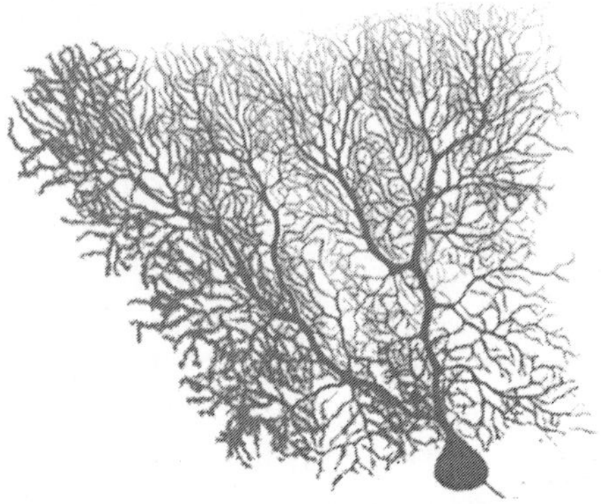

Figure 6.30 Image of a cerebellum Purkinje cell.

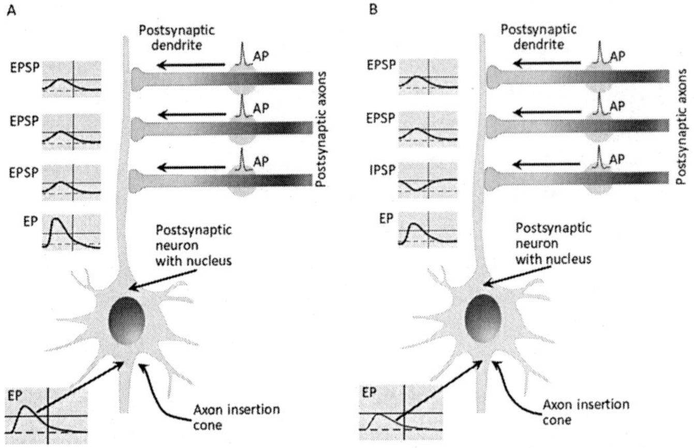

Figure 6.31 Schematic representation of (A) the spatial summation, in the central nervous system, of three distinct excitatory postsynaptic potentials and (B) the spatial summation of two excitatory postsynaptic potentials and one inhibitory postsynaptic potential. AP = action potential; EPSP = excitatory postsynaptic potential; IPSP = inhibitory postsynaptic potential; EP = electrotonic potential.

The neurons of the central nervous system generally have many hundreds of dendrites that emerge from the cell body or from the dendrites themselves, organized in complex branches (Figure 6.30) on which hundreds of synapses are formed. On each branch are both excitatory and inhibitory contacts, each formed by axons of presynaptic neurons.

This complex system offers a notable evolutionary advantage to the communication between nervous cells as it allows the modulation of the signal, as schematized in Figure 6.31, with two different modalities, spatial summation and temporal summation.

Spatial summation. Figure 6.31A shows a schematic of a postsynaptic neuron with numerous dendrites and an axon that emerges, as is known, from the insertion cone of the cell body. Three excitatory synapses have formed on one of the dendrites, inducing three excitatory postsynaptic potentials (EPSPs). The EPSPs combine to create a single potential that propagates electrotonically through the cell body to the insertion cone, decreases as a function of space, and can sum to other electrotonic potentials (EP) from other dendrites. If the electrotonic potential at the insertion cone reaches the threshold value, it generates an action potential. If one of the three synapses were of the inhibitory type (IPSP of Figure 6.31B), it would induce a postsynaptic potential that would subtract from the sum of the other two, reducing the amplitude of the electrotonic potential and making the formation of the action potential in the axon less likely.

The amplitude of the potential at the insertion cone depends not only on the number of synapses that generated it, but also on the characteristics of each. In addition to the type of synapse, excitatory or inhibitory, its distance from the cell body is important, due to multiple variables: the decrease in space of the electrotonic currents that generate the potentials, the amount of neurotransmitter bound to postsynaptic receptors, which determines the amplitude of the postsynaptic potential, the amount released proportional to the frequency of action potentials in the axon and the effectiveness of the removal of neurotransmitter from the intersynaptic space.

Temporal summation. As schematized in Figure 6.32, temporal summation occurs when a series of action potentials, three in the figure, arrive at the synaptic terminal in the synapse of a neuron with a dendrite. Each action potential induces the release of the maximum quantity of neurotransmitter, which generates an excitatory postsynaptic potential in the postsynaptic dendrite.

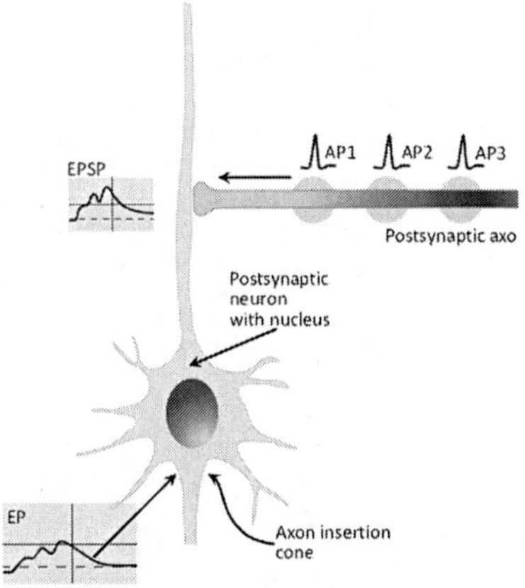

Figure 6.32 Schematic representation of temporal summation in the central nervous system. AP = action potential; EPSP = excitatory postsynaptic potential; EP = electrotonic potential.

Considering the time separating each action potential from the next, if their frequency is high enough for the dendrite to produce a second excitatory potential before the previous one has returned to its resting value, there is a progressive increase in the peaks and a potential value reached in the postsynaptic dendrite that is significantly greater than that induced by a single action potential.

Thus, the excitatory postsynaptic potential is electrotonically conducted to the insertion cone of the postsynaptic axon and, if the threshold is exceeded, an action potential is generated. The resulting potential is not itself possible to regulate with respect to time, but it is subject to the conditions described for spatial summation, i.e., it adds algebraically to other electrotonic potentials from other areas of the postsynaptic neuron.

6.3.4 Neuromuscular synapse

The synapse of the motor unit between the motor neuron and the muscle fiber (Data Sheet 7.1) has a number of special properties that make it unique and therefore in need of further examination. This particular

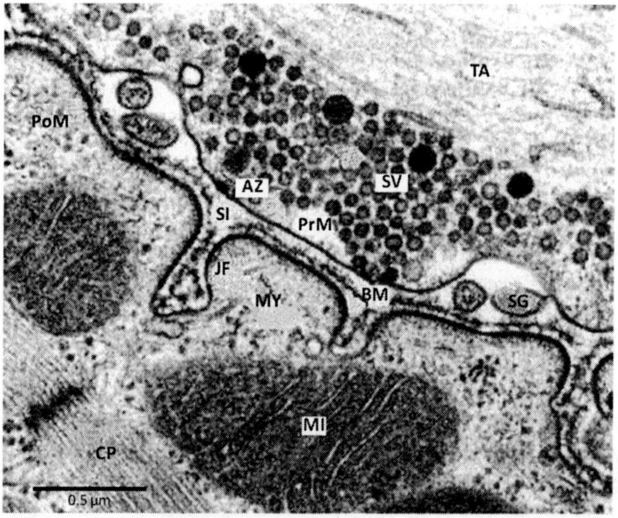

Figure 6.33 Electron microscope image of a neuromuscular synapse fragment. TA = terminal axon; SG = general secretory granules; PrM = presynaptic membrane; BM = basement membrane; PoM = postsynaptic membrane; MI = mitochondrion; MY = myofibril; CP = contractile proteins; JF = junctional fold; IS = intersynaptic space; SV = synaptic vesicles; AZ = active zone.

synapse is called the **neuromuscular junction**, **neuromuscular plate** or **motor plate**. It connects two cells of different origins, where the motor neuron acts as a stimulus while the muscle fiber transforms the excitation into mechanical work. It is a typical example of a chemical synapse that could in no way be replaced by an electrical synapse due to the obvious disparity in size and biophysical properties between the nerve cell and the muscle cell.

The typical structure of the neuromuscular synapse is evident in electron microscope images (Figure 6.33). Close to the presynaptic membrane, and especially in the active zones of the terminal axon, rich in voltage-dependent calcium channels (Figure 6.16), numerous vesicles ready to release the neurotransmitter acetylcholine (Paragraph 6.3.2.3) are grouped together. However, the structures that most characterize the neuromuscular synapse are the junctional folds, invaginations of the postsynaptic membrane juxtaposed to the active zones (Figure 6.33).

The Schwann cells, which form the myelin of the motor neuron, are also present in the axon of the presynaptic terminal (Figure 6.34), for which they perform the important function of mechanical protection from continuous shortening and lengthening during the contraction and relaxation of the muscle fiber.

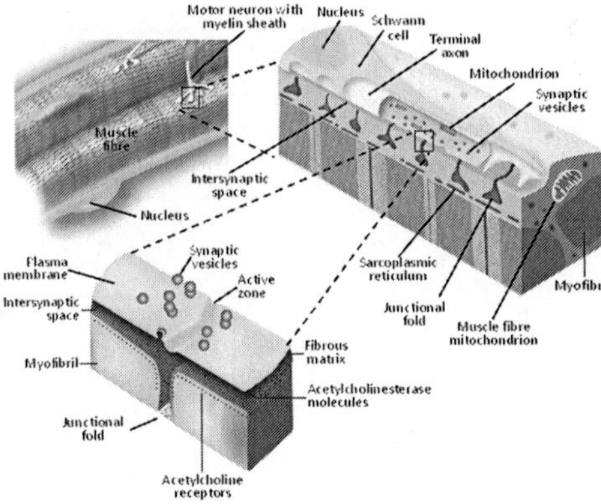

Figure 6.34 Structural details of the neuromuscular synapse.

In the synaptic terminal, much of the acetylcholine is synthesized using **choline and the enzyme choline acetyltransferase (both transported from** the cell body along the axon of the presynaptic neuron by the molecular motors) and acetyl-coenzyme A, produced by the numerous mitochondria present in the synaptic terminal, according to the reaction:

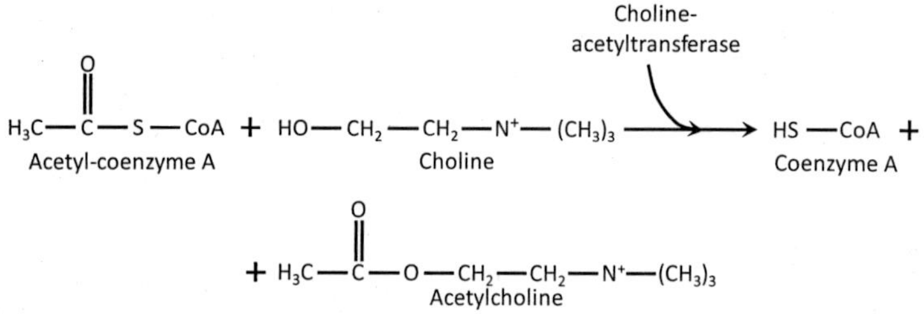

Between 5000 and 7000 molecules of acetylcholine are transferred into each vesicle. At this point, the vesicles begin the cycle that brings them to the numerous active zones (Figure 6.18), each zone ready to release by exocytosis a quantum of neurotransmitter, that is the minimum package of acetylcholine molecules. In the absence of action potentials, generally 1 quantum per second is released spontaneously, a quantity that is not able to induce a response in the muscle fiber. In the presence of an action potential, the number of quanta released becomes approximately 150,000 per second.

The postsynaptic muscle fiber has nicotinic ionotropic receptors for acetylcholine, at an estimated density of 10,000 per μm^2, embedded throughout the membrane of the neuromuscular junction and in the first section of the membrane forming the junctional fold. Acetylcholinesterase enzyme molecules are present on the entire outer face of the postsynaptic muscle fiber and throughout the extent of the junctional folds (Figure 6.34). In the deep areas of the junctional folds, and especially in the membrane outside the neuromuscular synapse, the voltage-dependent sodium channels responsible for the action potential are inserted; in the membrane outside the synapse, the voltage-dependent potassium channels responsible for the repolarization of the action potential are also present (Figure 6.35).

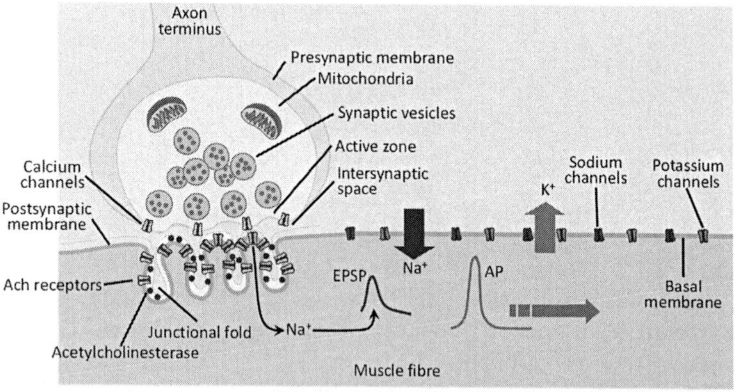

Figure 6.35 Schematic representation of the function of a neuromuscular synapse. EPSP is the excitatory postsynaptic potential, AP is the action potential, the large black arrow and large grey arrow represent the sodium and potassium currents, respectively, the thin black arrows represent the sodium ion flows and the thin grey arrow represents the propagation of the action potential along the muscle fiber.

Acetylcholine released into the intersynaptic space rapidly binds to its membrane receptors, which change conformation and become specific channels for the sodium ion. Sodium enters the muscle fiber (Figure 6.35), an excitatory postsynaptic potential is generated that depolarizes the membrane to the threshold value for sodium and potassium currents and an action potential is generated. The action potential propagates towards the triads (Paragraph 7.2.1) and activates the mechanism of muscle contraction.

The high affinity of the enzyme acetylcholinesterase for the neurotransmitter, estimated to be 10^5 times greater than the affinity of the receptors, leads to the rapid enzymatic elimination of acetylcholine according to the reaction:

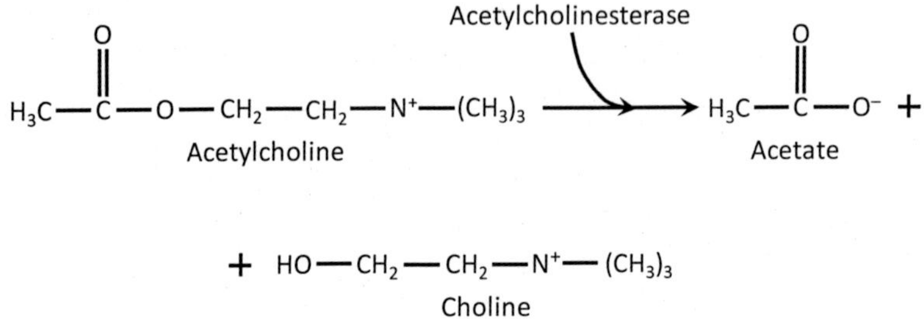

While acetate is eliminated from the circulatory system, choline is recovered from the synaptic terminal using the endocytosis vesicles formed during the synaptic release process (Figure 6.18) and is used for the synthesis of new acetylcholine molecules.

The percentage of acetylcholine that manages to bind to its receptors is actually quite low, but still in excess of that required for impulse transmission.

The two events just described, massive release of acetylcholine and massive and rapid intervention of the enzyme acetylcholinesterase, are apparently at odds, but in reality, they guarantee that synaptic transmission functions correctly. In fact, the massive release of the neurotransmitter guarantees safe stimulation of the muscle fiber, while the rapid elimination of the neurotransmitter guarantees absolute control of the contraction by the motor neuron, avoiding the risk of overstimulation due to unwanted residual neurotransmitter, yet ensuring that at and only at each action potential of the motor neuron is there a contraction.

6.4 Neuronal firing as a language in the nervous system

The single nerve action potential is essentially produced by TTX-sensitive sodium and delay rectifier potassium currents. Functionally, this means that the amplitude of the excitable wave lies between the sodium and potassium equilibrium potentials. In addition, the duration of the nerve action potential is determined by the kinetics of the currents of the two cations. Ultimately, the nerve action potential is an **all-or-none** phenomenon and is impossible to modulate, except by drastic intervention from outside, for example by changing ion concentrations.

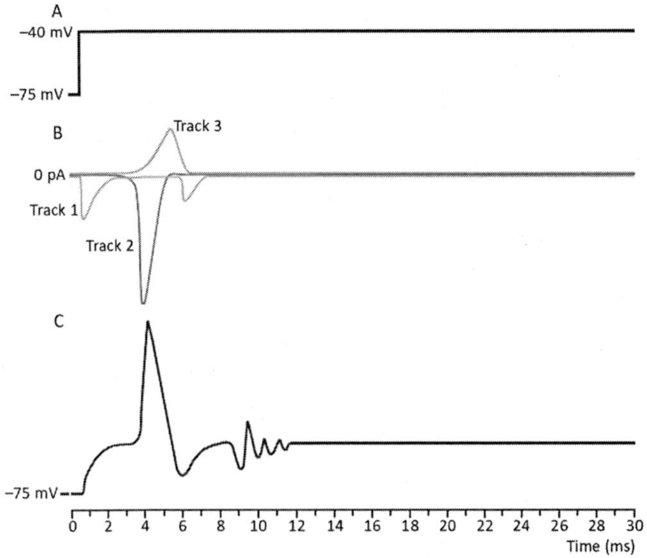

Figure 6.36 A large and prolonged stimulus (A) induces the onset (B) of a capacitive current (track 1), a fast sodium ion current (track 2) and a delay rectifier potassium ion current (track 3), which are capable of generating (C) an action potential, followed by oscillations that dampen to a constant potential value intermediate between the equilibrium potentials of sodium and potassium.

The neuronal cell, in order to send different information depending on the stimuli it receives, has mechanisms that modulate its ability to produce firing, bursts of action potential at different frequencies. The mechanism of neuronal firing requires changes in the structure of membrane permeabilities with respect to what was described in Chapter 5. Following a stimulus maintained over time, the response of the neuronal cell, in which there are TTX -sensitive sodium and delay rectifier potassium ion channels with comparable conductances, behaves as shown in Figure 6.36. The stimulus that raises the membrane potential from -75 to -40 mV (Figure 6.36 A) causes the onset of a capacitive current (trace 1 of Figure 6.36 B) that depolarizes the membrane (Figure 6.36 C) until the potential is reached at which the sodium currents (trace 2 of Figure 6.36 B) and potassium currents (trace 3 of Figure 6.36 B) are activated.

Following the fast depolarization, the cell tries to return to the resting values through the action of the potassium current, assisted by the inactivation of the sodium current. However, repolarization is hindered by continuation of the stimulus; the capacitive current forces the membrane to depolarized potentials, counteracting complete repolarization. The membrane potential reached may be permissive for the sodium channels to open. However, in the absence of complete repolarization, a large percentage of sodium channel remain inactivated. In order to open, the sodium channels must first close, which can only occur after complete repolarization of the membrane. For complete membrane repolarization, the potassium current must be much more intense. Figure 6.37 illustrates this situation, in which there is a considerable increase in the potassium repolarizing current (trace 3 of Figure 6.37 B). A larger potassium current has no influence on the onset of the action potential (Figure 6.37 C), but has a great impact on repolarization. As a consequence, the membrane potential reaches values close to the equilibrium potential of potassium and the potassium current deactivates. By the time the deactivation occurs, inactivation of the sodium current is complete. At this point, the maintained stimulus depolarizes the membrane via the capacitive current (trace 1 in Figure 6.37 B), the sodium channels open (those that are closed and not inactivated) and, as long as the stimulus remains, a cycle is initiated in which the sodium current is reactivated (trace 2 in Figure 6.37B), which regenerates the action potential.

It is important to understand what regulates the frequency at which the neuronal cell is able to produce action potentials. The action potential is an all-or-none event. This means that once triggered, it is impossible to modulate its development. In neuronal cells, it lasts 1-1.5 ms, and this is a fixed time. The only possible intervention is through the capacitive current. This is because membrane capacitance is a passive element induced by the external stimulus (Paragraph 5.13). For neuronal cells, the capacitor charge is extremely fast (50-100 μs). The kinetics of the capacitive current depend on the resistance and capacitance of the membrane (tab 2.1), which, in turn, depends on the surface area of the membrane itself (equation S2.3): smaller cells charge the membrane capacitance faster. Thus, smaller cells can produce bursts of action potentials at higher frequencies.

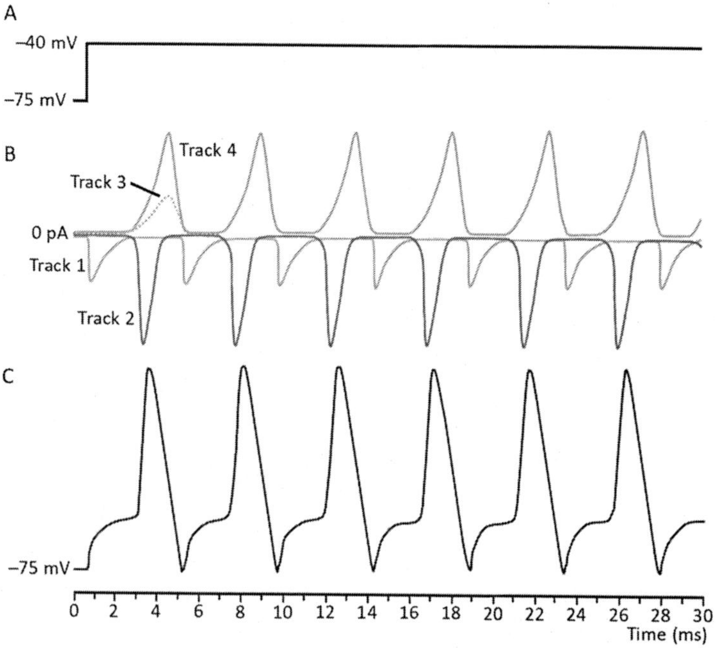

Figure 6.37 If the cell, in the presence of a large and prolonged stimulus (A), is able to produce (B) a large ionic current of delay rectifier potassium (track 4) compared to that in Figure 6.36 (track 3), there is the complete repolarization of the cell, which causes full deactivation of the sodium current (track 2). Under these conditions, the capacitive current, which is reproduced cyclically for the duration of the depolarizing stimulus (track 1), is able to induce the firing of action potentials (C) in the cell.

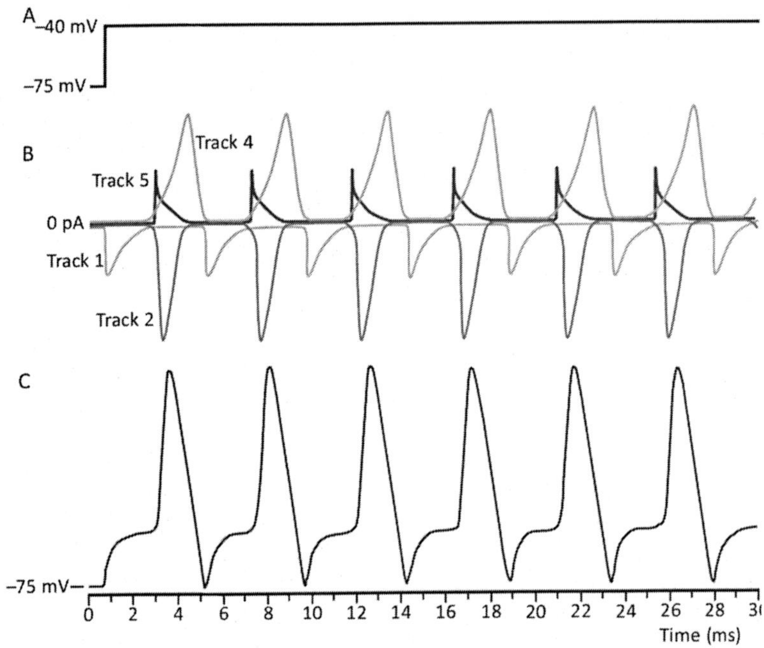

Figure 6.38 The hyperpolarizing $I_{K(A)}$ current (track 5) has no effect on the frequency of firing in a depolarized cell with a large, time prolonged stimulus (A) due to the prevalence of the large sodium current (track 2), the reduced amplitude of the $I_{K(A)}$ current and its rapid inactivation. Track 1, track 4 and panel C as in Figure 6.37.

From these considerations, it can be deduced that with a minimal, but appropriately proportioned set of membrane currents, the cell is able to produce firing. It can also be stated that each individual cell, reached by the appropriate stimulus and maintained over time, will produce action potentials at a maximum frequency related to its size. In general, neuronal cells belonging to homogeneous systems have comparable size and membrane resistance, so that they have the capacity to generate firing at similar frequencies. However, under the conditions just described, the neuron will only be able to respond to different stimuli with a single frequency, regardless of the intensity of the stimulus itself. The goal is to have an excitable cell that responds to stimuli of different intensities with different frequencies. For this to happen, the system must be modified. Changing the cell capacitance means changing the cell size. This is not possible. Sodium and potassium currents cannot be modulated during the progression of the action potential. As for many other cell functions, the only possibility is to activate an additional mechanism. The mechanism responsible for modulating the response to stimuli of different intensities must be voltage-dependent. Not only that, but it must be activated by a mechanism already present during the transitory imbalance of the resting membrane potential. The whole cycle consists of the membrane capacitor charging due to the onset of the stimulus, the fast depolarization, the repolarization and finally the return to resting membrane potential. The additional mechanism must be able to intervene at a suitable time during the excitability event, excluding during fast depolarization and repolarization, which are all-or-none phenomena. The only feasible time is during the early capacitive depolarization caused by the capacitive current (I_C, Equation S2.2). As has been stated previously, without an additional control, a constant stimulus triggers the maximum firing frequency for that particular cell (Figure 6.37). Thus, the only possible modulation is to increase the interval between two consecutive action potentials. The idea is to add a membrane current – voltage- and time-dependent – that acts between two consecutive action potentials. The current must have the opposite sign of the capacitive current. These requirements point to a fast-activated potassium current, triggered during the fast depolarization of the action potential with a voltage-dependent inactivation. These are the main characteristics of the A current ($I_{K(A)}$). Inactivation time is inversely proportional to the stimulus intensity. For strong stimuli, the current is activated and inactivated within the action potential without any effect on the frequency (Figure 6.38). On the contrary, if the stimulus is weak (Figure 6.39 A), the current $I_{K(A)}$ (trace 4 of Figure 6.39 B) is inactivated very slowly. The remaining current overlaps the capacitive current of the following action potential.

The effect on the membrane potential is strong even if the current is small. In this region of the action potential cycle, I_c (trace 1 of Figure 6.39 B) and $I_{K(A)}$ (trace 5 of Figure 6.39 B) are the only two active currents. Thus, membrane resistance is high and a small difference in the permeability has a large effect on membrane potential. If the cell is able to act on the inactivation of the $I_{K(A)}$ current, it can finely modulate the firing frequency, which at this point is directly related to the intensity of the stimulus.

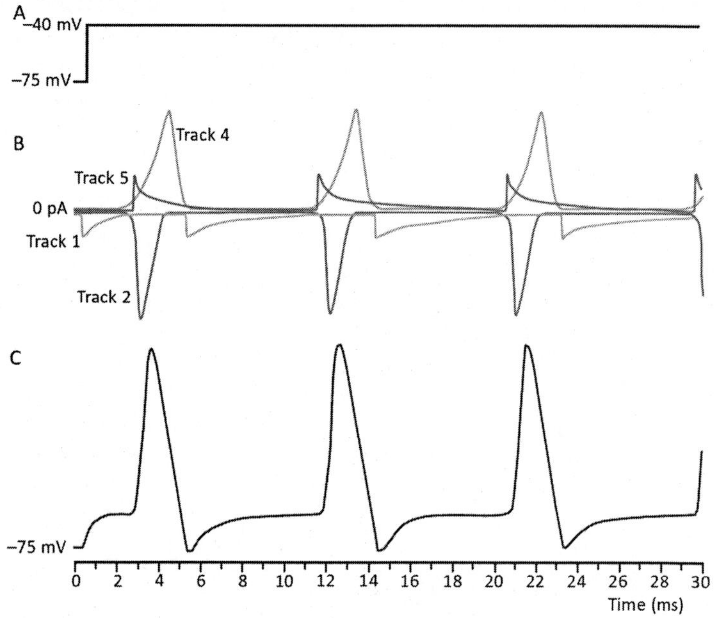

Figure 6.39 If we depolarize a cell with a prolonged weak stimulus (A), the $I_{K(A)}$ current (track 1) is inactivated slowly and remains sufficiently large during the depolarizations due to the capacitive current (track 5) slowing them down. This decreases the firing frequency (C). Track 2 and track 4 are as in Figure 6.37.

6.5 Autorhythmicity and pacemaker activity

The clearest example of the mechanism of autorhythmicity in excitable cells is certainly the sinus node's pacemaker activity in the heart. The ionic current responsible for the rhythmic change in resting membrane potential is the "funny current" (I_f). More recently, the I_f current has been described in the central nervous system and is denominated I_h.

6.5.1 The heart pacemaker

The pacemaker excitability cycle is composed of two distinct phases. In the first, the membrane voltage is slowly depolarized for tens to hundreds of milliseconds. The action potential takes place in the second phase, once the membrane voltage reaches a threshold (Figure 6.40). Compared with

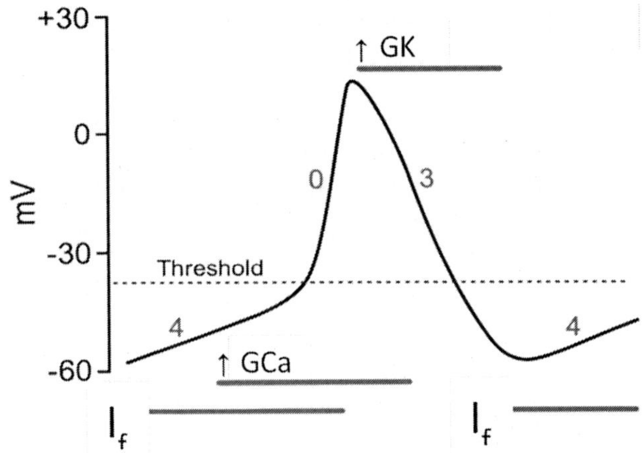

Figure 6.40. Example of pacemaker action potential Four major currents are indicated. The I_f current, responsible for the early part of the diastolic (pacemaker) depolarization; the T-type (I_{CaT}) and L-type (I_{CaL}) calcium currents (GCa), responsible for part of the late fraction of diastolic depolarization and for the upstroke; the potassium current (I_K), responsible for the repolarizing phase (GK).

the neuron excitability wave, the heart pacemaker is slower. This is because the action potential depolarization phase is governed by calcium permeability. The upstroke of the pacemaker action potential is not a sodium-induced fast depolarization. During the first phase of the excitability cycle, the slow voltage change is responsible for inactivating the sodium channels, compromising the development of an action potential. Calcium current has a slower activation and an even slower inactivation. Moreover, calcium ions in this context are both membrane voltage modulators and second messengers. The divalent ion responsible for cardiac cell contraction, differently from skeletal muscle, does not come from the external compartment and is not released from the internal store (Cap. 7.). The transition between the systolic (maximum depolarization) to the diastolic potential (maximum hyperpolarization) is realized by the potassium current (Figure 6.40).

The pacemaker current I_f deserves particular attention. I_f current has been molecularly characterized and defined as occurring via hyperpolarization-activated cyclic nucleotide-gated (HCN) cation channels. I_f is a sodium/potassium permeability. Its peculiar characteristic is that it is the only membrane ionic current activated in hyperpolarization. The I_f activation, occurring during the repolarization phase of the action potential, allows mainly sodium to flow in. The typical I_f current reversal potential is around -30 mV. The kinetics of the current are extremely slow. This is evident from patch-clamp, whole-cell current recordings (Figure 6.41). More information about the biophysical characteristics of I_f channel(s) can be found in DiFrancesco and Borer 2007 (DiFrancesco, D., Borer, J.S. The Funny Current. *Drugs* **67,** 15–24 (2007). https://doi.org/10.2165/00003495-200767002-00003)

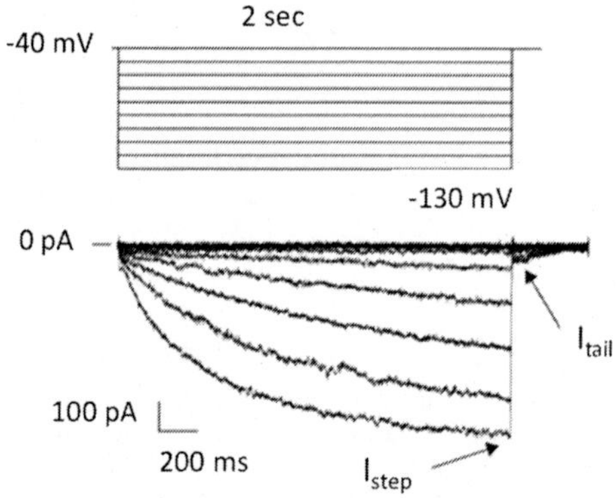

Figure 6.41. Patch-clamp whole-cell If current recorded from isolated sinus node myocyte during hyperpolarization.

The natural, spontaneous pacemaker rhythm is produced by the membrane enrichment of the I_f channel. The current is modulated by the intracellular level of cAMP. The nervous system control of the cardiac rhythm is mainly due to sympathetic and vagal innervation (Figure 6.42A).

Increased cAMP shifts the voltage dependence of the funny channel activation curve to the right, thus increasing current availability during diastole, hence the diastolic rate. This can be mimicked *in vitro* by perfusing isolated sinus node cells with noradrenaline (Figure 6.42 B). On the contrary, cAMP deprivation shifts the activation curve of the current to the left, lengthening the slow depolarization time and slowing the heart rate (Figure 6.42 B). It is possible to verify this in *in vitro* experiments using acetylcholine stimulation of muscarinic receptors.

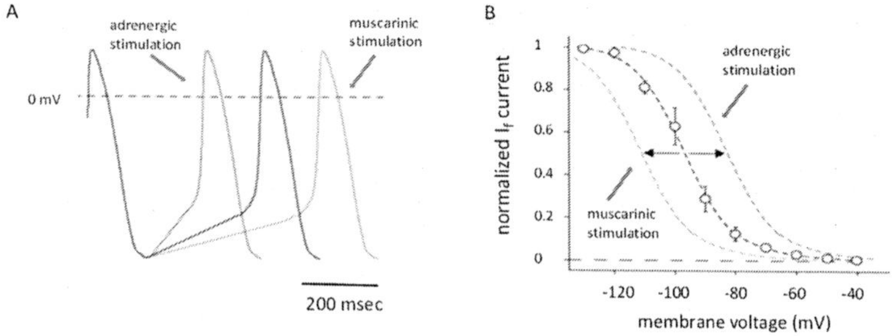

Figure 6.42. A. Modulation of pacemaker activity by noradrenaline (adrenergic stimulation) and by acetylcholine (muscarinic stimulation). B. If current activation curves in control conditions (o) and in the presence of noradrenaline (shift the curve at depolarized potentials) and acetylcholine (shift the curve at hyperpolarized potentials).

6.5.2 Neuronal spontaneous action potential rate and autorhythmicity

Neurons are capable of spontaneous activity. However, basic spontaneous neuronal excitability derives from the instability of the membrane resting potential. Because random action potential occurrences do not represent signals for the nervous system, neuronal cells have weak control of the resting potential. Depolarizing inputs are able to produce action potentials whose frequency is proportional to the amount of depolarization (Figure 6.43). The mechanism is linked to the complex set of ionic permeabilities present on the neuronal membranes (Cap. 6.4). Involvement of the pacemaker current I_h and of the T-type calcium channels is also possible.

Where I_h is essential is in the different automatic pacemaker activities of the central nervous system.

Figure 6.44 depicts a sequence of action potential bursts in bulbar respiratory center neurons. Action potential bursts are triggered by slow depolarization due to the I_h current.

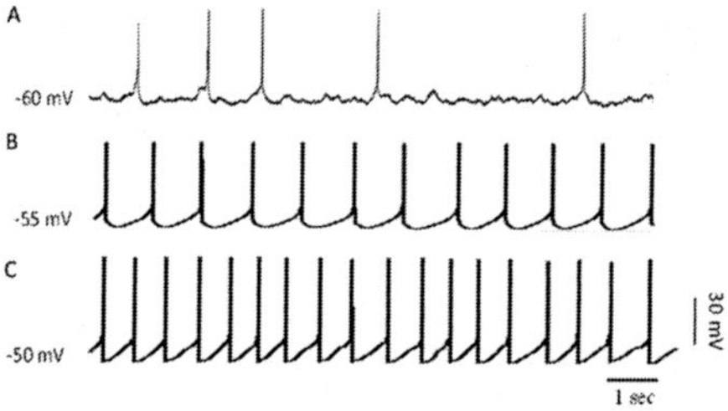

Figure 6.43. Spontaneous firing of a neuronal cell. Depolarization is responsible for increased frequency of firing, also producing a regular action potential sequence (B,C).

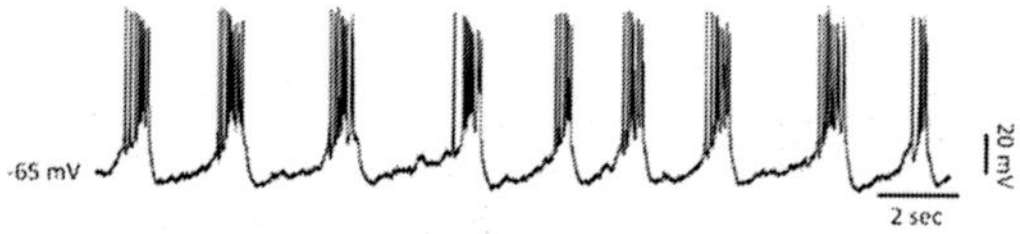

Figure 6.44. Respiratory neuronal pacemaker activity. Slow depolarizations between action potential bursts are due to I_h current activation. Action potential burst are due to fast sodium and delay rectifier potassium currents. Action potential firing ends after the increase of intracellular calcium. Repolarization is due to the calcium-activated potassium channels.

Data Sheet 6.1

Junctions between cells

The junctions between cells are specializations of the plasma membrane that enable and control the processes of adhesion and communication between two cells. They consist of specialized and functionally well-organized adhesion complexes. In vertebrates, the types are: gap junctions, anchoring junctions, tight junctions and desmosomes (Figure S6.1).

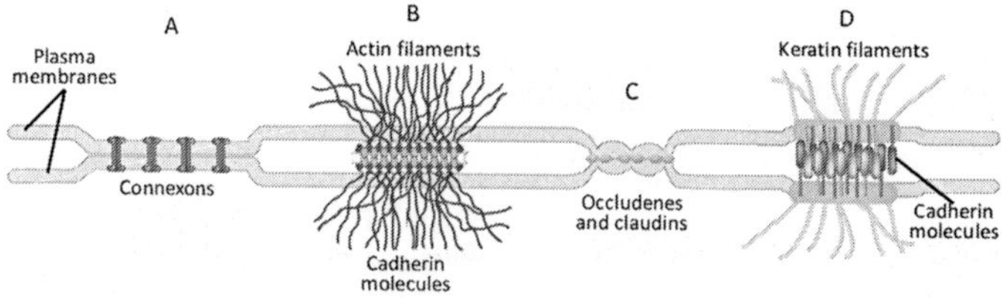

Figure S6.1 Diagram comparing the main junction structures between cells. A) Gap junction or communicating junction. B) Anchoring junction, adherent junction or zonula adherens. C) Tight junction, occluding junction or zonula occludens. D) Desmosome or macula adherens.

Gap junctions. Gap junctions (Figure S6.2), varying in number from a few dozen to a few hundred and with a regular arrangement, are formed by two identical structures called connecxons, each 7-7.5 nm thick, synthesized by each of the two cells involved and positioned in the membranes to form a single structure that reduces the space between the cells to about 1-2 nm. Connections are formed by a regular ring of six monomers of integral membrane glycoproteins called connectins, arranged to form a central pore (Figure S6.2C). In response to certain chemical signals, such as changes in pH or calcium ion concentration, the connectins slide radially. They form a protein channel approximately 2 nm in diameter that allows the passage of molecules of up to 1 kDa in molecular weight, such as in the intestinal epithelium (Paragraph 4.2.5), and of ions, thereby also creating electrical coupling between two cells, such as in smooth muscle (Paragraph 7.3) and heart muscle (Paragraph 7.4). Connessins can also move in the opposite direction to close the channel, in the presence, for example, of an excessive concentration of calcium ions, to prevent their passage into the other cell, whose survival would otherwise be at risk.

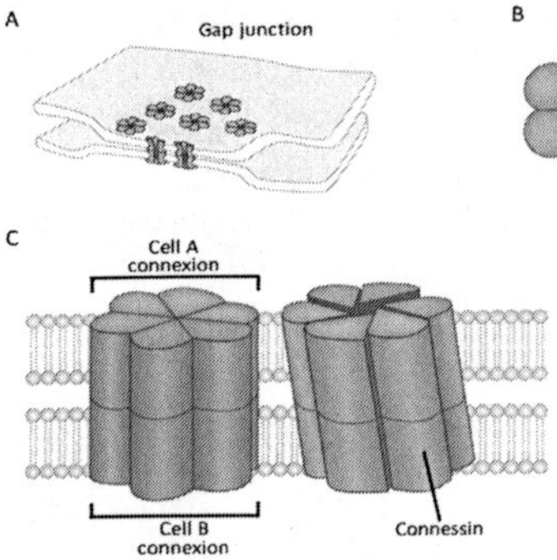

Figure S6.2 Gap junction. A) Diagram of the arrangement in the membranes of two contiguous cells. B) Detail of the pore opening. C) Detail of the juxtaposition of the two junctions.

Anchoring junctions. Anchoring junctions (Figure S6.3) are anchoring points between two cells, spaced 15-25 nm apart, or between a cell and the extracellular matrix. They form a continuous adhesion zone close to tight junctions and, because of the particular arrangement and complexity of the proteins used, they give structural consistency to tissues such as muscle or epidermis, dissipating forces from a number of directions. The two membranes are joined by structural proteins, the cadherins, which protrude outwards and join by intersecting with each other; the cadherins are bound to the actin filaments of the cytoskeleton by the intrinsic membrane proteins vinculin, β-catenins and α-actinins.

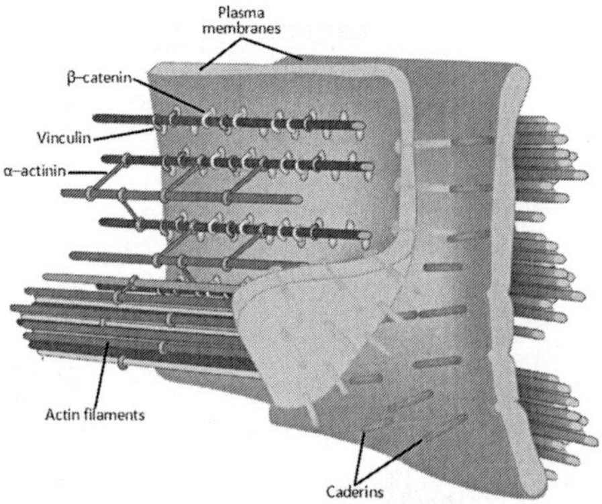

Figure S6.3 Anchoring junction: schematic representation.

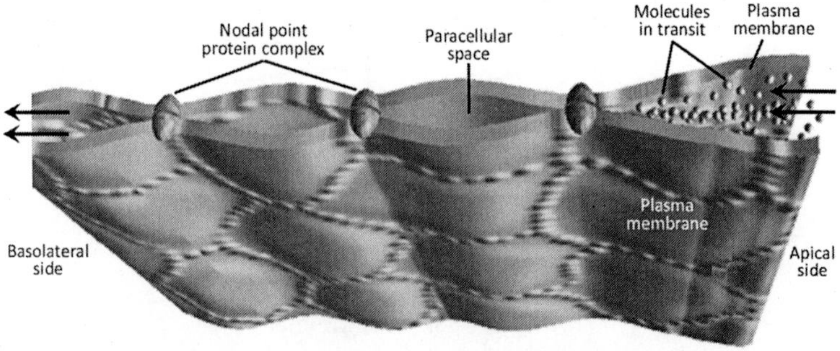

Figure S6.4 Tight junctions. The protein complex consists mainly of claudin-1 and occludin, plus others components lesser common such as E-cadherin, ZO-1, JAM-1, catenin, cingulin and actin. The black arrows indicate the direction of flow in the polarized tissue.

Tight junctions. Tight junctions (Figure S6.4) are typical of lining epithelia and intestinal epithelia, where they form along the entire perimeter of the cell (generally at the apex of polarized cells) a continuous belt called a zonula, which prevents the passage of fluids between the cells. The interstices between the cells are always extremely small and disappear completely at the so-called nodal points, where certain integral membrane proteins emerging from the plane of the outer face, mainly claudin-1 and occludin, form solid, non-covalent bonds with each other.

Desmosome. Desmosomes (Figure S6.5), which play a predominant role unloading the mechanical stresses experienced by tissues, are the best known cell-cell junction structures. They are characterized by the presence of a cytoplasmic thickening of protein material called the adhesion plate, formed by desmoplakins and placoglobulin, to which the fork-shaped intermediate filaments of the keratin cytoskeleton bind.

In the interstitial space, about 20 nm wide, the calcium-dependent integral membrane proteins known as cadherins, in particular prodesmocollin and desmoglein, emerge for each cell and bind to the adhesion plate.

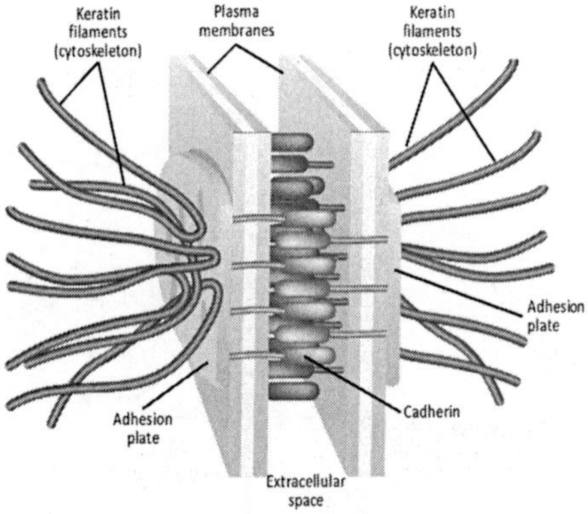

Figure S6.5 Desmosome.

Data Sheet 6.2

Peristalsis in the digestive tract

The perfect match between the structure of the digestive tract and the contractile function produces effective peristalsis, the important physiological function that allows the contents of the tube to travel continuously through the small and large intestines, uni-directionally from the duodenum to the rectum and anus. This function uses the gap junction mode of communication to synchronize the primary electrical signal.

The complex overall structure of the digestive tract of the mammalian intestine is schematized in Figure S6.6. The lamina propria faces the lumen and is bound by an epithelium covered with mucus to prevent damage from harmful substances; in the small intestine, it comprises the villi, the submucosa, the mesentery, various types of glandular tissue and numerous blood vessels. There are two layers of smooth muscle tissue (Paragraph 7.3): the muscularis externa and the muscularis mucosae. The muscularis externa consists of an outermost layer of longitudinal fibers and

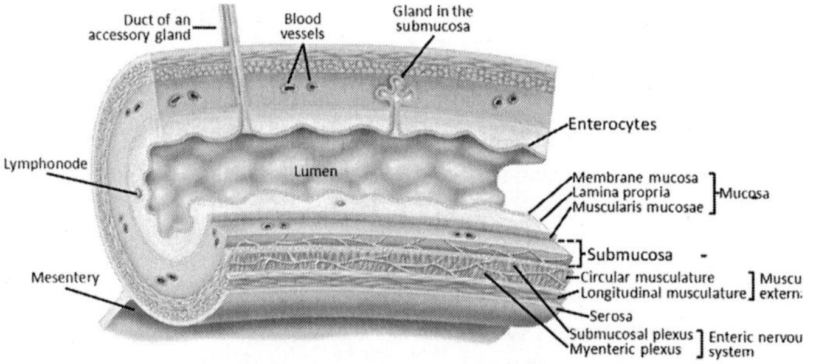

Figure S6.6 Sketch of the general structure of the mammalian intestine. Among the structures shown is the dual muscular tissues of the muscularis externa and muscularis mucosae. "Accessory gland' means a gland outside the intestine, such as the liver or pancreas.

an innermost layer of circular fibers in relation to the lumen of the tube.

The function of the muscularis mucosae layer, on the other hand, seems mainly to control the lamina propria and contribute to the formation and maintenance of the villi. The function of the stratum muscularis externa is twofold: to ensure that the bolus moves along the digestive tract and to mix and segment it into smaller portions so that the enzymes produced by the external glands of the lamina propria and submucosa can digest it more easily.

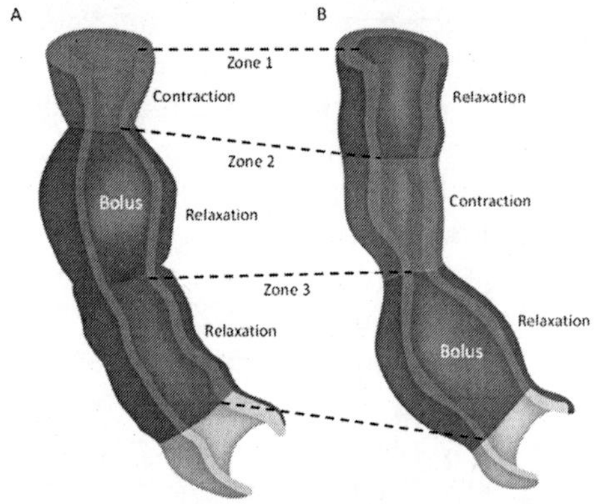

Figure S6.7 Diagram of the mechanism of peristalsis in the mammalian intestine, which continuously pushes the bolus, the result of the digestion of food by the stomach, towards the anus. Figures A and B represent two successive

The coordination between the contraction of the longitudinal fibers and the circular fibers generates the peristaltic waves that make the bolus continuously progress from the duodenum to the anus. The contraction of zone 1 (Figure S6.7A) pushes the bolus into zone 2, which is in the relaxation phase. At this point, there is a contraction of the circular fibers first in the band next to zone 1, to block backward movement, and then of the whole of zone 2 (Figure S6.7B). In this way, the lumen of the intestine is reduced and the bolus slides towards zone 3, favored by the simultaneous contraction of the longitudinal fibers. When necessary, there is also contraction of the circular fibers in the area with the bolus to fragment it and form smaller masses (arrows in Figure S6.7B).

This alternation between contraction and relaxation is produced by the slow oscillations of the membrane potential of the smooth muscle tissue (Figure S6.8), to which trains of action potentials can be added, if the potential exceeds the threshold value. With each depolarizing oscillation, there is the entry of calcium, which activates calcium release from the sarcoplasmic reticulum and the contraction (paragraph 7.3.1). The frequency of slow oscillations varies from 6 per minute in the cecum and proximal colon to 16-18 per minute in the duodenum.

Nervous control of intestinal peristalsis is provided by the submucosal plexus or Meissner's plexus, located below the muscolaris externa, and the myenteric plexus or Auerbach's plexus, located between the longitudinal and circular fibers of the muscolaris externa (Figure S6.6).

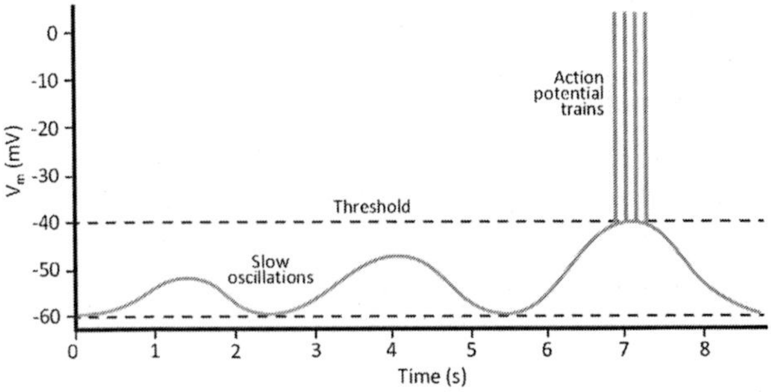

Figure S6.8 Potential oscillations due to incoming calcium currents produce peristaltic contractions of intestinal smooth muscle and, occasionally, trains of action potentials.

The submucosal plexus controls the secretory activity of the gastrointestinal tract, local blood flow, local absorption and the degree of contraction of the muscularis mucosae. The myenteric plexus controls the tone of the intestinal wall, the intensity of rhythmic contractions, the conduction velocity of excitation waves and the propagation velocity of peristaltic waves.

CHAPTER 7

Transformers 1: motors

In a recent review article on muscle contraction, Herzog (2017) of the University of Calgary (Canada) states that: "*Skeletal muscle mechanics have been studied ever since people have shown an interest in human movement. However, our understanding of muscle contraction and muscle mechanical properties has changed fundamentally with the discovery of the sliding filament theory (J. Hanson & HE Huxley, 1953; AF Huxley & R Niedergerke, 1954) and associated cross-bridge theory (A. F. Huxley, 1957). Nevertheless, experimental evidence suggests that our knowledge of the mechanisms of contraction is far from complete, and muscle properties and muscle function in human movement* ***remain largely unknown.***" Thus, despite decades of intensive research, the mechanism, at the molecular level, of how muscle contraction occurs in biological tissues has yet to be clarified.

This chapter will attempt to reconstruct the mechanism of action of contractile proteins on the basis of experimental evidence and hypotheses thereby derived. The principle that has been adopted until now will continue to be followed: physiological functions happen by taking advantage of the potential energy stored by the system and released following a suitable stimulus. In the specific case of muscular contraction, and in particular that of skeletal muscle, a role of fundamental importance is played by the structure/function relationship.

The idea is to first give a general overview of the problem of muscle contraction, the different types of tissues and their organization. This will be followed by an initial description of the structure of the skeletal muscle, as 95% of the muscle tissue of the human body belongs to this type, and its functional organization. Then, with the presentation of biomechanical experiments, we will try to understand how their interpretation led to the model of the sliding filaments and the physical interaction between the myosin heads and the actin filament. Finally, an attempt will be made to hypothesize a molecular model of force generation based on the morpho-functional studies carried out, in particular, by A. F. Huxley (1974).

The contractile apparatus can be defined as an energy transformer. Potential energy, accumulated through cellular metabolism and thus derived from chemical energy, is transformed by the muscle tissue into **mechanical energy,** with which it is able to physically interact with its surroundings. This operation is one of the best examples of the synergy between structure and function that is often encountered in physiological mechanisms. Based on the hypothesis of sliding filaments and the interaction between myosin and actin heads, an attempt will be made to produce a hypothetical model of force generation at the molecular level that exploits the structural organization of skeletal muscle and stored potential energy.

The development of a system that allows movement, an evolutionary process whose beginnings go back a long time, is the most important of the factors that have given living organisms the ability to interact with the external environment and draw greater evolutionary advantages from it. In many cases, during the course of evolution, plants have acquired the ability to move in order to continually find, through the root system, substances useful for energy production in an environment with a low concentration of molecules, which are additionally poor in energy. At the same time, however, plants did not develop other fundamental apparatuses, such as a coordination center or a flexible system that would allow the accumulation in the organism of these recovered substances where they were present in abundance. Furthermore, movements are limited and in response to simple external stimuli, such as changing the direction of growth in the presence of an obstacle or adhering to one support rather than another in another position. Probably the first step towards the development of organisms with greater autonomy, less and less dependence on the distribution in the environment of the nutrients needed to have a sufficient amount of energy, free from the constraint of a root system and the need for a constant supply of energy, was the selection of structures that gave animals the possibility of movement. The next step was, by necessity, to select and organize more and more effective structures for the transformation of natural substances into metabolic energy and for the accumulation of this energy in different forms. Subsequently, as organisms grew in size and complexity, control systems were developed both for receiving external stimuli and for responding effectively to those stimuli.

Over the course of evolution, the first complex control system was the radial network present in coelenterates (Figure 7.1A), which allows rapid activation of vast portions of the organism. Then, with increasing complexity, the systems became multicentric with the metameric type, essentially composed of chains of ganglia organized in metameres. This was the advent of different tasks for different ganglia. The cervical ganglia perform a function of relating to the surrounding environment. The metameric ganglia have the same structural complexity and great functional autonomy (Figure 7.1B).

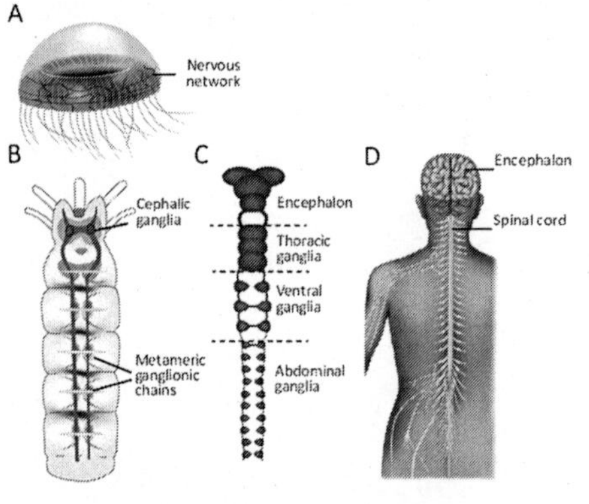

Figure 7.1 Diagram of the nervous system of the (A) radial coelenterate type, (B) metameric annelid type, (C) predominantly insect cephalic type and (D) mammalian cephalic type.

The evolution of the metameric system towards structural specialization of the different areas of the organism led to the development of the encephalon, the single center of coordination and elaboration, although some specific functions remained in the periphery (Figure 7.1C). Finally, the encephalon became the true control center that receives all of the information from the surrounding environment through sensory systems, processes it and sends appropriate signals to the periphery through the spinal cord, which nevertheless retains a metameric organization (Figure 7.1D).

The acquisition, in the course of evolution, of an autonomous movement apparatus in increasingly complex organisms, enabling animals to search for food, has led to the selection of increasingly sophisticated sensors for position, balance and orientation (Chapter 8). In parallel, a complex sensory organization has been instrumental to develop more precise control and stimulation systems to ensure the survival of the individual. Control systems function to evaluate the stimuli that reach the body from outside through the sensory systems; they also control the stimuli coming from inside the organism. The control systems receive messages from the sensory systems and generate the appropriate responses. Responses are based on physical activity like movement, sphincter control, defense or attack strategies. It could be said that organisms with autonomous movements have selected the appropriate structures for these functions to respond adequately to external and internal stimuli. However, there has not always been a parallel and harmonious development of sensory, control and physical response apparatuses. During the course of evolution, the Earth has been colonized by large reptiles that were not competitive on a evolutionary point of view when food resourses were not easily accessible. In fact, the effector system, essentially the movement system, was not proportionate with sensory and control systems. Animals such as the brontosaurus were gigantic, needing to move constantly in order to obtain the large amounts of food they required, but were incapable of responding adequately to external stimuli, even in the case of danger (commentary on Figure 6.1). Conversely, from a strictly physiological point of view, it could be argued that the success of small and medium-sized animals is due to a balance between energy supply, movement and control systems.

The ability to respond ever more quickly to stimuli from the external environment became of greatest value to homoeothermic animals, which won the competition with the large reptiles and were able to colonize the

Earth with unprecedented success. the evolutionary pressure was a key factor in the development of a sofisticated control system of movements and was certainly essential in directing evolution up to the appearance of *Homo sapiens.*

Although evolution led to a downsizing and harmonization of control systems and movement structures, the latter are more frequent in known organisms because of their special functions. While the mechanisms that distinguish phenomena such as excitability or the absorption of nutrients are essentially cellular and use the potential energy accumulated in the form of charge separation across cell membranes, the structures for movement can be defined as true transformers. The set of cells that constitute them convert the potential energy, accumulated in various forms in muscle cells, into mechanical energy capable of performing macroscopic and measurable work. This is because, in the course of evolution, muscle cells changed and organized themselves into macrostructures capable of having sufficient energy and force to carry out work on their surroundings.

7.1 The structures for movement

Muscle contraction is indispensable for generating all types of movement. Three main types of muscle tissue are known (Figure 7.2). Smooth muscle, cardiac muscle and skeletal muscle, while sharing the molecular mechanism of the contractile process, have very different macroscopic functions, tissue organization and cellular organization. While in smooth muscle and cardiac muscle, the cells remain distinct, in skeletal muscle, embryonic muscle cells fuse into a

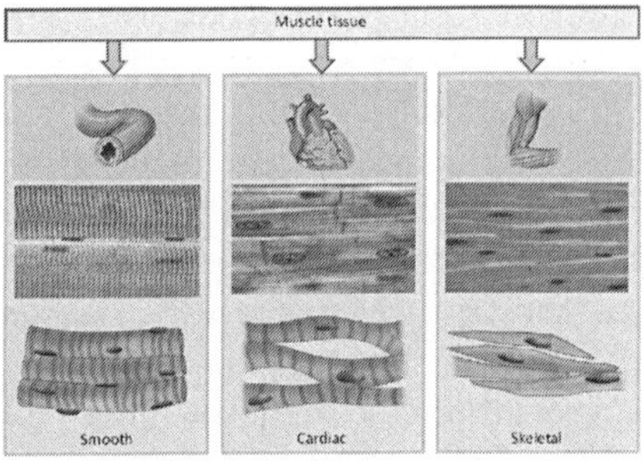

Figure 7.2 The three types of muscle tissue in humans. From top to bottom: diagrams of the macroscopic organization of the three muscle tissues, medium magnification light microscope images and diagrams of cellular organization.

multinucleated syncytium forming cylindrical structures that can reach several centimeters in length, with diameters ranging from 50 to 250 µm. Another fundamental macroscopic distinction lies in the accessory structures. Skeletal muscle adheres to the skeleton through tendons in order to perform actions on the surrounding environment. Smooth muscle and cardiac muscle, on the other hand, are cellular organizations that essentially provide functions within the body, such as the mechanical handling of ingested food and blood circulation. In the cytoplasmic organization of the proteins directly involved in the mechanism of contraction, cardiac and skeletal cells are much more similar, both presenting characteristic striations that are absent in smooth muscle tissue. Finally, from a functional point of view, smooth muscle cells and cardiac cells are linked by cytoplasmic bridges, the gap junctions (Figure 6.1), which ensure radial diffusion of excitation and consequent contraction. In contrast, skeletal muscle fibers are stimulated by chemical synapses (section 6.3) and perform essentially longitudinal contraction.

7.1.1 The structure of skeletal muscle

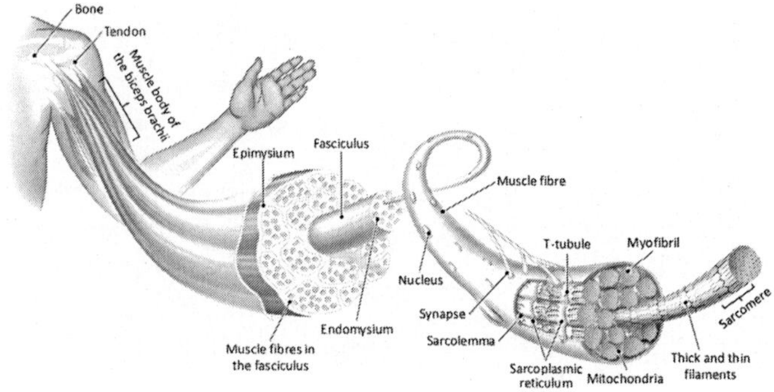

Figure 7.3 The complex organization of a skeletal muscle: the body, fascicles, muscle fibers and myofibrils can be distinguished.

Skeletal muscle (Figure 7.3) consists of a spindle-shaped body firmly connected to visco-elastic non-contractile structures, the tendons, which in turn are firmly attached to the skeletal system. In rare cases, the skeletal muscle is connected directly to the skin, such as facial muscles, to cartilage, such as the muscles of the larynx, or to other muscles.

The muscular body, with a diameter ranging from a few millimeters to several tens of centimeters, depending on the size of the muscle, is made up of many thousands of fascicles, which in the most voluminous muscles are grouped into muscle bundles, with a diameter of several tens of micrometers. The fascicles are arranged longitudinally relative to the muscle body and parallel to each other, each with its own connective sheath, the endomysium, with blood vessels and nerve endings. The fascicles are made up of many thousands of cells, the muscle fibers. Each muscle fiber is wrapped in a sheath of connective tissue, runs the length of the entire muscle body, from a few tens of micrometers to tens of

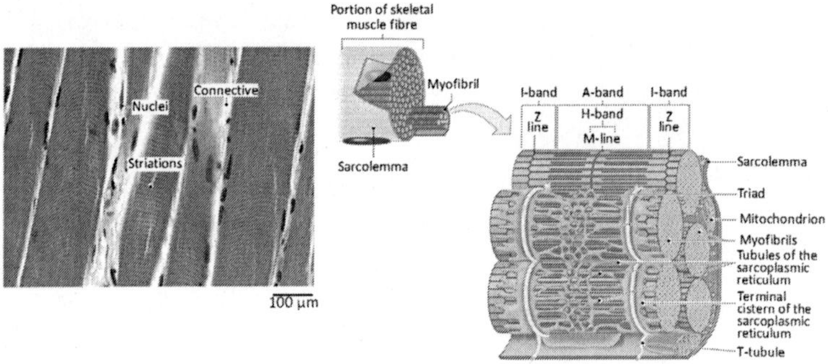

Figure 7.4 A) Light microscope image at medium magnification of muscle fibers in a fragment of striated muscle. Note the numerous nuclei arranged at the periphery of the cell, the connective tissue arranged around each cell and the striations parallel to each other and transversal to the larger structure. B) Additional cellular details of the complex organization of a skeletal muscle such as that sketched in Figure 7.3.

centimeters, and has a diameter that can vary from a few tens of micrometers to a few hundred.

The fully differentiated muscle fiber is a single cell obtained from the fusion of hundreds of myoblasts, the mononuclear muscle stem cells that are derived from the satellite cells by differentiation, to form a polynuclear syncytium with a few hundred nuclei placed in a peripheral position (Figure

7.4A). The skeletal muscle is also called striated muscle because in light microscopy images, there are striations transversal to the largest dimension of the muscle and parallel to each other. The striations are due to the regular alternation of refractive bands and less refractive bands, about 2 μm apart when the muscle is at rest. They are produced by the special arrangement of the contractile proteins in which the cell is rich (Figure 7.4A and paragraph 7.1.2). Inside the fiber, the space is almost completely occupied by the contractile proteins organized in myofibrils, which in turn are composed of the functional units, the sarcomeres, placed in series with each other. Each sarcomere is completely surrounded by the fenestrated sarcoplasmic reticulum that extends from one T-tubule to the other and forms, at both ends, the terminal cisternae. The T-tubules, invaginations of the plasma membrane, run perpendicular to the main axis of the muscle fiber and come into functional contact with the terminal cisternae of the reticulum (Figure 7.4B).

The electrical stimulus for muscle contraction is carried by the axon of a motor neuron that originates in the spinal cord and branches out. Each branch makes contact with its muscle fiber through its own neuromuscular synapse (section 6.3.3). The motor neuron axon, its branches and the muscle fibers connected to it make up the motor unit (section 7.1).

From a biomechanical point of view, the muscle structure has active and passive elements. The passive elements may contain an undamped (elastic) and a damped (viscoelastic) component; all biological tissues, when subjected to mechanical traction, show an elastic and a viscoelastic component that are difficult to separate. The structures with an elastic component include the tendons, the collagen bands that surround the fibers and many of the accessory structures that contain the contractile proteins organized in fibrils. There is also an elastic component associated with each acto-myosin complex when they are formed during contraction. The viscoelastic components are all the structures that make up the contractile machine. Of these, some are firmly connected to each other. Others can only resist traction due to the friction that develops between the different structures of the tissue; they are separated by compounds in the form of gels or ionic solutions. The biomechanical behavior of skeletal muscle will be discussed in more detail in section 7.2.4 and in subsequent sections.

7.1.2 The sarcomere

All of the myofibrils that form the striated muscle fiber are formed by pseudo-cylindrical subunits, the sarcomeres, which are identical, are arranged in series and repeat along the whole length of the myofibril (Figure 7.5). Each sarcomere is delimited by two T-tubules, introflexions of

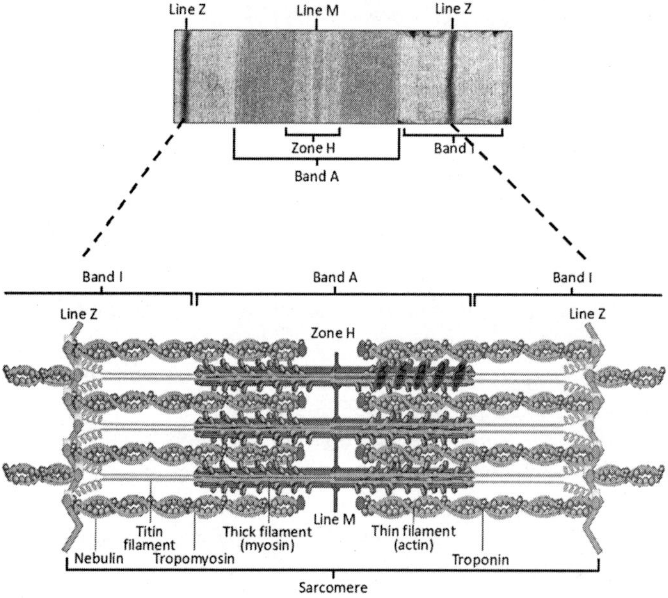

Figure 7.5 A) Electron microscope image of a mammalian muscle fiber sarcomere. Areas of different density are highlighted (two Z lines, the I band, the A band, the H zone and the M line). B) Schematic representation of the structure of the sarcomere, constructed with information provided by images such as that in (A) and a series of other experiments carried out over the years using different techniques.

the sarcolemma involved in the transmission of the excitatory wave. Each also has a developed sarcoplasmic reticulum (Figure 7.4) implicated in the control of the intracellular calcium concentration, as it has the task of storing calcium ions and releasing them during the process of contraction.

Figure 7.5A shows an electron microscope image of a resting mammalian muscle fiber sarcomere. Areas of different density are evident: two Z lines, the I or isotropic band, the A or anisotropic band, the H zone and the M line. Figure 7.5B is the schematic representation of a sarcomere assembled with the information provided by images like that of Figure 7.5A and by a series of other experiments carried out over the years with different techniques. In detail, inside the sarcomere there are two types of filiform protein complexes, the thin filaments and the thick filaments, oriented parallel to the major axis of the fiber and disposed longitudinally in an extremely ordered way. The filaments transform energy into work for muscular contraction. Further, there are the two Z lines, which delimit the sarcomere and bind the thin filaments. Band I, which involves two contiguous sarcomeres, is created by the presence of only thin filaments and is divided into two parts by the Z line. Band A, which occupies most of the sarcomere, is as wide as the thick filaments, about 1.2 μm, and is created by the simultaneous presence of thin and thick filaments. Finally, in the middle of the A band, the M line can be recognized, which is located approximately in the middle of the thin filaments, in the area where the two myosin molecules join.

The striations that are observed in the muscle under the light microscope and that characterize skeletal muscle fibers (Figure 7.4A) are produced by the Z lines of each myofibril, which are positioned in phase with those of the other myofibrils that form the muscle fiber. Other accessory proteins are titin and nebulin, which are involved in stabilizing the structure of the sarcomere.

The ultrastructure of the contractile proteins has an extremely precise organization. Figure 7.6A shows a high-resolution electron microscope image in which the transverse bridges between the thin and thick filaments are evident. The thin filaments (Figure 7.6B), which are about 1 μm long, consist of two spiral chains of G-actin, a globular protein, two filiform tropomyosin molecules and the troponin complex. The troponin complex consists of troponin I, which binds F-actin, troponin C, which binds Ca^{2+} ions, and troponin T, which binds tropomyosin. Tropomyosin and the troponin complex are involved in the regulation of contraction.

The thick filaments consist mainly of myosin and are formed by two α-helical protein filaments of light meromyosin wound in a spiral, which form the so-called tail. Each filament has a heavy meromyosin head with a specific attachment site for actin and a specific attachment site for ATP (Figure 7.6C). About 200 pairs of myosin molecules, joined together at the

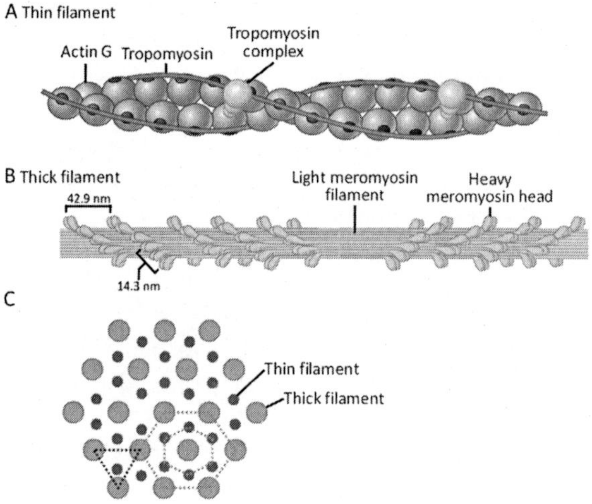

Figure 7.6 A) Schematic of the structure of a thin filament of the sarcomere with its main constituent proteins. B) Schematic of the structure of a thick sarcomere filament. C) Reconstruction of an electron microscope image of a section of myofibril. Each myosin molecule is surrounded by six actin molecules (small hexagon outlined in grey) and six myosin molecules (large hexagon outlined in grey). Each actin molecule is surrounded by three myosin molecules (dashed blue triangle).

tail end, form the thick filament. They are arranged in such a way that the various head pairs emerge from the structure at regular intervals, every 14.3 nm following a spiral line with a pitch of 42.9 nm (Figure 7.6C).

In order to have maximum efficiency in contraction, the thick filaments of myosin and the thin filaments of actin are distributed in a very regular manner: each thick filament is surrounded by six thin filaments and each thin filament is surrounded by three thick filaments. In this way, each thick filament is in the center of one hexagon, with six thin filaments at the vertices, and of a second hexagon with six thick filaments at the vertices, surrounding six thin filaments (Figure 7.6D). This structure allows maximum useful contact between the two contractile proteins when they are interdigitated during muscle contraction.

7.2 Skeletal muscle contraction

Skeletal muscle is activated by a motor neuron via the neuromuscular junction (section 6.3.3), which is usually located at the proximal third of the fibers that make up the muscle. In an experiment, such as that shown in Figure 7.7A, the motor neuron is stimulated to produce an action potential that causes the release of acetylcholine into the intersynaptic space. The neurotransmitter molecules interact with nicotinic receptors

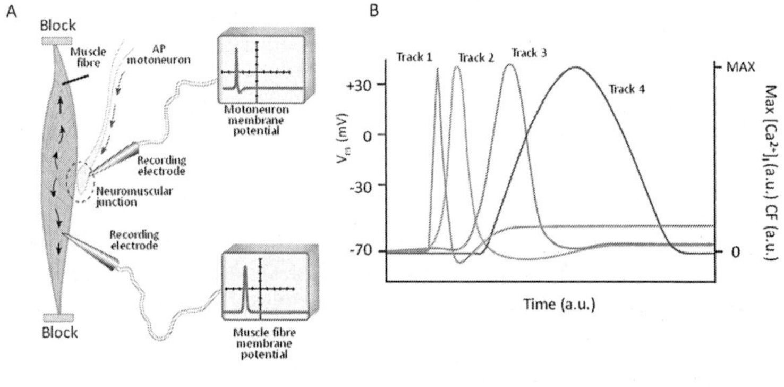

Figure 7.7 A) In the diagram, a motor neuron "propagates" an action potential (AP) from the central nervous system (downward arrows) to the neuromuscular junction (dashed ellipse). The action potential then spreads throughout the muscle (upward arrows), which contracts. B) The various delays between the onset of the action potential in the neuron, the action potential in the muscle, the calcium transient and muscle contraction. Track 1: action potential of the neuron, Track 2: action potential of the muscle, Track 3: calcium transient, Track 4: muscle contraction. V_m = membrane potential. $[Ca^{2+}]_i$ = intracellular concentration of the calcium ion. CF = contraction force. Time axis, $[Ca^{2+}]_i$ and CF are expressed in arbitrary units (a.u.).

located on the postsynaptic membrane of the muscle and an action potential is generated in the muscle fibers, which are induced to contract. As we are now able to record simultaneously the action potential of the

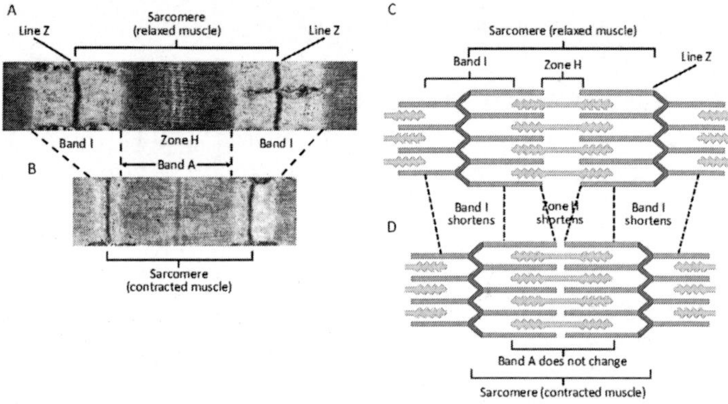

Figure 7.8 Electron microscope images (A, B) and model (C, D) of a sarcomere from a relaxed muscle (A, C) and a contracted muscle (B, D).

neuron movement, the action potential of the muscle, the intracellular calcium of the muscle fiber and finally the force at the ends of the muscle, we can produce a graph such as that in Figure 7.7B. The temporal sequence of events is important. The nervous stimulus, followed by a delay given by the mechanism of the chemical synapse, induces an electrotonic current in the muscle fiber, which triggers an action potential. Before contraction occurs, the excitable wave must invade the entire fiber. In fact, the action potential, by invading the T-tubules, manages to generate a uniform increase in intracellular calcium. The mechanism of intracellular calcium release is slower than the propagation of the action potential into the T-tubules. The speed with which the action potential is transmitted throughout the muscle manages to make the release of calcium, a much slower phenomenon, homogenous. In turn, calcium increases in all fibers with timescales that precede force generation at the ends of the muscle. In the end, the entire muscle fiber has a synchronous contraction, since the action of the contractile proteins is much slower than the mechanism of calcium release from the reticulum. The delays in the various stages have a physiological value in this case.

The ultimate goal of this sequence of events is to bring the Z lines of each sarcomere closer together (Figure 7.8B and D). As the sarcomeres are all in series with each other and in close contact with those of the adjoining fibers, the shortening must take place synchronously, to avoid breaks and inhomogeneity in the contractile process. Once the action potential is produced in the muscle cells of the neuromuscular junction (Figure 7.9), the excitation must be transformed into a massive intracellular release of

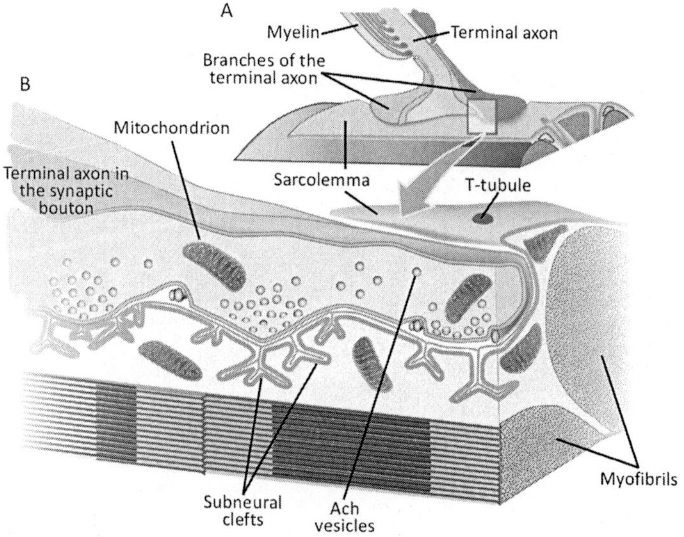

Figure 7.9 A) Schematic of the terminal zone of a neuromuscular junction between a myelinated axon and a muscle fiber. B) Schematic drawing of the detail of a synaptic bouton. The synaptic neurotransmitter vesicles are highlighted. Subneural clefts are invaginations of the muscle fiber membrane, rich in receptors for the neurotransmitter: they greatly increase the probability of contact between receptor and neurotransmitter.

calcium ions and therefore into the interaction between actin and myosin with the generation of force at the ends of the muscle. This mechanism is called excitation-contraction coupling.

7.2.1 Excitation-contraction coupling.

The physiological mechanism by which the action potential (the electrical signal conducted to the muscle fiber by the axon of the motor neuron) is transformed into a chemical calcium signal that commands the contraction of the sarcomere takes place in the triad. The triad is formed by two cisternae of the sarcoplasmic reticulum belonging to two contiguous sarcomeres and the ction coupling corresponding T-tubule, an invagination of the sarcolemma. The action potential propagates, in a manner similar to that seen along the axon (paragraph 6.2.4.2), through the depth of the muscle fiber, through the system of T-tubules (Figure 7.10).

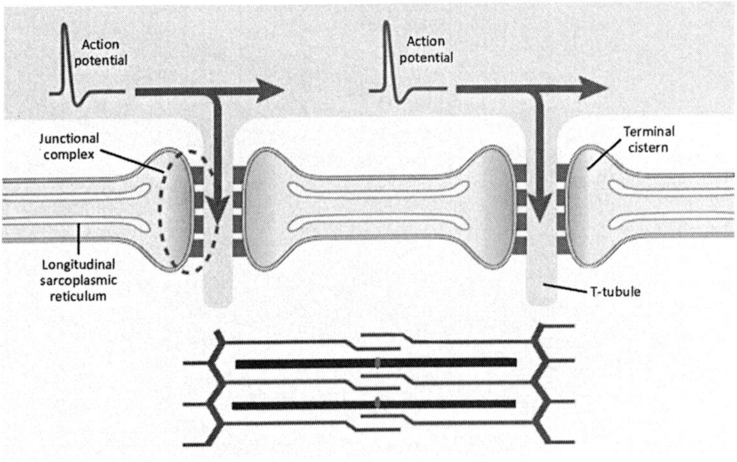

Figure 7.10 The triad, formed by the sarcoplasmic reticulum of two contiguous sarcomeres and the T-tubule, allows the action potential to reach deep into the muscle fiber, close to the contractile actin and myosin filaments. The junctional complex is highlighted (dotted oval).

The structure of the T-tubule, adhering to the terminal cisternae of the sarcoplasmic reticulum, is illustrated by Figure 7.11A, which shows the organization of several myofibrils surrounded by the reticulum. Electron microscopy showed the presence of electron-dense particles, called junctional complexes (Figure 7.10), which are interposed between the membrane of the T-tubule and the membrane of the sarcoplasmic reticulum (Figure 7.11B).

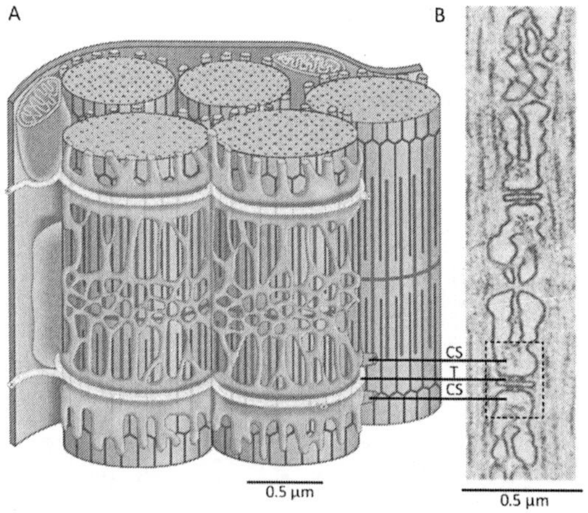

Figure 7.11 A) Reconstruction of the organization of different myofibrils. B) Electron microscope image of the longitudinal section of a myofibril. The triad, formed by two sarcoplasmic lattices of contiguous sarcomeres (CS) and the T-tubule interposed between the two sarcomeres (T), and the electron-dense particles interposed between the two lattices and the T-tubule (dashed rectangle) are evident.

Electron microscopy images (Figure 7.12A) show that the contacts between the membrane of the T-tubule and that of the sarcoplasmic reticulum are several and in variable numbers depending on the section of tissue analyzed (Franzini-Armstrong, 2004). The protein complex underlying the electron-dense particles consists of two proteins, illustrated in Figure 7.12B. Embedded in the membrane of the T-tubule are dihydropyridine receptors. Contiguous with dihydropyridine but inserted in the sarcoplasmic terminal cisterna of the sarcoplasmic reticulum is the ryanodine protein complex (RyR1). The dihydropyridine receptor is nothing more than an L-type calcium channel, but it lacks the ability to conduct calcium ions. In fact, it functions only as a voltage sensor, changing conformation when it is affected by the action potential. The ryanodine

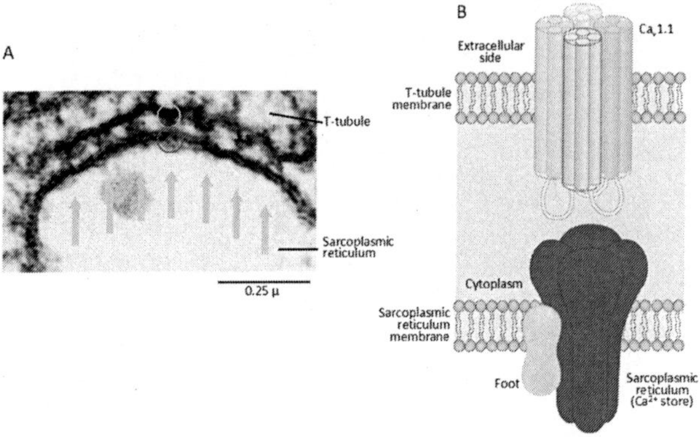

Figure 7.12 A) High-magnification electron microscope image of a portion of the skeletal muscle triad. Numerous electron-dense particles (arrows) are evident, consisting of a part embedded in the membrane of the T-tubule (upper circle) and a part embedded in the membrane of the sarcoplasmic reticulum (lower circle). B) The two components of the electron-dense particles are the $Ca_V1.1$ dihydropyridine receptor and the RyR1 protein complex.

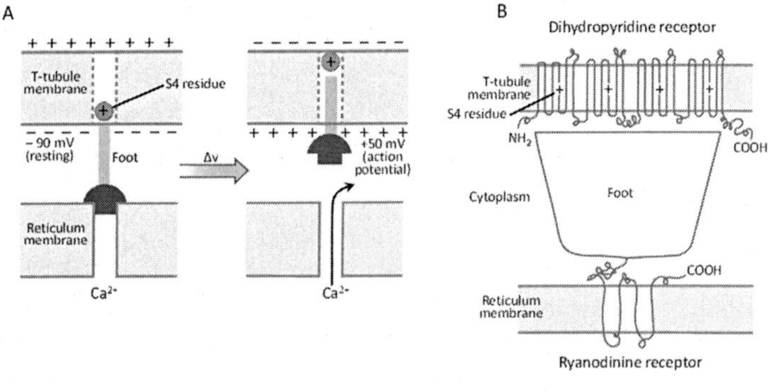

Figure 7.13 Biomechanical model (A) and biomolecular model (B) of the ryanodine receptor, which is sensitive to variation in membrane potential (ΔV) in a skeletal muscle cell.

protein complex, on the other hand, is a calcium channel whose task is to release the ions stored in the sarcoplasmic reticulum. The two proteins physically connect through a structure belonging to the ryanodine protein complex called the foot, which functions as a mechanical transducer. In practice, simplifying the mechanism as illustrated in the biomechanical model in Figure 7.13A, following depolarization of the T-tubule membrane by the onset of the action potential, the dihydropyridine receptor activates voltage-sensing S4 residues, changing its conformation. The change is transmitted via the foot to the ryanodine receptor, which is activated and releases calcium from the sarcoplasmic reticulum. The biomolecular model is also shown in Figure 7.13B. Since the ryanodine receptor is only present at the terminal cisternae facing the T-tubules, the release of calcium is spatially confined. For this reason, the rest of the sarcoplasmic reticulum is rich in cytoplasmic calcium-sensitive calcium channels (CICRC, *Calcium-Induced Calcium Release Channels*) that release calcium following calcium entry through the ryanodine receptors activated by action potential depolarization. In the case of skeletal muscle, one can speak of the DICR (*Depolarization-Induced Calcium Release*) mechanism, followed by the

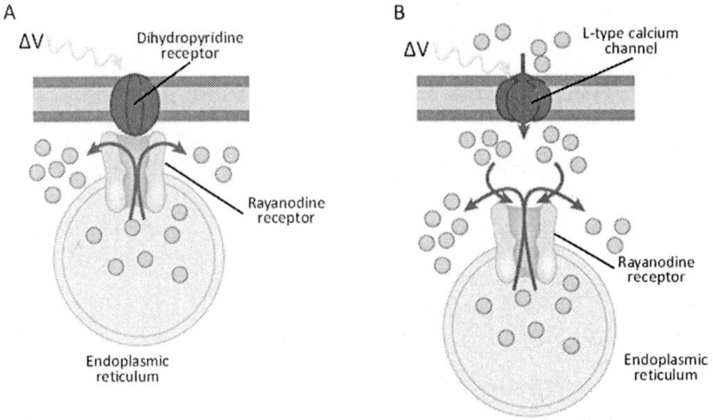

Figure 7.14 Schematic of the mechanism of calcium ion release from the endoplasmic reticulum by activation, due to a change in membrane potential (ΔV), of the dihydropyridine receptor, which in turn activates the ryanodine protein complex, in a skeletal muscle cell (A) and in a cardiac muscle cell or nerve cell (B).

CICRC-type mechanism. This mechanism differs between skeletal muscle and cardiac muscle. In cardiac muscle, there is no mechanical coupling between the dihydropyridine receptor and the ryanodine receptor. In heart cells, the dihydropyridine receptor is, for all intents and purposes, an L-type calcium channel. Calcium, entering through the dihydropyridine receptor, diffuses towards the reticulum and activates the ryanodine receptor, which in turn releases calcium ions (Figure 7.14). In cardiac cells, this mechanism is effectively a CICRC-type mechanism. An overview of the events involved in the excitation-contraction coupling mechanism is shown in Figure 7.15.

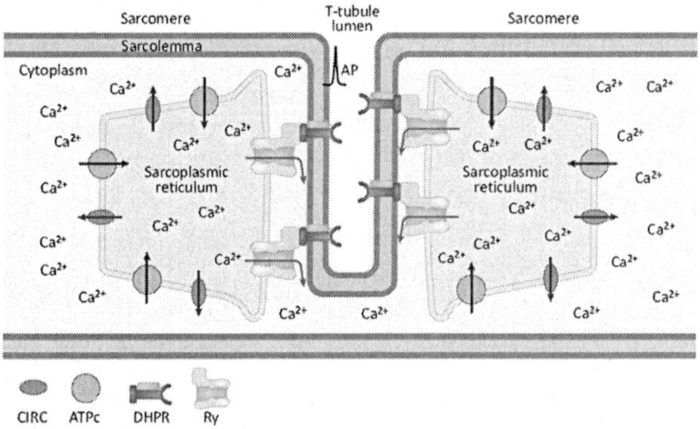

Figure 7.15 Schematic of the mechanism for the action potential (AP)-induced increase in intracellular calcium concentration in a skeletal muscle triad. Dihydropyridine-sensitive receptors (DHPR), ryanodine-sensitive calcium channels (Ry), calcium-activated calcium channels (CICR) and Ca^{2+}-dependent ATPases (ATPc) are represented.

7.2.2. A theoretical model of muscle contraction

Having analyzed the close link between structure and function of all of the elements that make up the complex mechanism of the transformation of an electrical signal into mechanical work (paragraph 7.2.1), it is necessary to address, at molecular level, the biomechanical event of muscular contraction. As mentioned, there is currently no agreed theory on how the interaction between the two major contractile proteins, actin and myosin, can generate force. In particular, it is not clear what energies are involved that allow the myosin head to rotate when it interacts with the actin attachment site. To try to explain this point, and in light of current mechanical, physiological, electrophysiological and molecular knowledge, the idea is to propose a logical model of what might be a mechanism of contraction at the molecular level. The hypothetical mechanism should take into account the principle that any functions take advantage of the potential energy previously stored, and must be compatible with a close morpho-functional relationship.

Consider a container of infinite size containing a solute in which there is a protein molecule M. The system is at a definite temperature above absolute zero and in thermodynamic equilibrium, i.e., there is no energy exchange with either the external environment or between different parts of the same environment. The system is in thermodynamic equilibrium. Let us suppose, moreover, that at time t = 0, the molecule is at point 1 of Figure 7.16. Under these conditions, the molecule under consideration is free to move in a random direction, with a high number of degrees of freedom. The kinetics of movement is directly proportional to temperature and inversely proportional to size. The molecule can move as it is charged with kinetic energy, which is derived from thermal agitation and collisions with solute molecules. It moves with uniform rectilinear motion until it collides with another solute molecule, say at point 2, then it moves again and in the same way to say point 3, and so on until point n, where it is at time t+δt. At each instant, M has a high probability of being at any point in space. The

Figure 7.16 The M-molecule moves in an infinite environment due to Brownian motion caused by the heat in the molecule itself and by collisions with solute molecules.

Figure 7.17 In a cylindrical vessel of limited volume, the degrees of freedom of the random motion of molecule M in solution are reduced.

molecules move according to Brownian motion kinetics, as they were first observed by the Scottish botanist Robert Brown (Montrose, 1773 - London, 1858) in 1827 while he was studying pollen particles under a microscope. They do this because they are bombarded by the other moving particles in the fluid. Brownian motion occurs at a speed and with changes in direction and frequency close to infinity. They were later explained and described quantitatively by Albert Einstein in 1905. Now, the molecule M's environment is transformed into a cylinder. This first constraint reduces the degrees of freedom of the random movements of M, which would make some displacements with modalities very similar to those of Figure 7.16, but in a limited number of directions and with at least partially predictable paths (Figure 7.17). Reducing then the diameter of the container, down to a value comparable to the dimensions of the molecule, the degrees of freedom are drastically reduced, so that the molecule itself will move, due to thermal agitation, almost exclusively according to the largest dimension of the container (Figure 7.18).

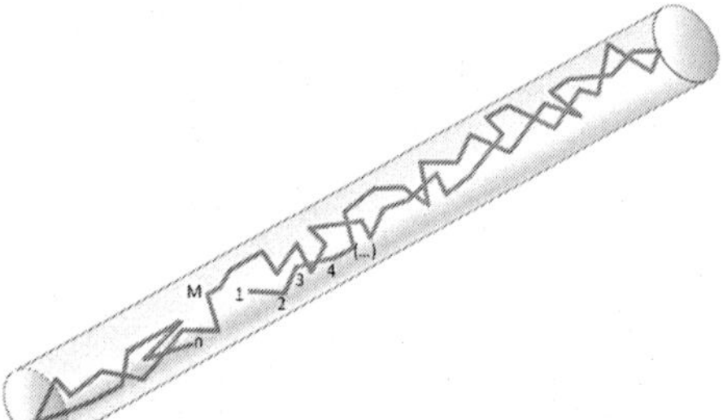

Figure 7.18 In a cylindrical vessel, with the diameter comparable to its size, the molecule M moves almost exclusively according to the largest dimension of the cylinder.

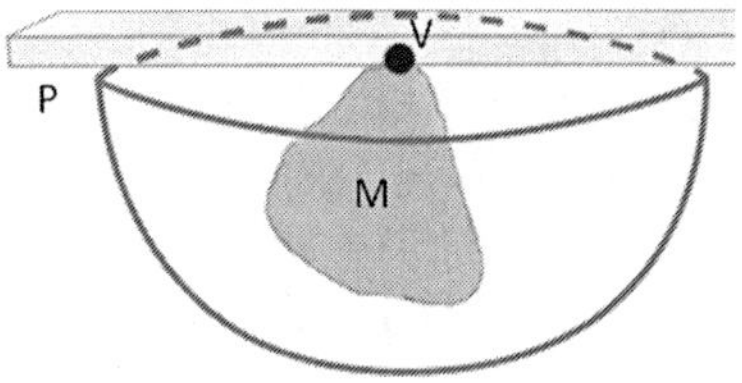

Figure 7.19 The point of constraint (V) of the structure M to the protein structure P drastically limits the degrees of freedom of its random motion (Brownian motion) to the yellow part of the space.

A further drastic reduction of the degrees of freedom of movement occurs if the molecule under examination is semi-rigidly bound to a support P (Figure 7.19). In these conditions, the molecule cannot move in the solution anymore, but is limited to oscillating around the point of constraint V in the portion of space in which it is immersed in the shape of a hemisphere (Figure 7.19). This is in small part due to the effect of the collisions with the solute molecules and for the most part due to thermal agitation.

There is further limitation of movement if the constraint V with P (Figure 7.20) is such that the molecule under examination can move, again almost exclusively due to thermal agitation, only along a plane. In these conditions, the protein is subject exclusively to a movement very similar to a pendulum. At time t = 0, the molecule M is in position -1, at the maximum potential energy, the minimum kinetic energy and oscillates due to thermal agitation. It has a high probability of oscillating only towards position 0 and from this towards position +1, given the limitations of movement imposed by the constraint at P. This oscillation, which, as said, is characterized

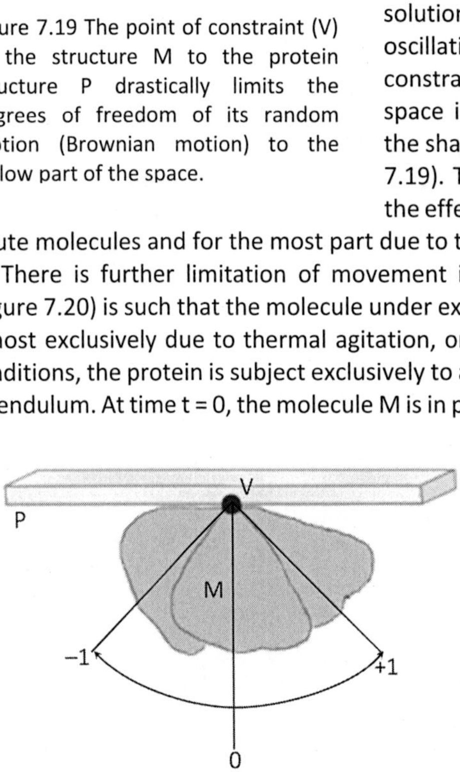

Figure 7.20 The molecule M bound in V in a suitable way to the fixed protein structure P can only move from position -1 to position +1 and vice versa along a trajectory passing through the 0 point.

by practically infinite speed and frequency, is made possible by the fact that potential energy is transformed into kinetic energy. The kinetic energy of M increases up to position 0, when it reaches its maximum, and then decreases up to position +1, where there is again the maximum of potential energy. From position +1 it starts to oscillate to -1 and vice versa. Each time, it alternately charges itself with potential energy at the expense of the kinetic energy provided by thermal agitation and with kinetic energy at the expense of potential energy. While M oscillates between position +1 and position -1 and vice versa, another protein molecule S, present in the environment of M, could come into contact with M. If S, of globular nature and present in position a in Figure 7.21A, were to be found, due to thermal agitation schematized by the broken grey arrow, in position b at the instant in which M oscillates from position -1 to position +1, the molecule S would be pushed, following a random trajectory, to position c in Figure 7.21A. Similarly, if the S molecule were to move from position a to position b in Figure 7.21B at the instant in which M oscillates from position +1 to position -1, it would be pushed to position c in Figure 7.21B following a random trajectory. The protein molecule S therefore moves in a different way depending on the direction of the oscillation of the protein molecule M. In order to be sure that the thrust of the oscillating protein M

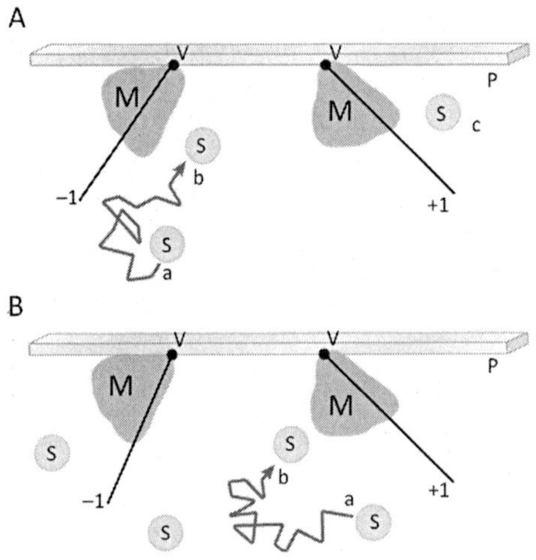

Figure 7.21 If the protein molecule S is added to the part of space in which M is moving, it moves, due to the kinetic energy of M, either in the direction from -1 to +1 (A) or in the direction from +1 to -1 (B), depending on its initial position with respect to M.

on the spherical protein S is not lost in random trajectories, it is necessary to bind in a semi-rigid way S to the protein structure P1 (Figure 7.22). P1 must be placed parallel to the protein structure P. In this way, the oscillations of protein M make it collide with the protein S, which, in turn, will move with the same trajectory as protein M and drag structure P1, moving on a single plane, parallel to the structure P. The direction of the movement depends on the position of S at the moment of the collision with the M molecule. If protein structure S is on the right of M (Figure

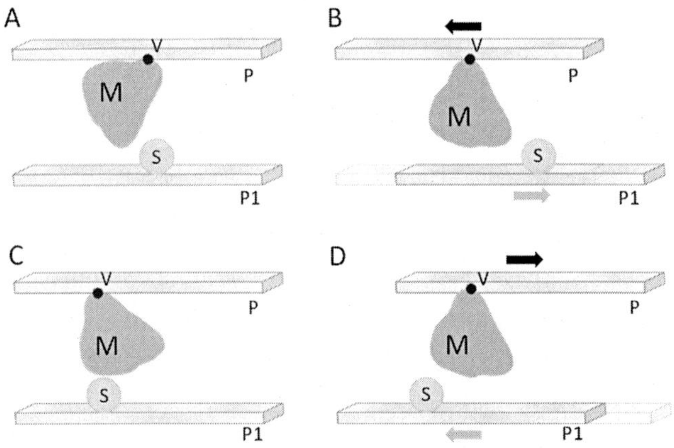

Figure 7.22 Due to the impact of the protein M, the protein molecule S moves, dragging the protein structure P1, to the right (B) if it is randomly located to the right of M (A); the spherical protein S moves, dragging the protein structure P1, to the left (D) if it is randomly located to the left of M (C).

7.22A), the protein structure P1 moves to the right (grey arrow of Figure 7.22B) and the protein structure P moves to the left (black arrow of Figure 7.22B). Vice versa, if the protein structure S is on the left of M (Figure 7.22C), the protein structure P1 moves to the left (grey arrow of Figure 7.22D) and the protein structure P moves to the right (black arrow of Figure 7.22D). In both cases, however, the potential energy of the M molecule is transferred as kinetic energy to the S protein and from S to the P1 structure, which transforms it into work and moves.

For the system to work effectively, however, it is essential to add a number of elements to the model that further reduce the degrees of freedom. This is to ensure that the protein S and the structure P1 move only when the protein M (which must be able to make complete

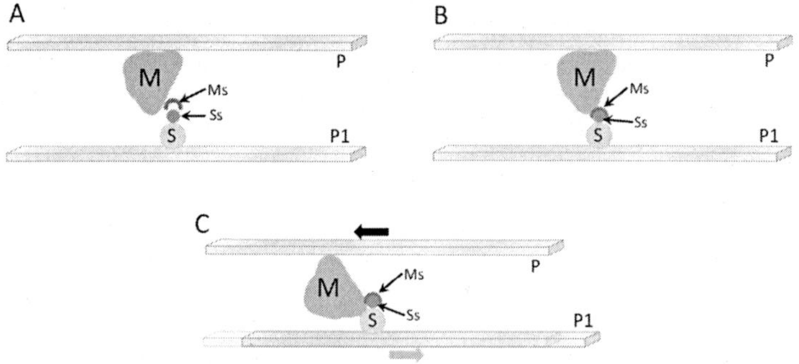

Figure 7.23 The Ms specific site of the M protein is able to bind to the Ss specific site of the S protein (B) and produce the shift of the P and P1 protein structures (black arrow and grey arrow) only when the left-to-right oscillation of the M protein brings the two specific sites into a favorable position (A).

oscillations) is, for example, oscillating from position -1 to position +1 as in Figure 7.21A and not from position +1 to position -1 as in Figure 7.21B.

The aim is achieved if an interaction site is added to the protein M at a suitable position in the molecule (Ms of Figure 7.23A) and an interaction site is added to the globular protein S (Ss of Figure 7.23A) so that the two proteins can interact only when, by random chance, the two sites are in a position that is mutually favorable to the formation of the bond between M and S. The bond must be strong and with high affinity only when the protein M oscillates in one of the two possible directions. Under these conditions, the kinetic energy of protein M will transfer almost completely to protein S and the P1 structure (grey and black arrows in Figure 7.23C).

Until now, all of the mechanisms described are not energy consuming if we consider thermal agitation a natural occurrence. However, because of the high-affinity binding between M and S, the system needs metabolic energy to separate the two proteins and to allow the M protein to repeat the sequence (Figure 7.23). Accordingly, the cycling sequence of events is: 1) transformation of potential energy into kinetic energy; 2) oscillation; 3) binding between the Ms and Ss sites of the M and S proteins respectively; 4) transfer of kinetic energy to the P1 structure; 5) shift of the P1 structure with respect to P. To result in a repeating sequence, timing is essential, and indeed fundamental in this model, as in all dynamic biological phenomena. Binding between the two proteins can only take place if the S protein is in the most favorable position at the precise moment in which the M protein reaches it during its oscillation from the -1 position to the +1 position or vice versa in Figure 7.21.

Timing is also important when the M protein reaches the maximum point of oscillation (Figure 7.23C) and must return as quickly as possible to its initial position (Figure 7.23A) to be ready to start a new cycle. It is therefore necessary to hypothesize the intervention of a molecule with a high energy level that is able to break the bond between the two proteins. This action becomes a functionally limiting factor, as the high-energy molecule must be present in sufficient concentration at the time when the two proteins must separate. It is necessary to add a high-affinity binding site for a high-energy molecule to the specific Ms site of the M protein (E in Figure 7.24A). The molecule binds to the M protein before the interaction of M with S (Figure 7.24 A). Once Ms and Ss interact and the transfer of mechanical energy occurs, the high energy molecule E becomes activated (Ea of Figure 7.24A) and provides the energy to cause the Ms and Ss complex to detach. Timing is essential: E activation (Ea) must occur when the Ms-Ss complex is at its maximum kinetic energy. Separation allows M to resume oscillation and to be ready for a new binding to another S protein.

All these sequences of events need a starting signal. This is provided by the release of C (Figure 7.25) from internal stores. C can be associated with calcium ions in a physiological scenario. Once the action potential activates the dihydropyridine/ryanodine complex, and consequently the CICR channel, calcium is massively diffused in the sarcomere. This high concentration increases the probability of interaction with a C binding site present on the S protein (Sc in Figure 7.25). Binding of C with S causes Ss to be available for the mechanical interaction with Ms.

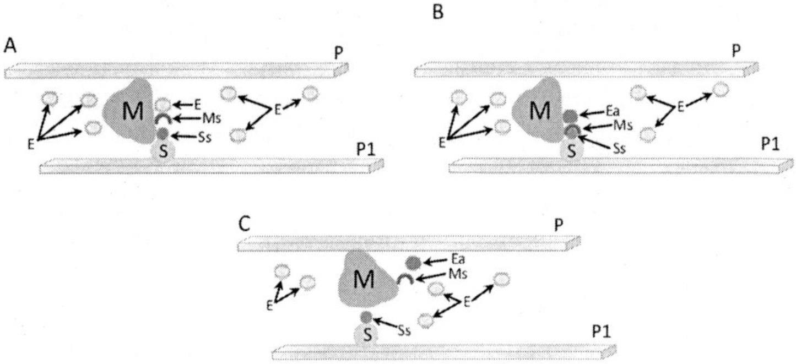

Figure 7.24 A high-energy molecule E always binds to its specific site Ms (A). Interaction of Ms with Ss activates E, becoming Ea (B). E activation causes protein M to detach from protein S (C).

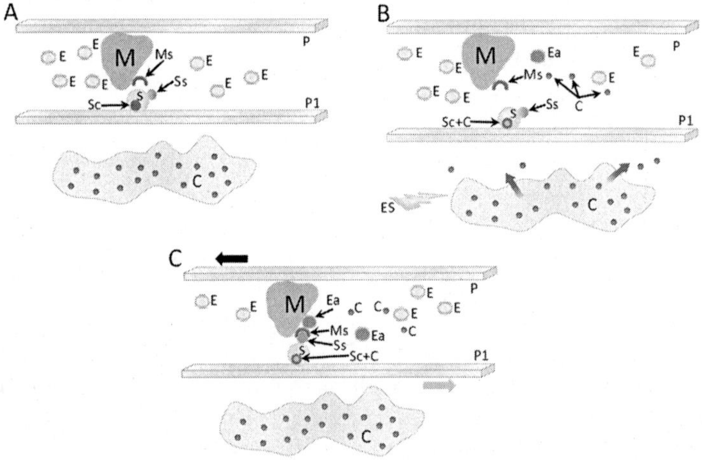

Figure 7.25 The release of compound C (thick arrows) following an external stimulus ES (B) makes protein S available for binding to protein M and reciprocal sliding of the P and P1 structures (C). E, Ea, Ms, Sc and Ss are

The model under consideration can represent the single basic contractile unit, whose structure is organized such that the function of transforming potential energy into work and the generation of force, both random phenomena that occur without the direct intervention of energy, are essentially based on Brownian motion, thermal agitation and conformational changes of the protein molecules. The outward production of work, represented in this model by the sliding of the protein structures P and P1, can only occur if all of the elements are in a useful position and with the exact timing necessary for progression.

7.2.3 A molecular model of muscle contraction

There are conditions in which the contractile unit might not be able to work because, by chance, not all of the elements involved have assumed the necessary configuration. This is despite the molecular phenomena involved being highly probable, given the very high frequency with which they occur. To overcome this problem, it is necessary to increase the number of units (Figure 7.26A). At any time, the M protein can randomly assume any of the infinite positions along the trajectory between position

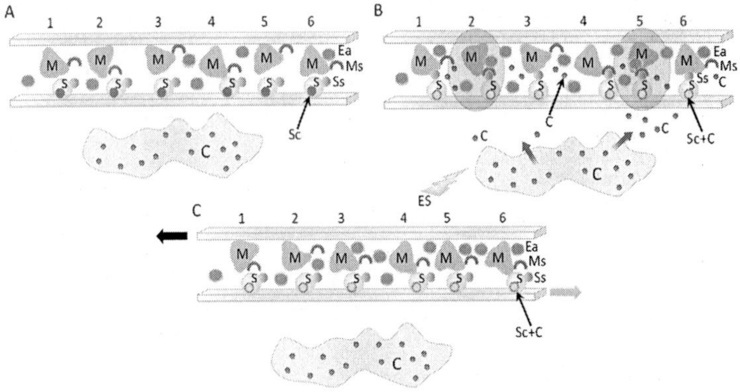

Figure 7.26 If there are numerous contractile units (A), following an external stimulus, ES, the units will release compound C (thick arrows in B) from its stores in sufficient quantity to activate the dual sliding process of the P and P1 protein structures (black and grey arrows in C). M, S, E, Ea, Ms, Sc and Ss are as in Figure 7.25.

-1 and position +1 (Figure 7.21). To simplify the picture, only three configurations are taken into consideration: two extremes at maximum potential energy and one at the peak of kinetic energy (-1, 0 and +1 respectively). Statistically, the maximum probability of the M protein having the right orientation to bind S is 33%. Increasing the number of M-S complexes results in an increase in the probability of useful interactions. In the schematic of Figure 7.26B, only the contractile units 2 and 5, highlighted in the Figure, are able to function correctly (grey and black arrows of Figure 7.26B). This is because all of the necessary conditions are present in the correct temporal sequence: 1) the external stimulus SE causes the release of compound C; 2) compound C frees the binding site of protein S for protein M; 3) the specific sites of protein M and protein S are correctly exposed; 4) protein M, already bound to the high-energy molecule Ea, interacts with the S site of S; 5) mechanical energy is transferred from M to S; 6) the Ea reaction provides the energy that, at the end of the cycle, causes the detachment of M from S to promote another cycle.

With a sufficiently high number of contractile units, at each instant there will be many links between M and S (Figure 7.26), distributed randomly along the protein structures P and P1. In the following instant, an equally high number of contractile units, but different from the previous ones, will be involved. And so it continues until C is freely diffuse. Cycling ends when electrical stimulation stops and C molecules are re-loaded in the stores. With no C molecules present, S returns to the resting condition and M returns to freely oscillate from position -1 to +1. (Figure 7.26C). This mode ensures continuity and high homogeneity of the sliding of the P and P1 protein structures.

The sliding process of P and P1, however, must serve to carry out outward work such as muscular contraction: for this to occur, one of the two types of protein structures must be firmly connected to a rigid structure (Figure 7.27). The effectiveness of the contraction, however, changes depending on how these filaments are actually arranged. If it is the protein structure P that is firmly connected to the rigid structure SR, the structural protein P1 moves to the right (grey arrow in Figure 7.27), creating less and less interaction between S and M. Vice versa, if the P1 protein structure is firmly connected to the rigid SR structure, the P protein structure moves to the left (black arrow of Figure 7.27B), increasing the possibility of creating more M-S complexes.

The structural organization of the model in Figure 7.27B presents two types of problems: 1) the risk of poor performance from the point of view of the work done, as there is no guarantee that the protein structure P runs parallel to the longitudinal axis; 2) the risk that the amount of work done is small due to the low number of bonds that can be formed between M and S, given the small number of contractile units that might be available at each instant. To ensure that the protein structure P1 slides neatly parallel to the longitudinal axis of the model and to the protein structure P (arrow of Figure 7.28), the number of protein structures P1 can be increased, each with numerous protein structures S with its own protein binding site M

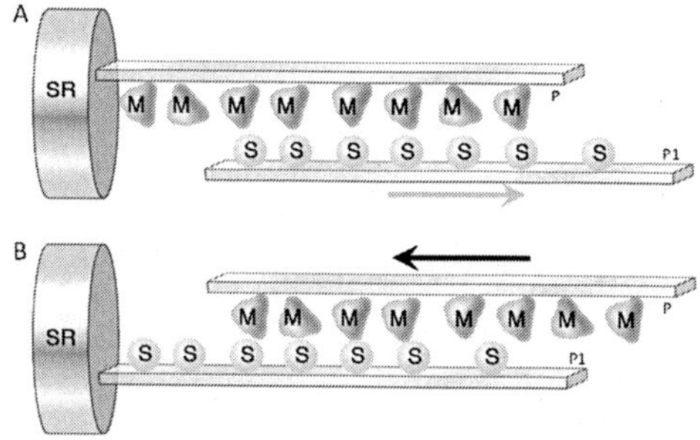

Figure 7.27 If we fix the protein structure P to the rigid structure RS, the mobile protein structure moves to the right according to the grey arrow (A). If we fix the protein structure P1 to RS, the mobile protein structure moves to the left according to the black arrow (B). M and S are as in Figure 7.24.

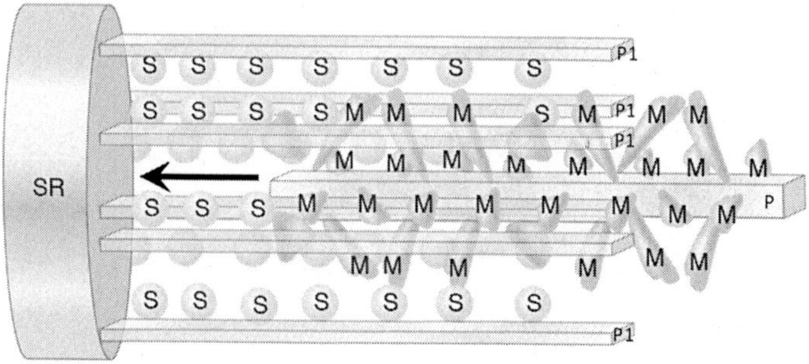

Figure 7.28 In order to increase the efficiency of the system, it is possible to have numerous protein structures P1 surrounding a protein structure P, rich in protein M. S is as in Figure 7.24.

(Figure 7.28) and rigidly connected to the same rigid structure SR, so that they can surround the protein structure P.

In order to guarantee a sufficient amount of work for the system, it is possible to arrange the M proteins on the P protein structure in such a way that they are all close to the S protein molecules of the P1 protein structure. For maximum functional performance, the numerous M molecules must be distributed in an orderly and compact manner on the P protein structure, just as the S protein molecules must be on the P1structure; they must emerge outwards in as many directions in space as possible.

The work proposed at the moment by the model under examination is to move the protein structure P with respect to P1. However, the goal of a contractile model is to generate force at the extremities by shortening the elementary unit of the contractile machine. To accomplish this task, it is sufficient to add a structure similar to Figure 7.28 in a mirrored position. Figure 7.29 shows the assembly of a hypothetical single contractile unit with the protein structures P connected between each other (Pa and Pb; Figure 7.29). Each cycle carried out by the M/S complex generates sliding of the protein structure P1 with respect to the protein structures Pa and Pb (grey arrows). This results in a shortening of the distance between the two opposite SR structures (black arrows). The different components of the single contractile unit can be easily associated with the diverse protein complexes of the sarcomere (Figure 7.5, 7.6 and 7.8). SR represent the Z line, P1 the actin filaments, P the myosin filaments, M the myosin heads and S the action filament carrying the attachment sites of the myosin

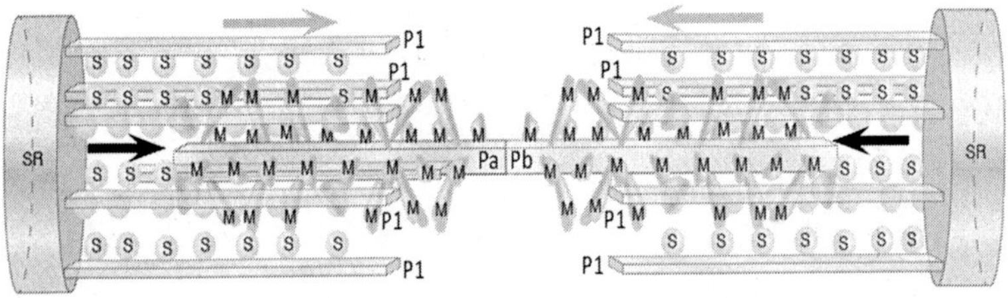

Figure 7.29 The dual sliding between the Pa protein complex, Pb and the P1 protein structure leads to the rigid structures RS moving nearer to each other. M and S are as in Figure 7.24.

heads.

Several units like the one in Figure 7.29, arranged in series with the SR rigid plates attached one to another, could form a practically infinite contractile fiber (myofibrils). Many myofibrils in parallel surrounded by a biological membrane could represent the single contractile fiber cell (myofiber).

In conclusion, a functional model of muscular contraction has been constructed, which may have retraced the steps selected by evolution to produce the sarcomere, and which takes into account the chaotic movement of Brownian motion and the effects of the thermal agitation of particular molecules in solution, to which the degrees of freedom are gradually limited. This model must be associated with a specific structure that is able to produce changes in the length of contractile structures or the development of tension. In both cases, the system performs work fueled by the potential energy accumulated in the cell, transformed into kinetic energy and heat. Performing a contraction is triggered and modulated in strength and in time by the nervous system. In this model, calcium ions and ATP molecules (the compound C and the energy molecule E in Figure 7.26) are decisive in coordinating the modalities and the time with which the contractile structure must function. ATP is constantly bound to the head of myosin so as not to be a limiting factor for the intervention of calcium when the action potential reaches the system. Both calcium and the action potential are required to initiate the contraction.

The whole organization, finally, makes it possible to take maximal advantage of the potential energy of the random oscillatory movement of the myosin head and to maintain it with a high momentum. This allows, at the moment of detachment of the myosin from the actin, the oscillatory movement to continue, to recover energy for a new oscillation and therefore a new cycle of contraction. It is important to note how a single event, as part of a highly numerous events, can be transformed from 'digital' to apparently 'analogue', such as the shortening of a muscle fiber. The contraction event remains a digital phenomenon as it is promoted by single elements that are all the same and that, by increasing the probability of interactions, transform the macroscopic event into a continuous phenomenon. We have already seen how a macroscopic signal produced by a high number of single microscopic digital events is perceived as "analogue" even if it is not (paragraph 5.12).

7.2.4 Biomechanics of contraction

Skeletal muscle biomechanics is concerned with gathering information about the mechanism by which contractile structures are able to activate and generate force. The functional experiments that will be described represent elements to support the theoretical model (section 7.2.2) and the molecular model (section 7.2.3) of skeletal muscle contraction.

A generic skeletal muscle (Figure 7.30A) can be functionally divided into two types of biomechanical elements (Figure 7.30B): elastic or undamped elements and non-elastic or damped elements.

An undamped element behaves according to Hooke's law (Robert Hooke, Freshwater 1635 - London 1703):

$$\vec{F} = -k \cdot \Delta\vec{x} \tag{7.1}$$

which states that the elongation Δx is directly proportional to the applied force F and the constant of proportionality k is the elastic constant, which depends on the intrinsic properties of the material from which the undamped element is made.

Damped elements, on the other hand, do not have a linear relationship between force and length. Indeed, in biological materials, and skeletal muscle in particular, it is difficult to divide elastic and non-elastic elements experimentally, as there is no biological tissue that is purely undamped or purely damped. As a first approximation, in the model in Figure 7.30, the undamped element is most closely identified with the tendons that attach the muscle to the skeletal structure. The damped elements, on the other hand, are more closely associated with the more contractile material.

In addition to the tendons, the collagen fibers that bind the myofibrils, the membranes of the myofibrils and the structures that make up the

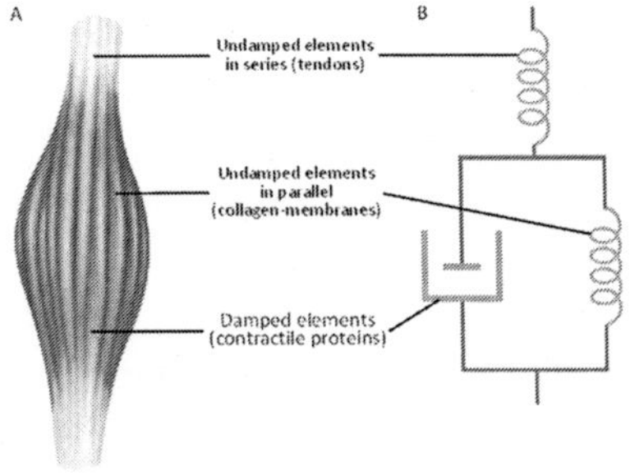

Figure 7.30 Schematic of a generic mammalian skeletal muscle (A) and a biomechanical model of its damped and undamped components (B).

myofibrils, such as the sarcomeres placed in series, show a certain degree of elasticity (Figure 7.3). Lastly, once the binding has taken place due to the presence of intracellular calcium, the bridges between actin and myosin show a non-negligible elasticity. The functionality of all of these elements is also partly influenced by the viscosity of the different tissues and the solutions in both the gel and liquid state.

7.2.4.1 Passive properties of skeletal muscle

If a stretch is applied to the unstimulated skeletal muscle, the passive response of the muscle can be ideally broken down in three different elements: the contractile engine, the undamped elements and the damped elements (Figure 7.30B).

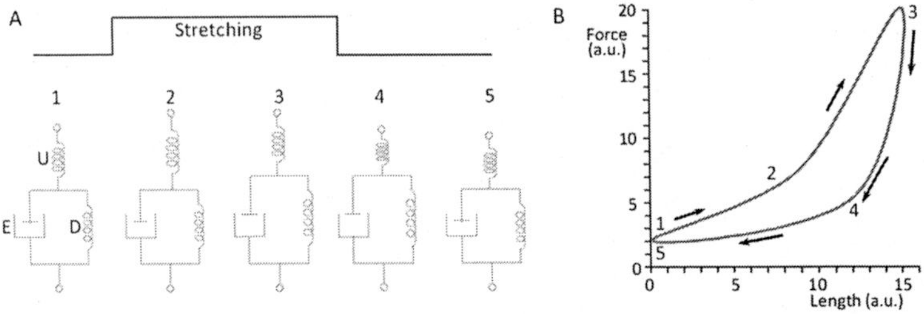

Figure 7.31 A) Biomechanical model of the responses in 5 phases of a skeletal muscle stretched for a certain time and not stimulated; the involvement, with different modalities and times, of the muscular motor (E), undamped elements (U) and damped elements (D). B) Experimental graph of the force developed by a skeletal muscle stretched for a certain time and not stimulated as a function of length. The phases from 1 to 5 corresponding to the biomechanical model in A are indicated. Force and length are expressed in arbitrary units (a.u.).

In Figure 7.31A (phase 1), the muscle is at rest. By performing a sudden stretch (phase 2), most of the muscle's elongation is absorbed by the undamped elements and the force rises moderately from 1 to 2 (Figure 7.31B). In a second step, the muscle motor and damped elements elongate (step 3 in Figure 7.31A and 2 to 3 in Figure 7.31B) at the expense of a partial shortening of the undamped elements. The force at the ends of the model increases because the resistance to elongation is no longer only that of the undamped elements, but also that of the muscle motor and the damped elements. The shortening after the stretching stops (phase 4 of Figure 7.31A) sees a sudden drop in the force (from 3 to 4 of Figure 7.31B) due to the shortening of the undamped part, followed by a more moderate drop in the force (phase 5 of Figure 7.31A and from 4 to 5 of Figure 7.31B) due to the shortening of the muscular motor and the damped elements.

The hysteresis of the force/length graph denotes the viscoelastic nature of the model, which represents very well the behavior of a striated skeletal muscle subjected to traction and then relaxed. The passive elements of the muscle resist traction by increasing the force at the ends of the muscle.

7.2.4.2 Active properties of skeletal muscle

The next test is done by stimulating the skeletal muscle in isometric mode, i.e., at constant length, with a stimulus capable of generating an action potential in the muscle fibers. Figure 7.32 shows the original experiment in which the rise in force is recorded following a single stimulus, at stimulus frequencies of 10 Hz, 20 Hz and at a high frequency (50 Hz). In the first case, the contraction is called a single twitch, the second and third are called clones or unfused tetanus, while the fourth trace is typical of a tetanic contraction. It must be kept in mind that the first three contraction modes do not exist in nature. They are only reproducible in the laboratory and on isolated muscle. All physiological contractions are tetanic, like trace 4 in Figure 7.32.

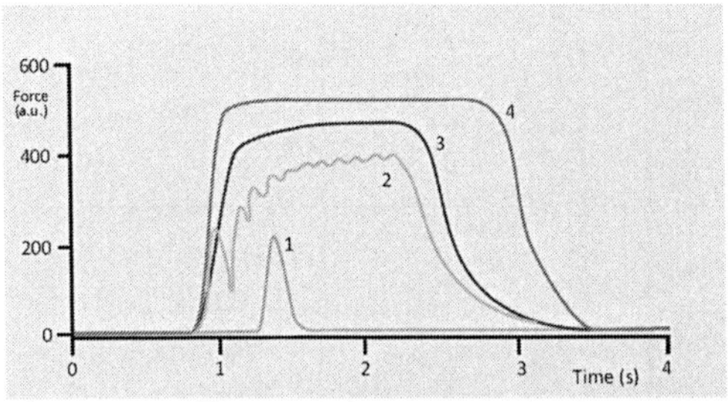

Figure 7.32 Example recording of the trend of contraction force as a function of time in a frog gastrocnemius muscle, maximally stimulated in isometric mode at different frequencies. Single stimulus (1), stimuli at a frequency of 10 Hz (2), 20 Hz (3) and 50 Hz (4).

The usefulness of the in vitro recordings is that they make it possible to investigate, both from a macroscopic and microscopic point of view, the mechanism of contraction. The first aspect that is interesting to study concerns the difference in force development of the muscle when stimulated by a single event (trace 1 of Figure 7.32) rather than by high-frequency stimuli (trace 4 of Figure 7.32). In a biomechanical model of a skeletal muscle (Figure 7.33A), a single stimulus, which makes the muscle motor contract, also shortens the damped elements. The total force developed by the contractile engine does not reach the muscle ends. Part of the total force is absorbed by the elongation of the undamped elements. The force rises (Figure 7.33B) with a delay with respect to the onset of the action potential and decays with a characteristic time trend (Figure 7.7B). In order for the muscle to develop full force at its ends, the undamped elements must be stretched and become practically rigid. From an operational point of view, this means that the time between one stimulus and the next must be shorter than the time in

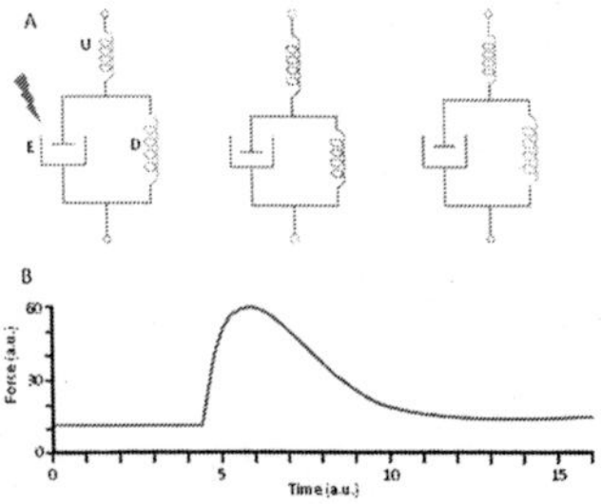

Figure 7.33 A) Biomechanical model of the effect of applying a single stimulus to a skeletal muscle and of the involvement, with different modalities and times, of the muscular motor (E), the undamped elements (U) and the damped elements (D). B) Graph relating the force developed, with the molecular mechanisms described in A, by a skeletal muscle stimulated with a single stimulus, and the length. Force and time are expressed in arbitrary units (a.u.).

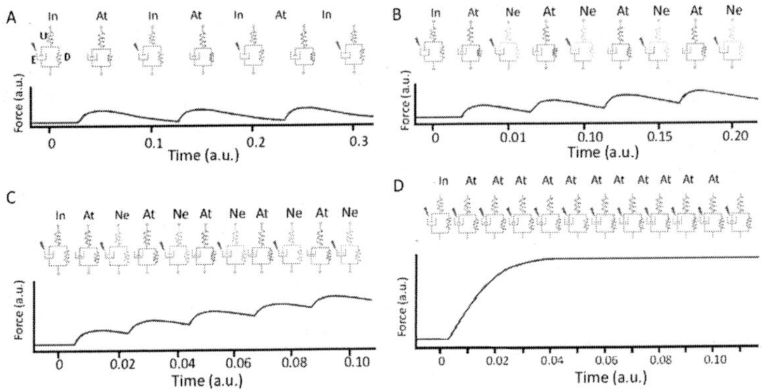

Figure 7.34 Biomechanical model of the effect of applying a series of stimuli (grey thunderbolts) to a skeletal muscle, at frequencies of 10 Hz (A), 20 Hz (B), 50 Hz (C) and 100 Hz (D). For each frequency, the upper part of each figure shows the behavior of the muscle motor (E), the undamped elements (U) and the damped elements (D). The lower part shows the force graph, expressed in arbitrary units (a.u.), as a function of time. The components of the muscle are shown as inactive (In), active (Ac) and active but not effective (Ne) for the purpose of contraction, due to the state of incomplete relaxation of the damped and undamped elements of the muscle.

which the undamped elements would relax. In the simulation carried out, still using the biomechanical model of the skeletal muscle (Figure 7.34), we can see how, as the stimulation frequency increases from A to D, the force rises more and more until reaching, at 100 Hz of stimulus frequency, a value that stabilizes in a plateau and the muscle develops its maximum force. All of the force developed by the contractile engine reaches the muscle ends because the undamped elements are stretched and rigid.

7.2.4.3 Isometric and isotonic contractions

The model we have just seen is instrumental in understanding the behavior of the different biomechanical components, such as the contractile engine, and the undamped and damped components of mammalian skeletal muscle. In skeletal muscle, these components are present to varying degrees in all the different muscle types, although prevalent in specific structures. Therefore, from now on, in order to

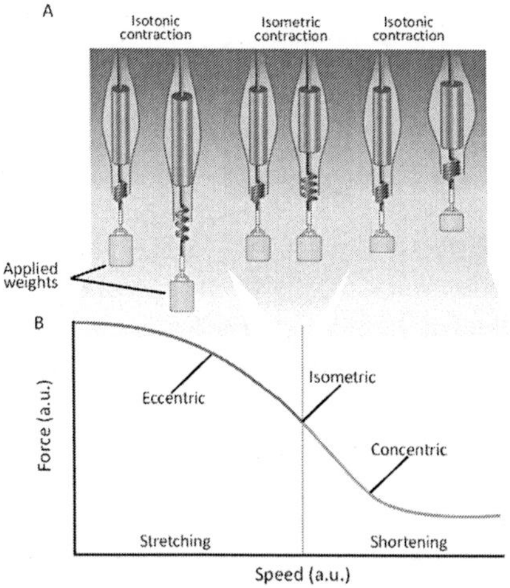

Figure 7.35 A) Comparison between the schematic behaviour of a skeletal muscle that, suitably stimulated with the application to one end of the muscle of decreasing weights, passes from the resting state (left scheme of each pair of muscles) to an isotonic contraction with lengthening (left part of the figure), to an isometric contraction (central part of the figure) and to an isotonic contraction with shortening (right part of the figure). B) Trend of the force, expressed in arbitrary units (u.a.), as a function of the rate of change of the length of the muscle, expressed in arbitrary units (u.a.), for the three types of experiments described in A.

address the biomechanics of muscle contraction, we will refer exclusively to active components, represented by the functional elements of contraction, and to passive components, which include all of the structural elements of the muscle.

Skeletal muscle has different ways in which it can contract when stimulated appropriately. A macroscopic distinction can be made between contractions with constant muscle length, or isometric contractions, and contractions in which the weight applied to the muscle is constant, or isotonic contractions (Figure 7.35A). In turn, isotonic contractions are divided into eccentric contractions, when the applied weight is greater than the maximum force that the muscle can develop, and concentric contractions, when the applied weight is less. In the force/velocity diagram in Figure 7.35B, it is evident how a muscle, subjected to different conditions, has the capacity to develop an active force. At the point indicated by the arrow, the muscle neither shortens nor lengthens, and has a length at which it develops maximum isometric force. This is the point at which isometric and isotonic contraction coincide. In eccentric contraction, the velocity of lengthening is directly proportional to the force: the greater the weight applied, the greater the speed with which the muscle lengthens. Conversely, in concentric contraction, the less weight applied, the greater the speed of shortening.

Experimentally, the maximum isometric force is found by constructing the force/length graph shown in Figure 7.36. This graph is an interpolation of several experimental points, where the total force during a tetanic contraction (T of Figure 7.36) is obtained in isometric conditions at different lengths of skeletal muscle. The muscle must reach a particular length before the force starts to increase. As the muscle length is increased, the tetanus force also increases. At a length that is specific to each muscle, the force reaches its maximum. After that, the force starts to decrease. However, further elongation results in a new increase of the contractile force. The second increase in force occurs at the same lengths where the muscle begins to resist elongation due to its passive structures. In the early stages, the total force (T), overlaps with the active force (A). When the active force starts to decrease, the passive forces (P) cause an exponential rise of the total force by opposing resistance at further elongation. To obtain the graph of the active force alone, the passive force P can be subtracted from the total force T. The gain in tensile force is limited. It is clear that, since we are dealing with biological tissues, further pulling would have the effect of irreparably damaging the muscle, to the point of destroying it. From the graph in Figure 7.36, we can deduce that there is

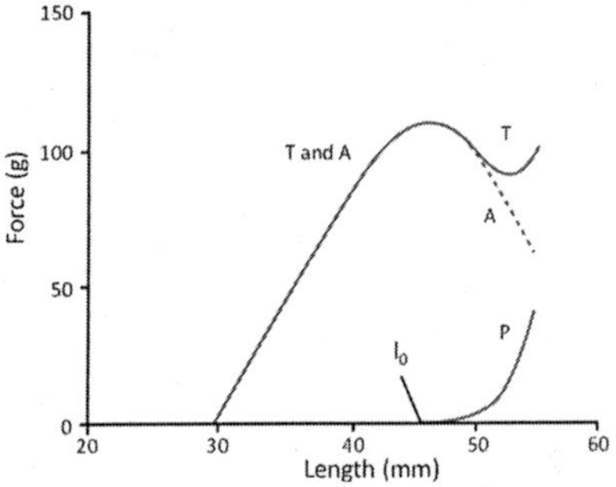

Figure 7.36 Force/length graph of toad sartorius muscle. The P curve is obtained by stretching the relaxed muscle beyond l_0, the maximum length of the muscle in the absence of tension in the passive elements. The T curve is obtained by stimulation of the muscle at tetanic frequency in isometric conditions at different lengths of the muscle. The C curve is obtained by subtracting the values of the P curve from the values of the T curve.

an optimum length at which each muscle typically develops maximum force. Functional experiments on skeletal muscle as a whole do not provide more detailed information on the molecular mechanism of contraction. In addition to an important passive component, the multitude of muscle fibers in parallel makes the individual events that occur at the level of the acto-myosin complexes when the muscle is stimulated largely homogeneous. To obtain more detail, it is necessary to make the biological system less complex and to reduce the preparation under observation as close as possible to the minimum known contractile unit, represented by the sarcomere. Huxley (1974) developed an experimental apparatus capable of recording the variation in force and distance between the Z-lines of individual sarcomeres from a group of isolated frog muscle fibers. Figure 7.37A shows the force/length diagram constructed from experiments carried out on a few muscle fibers.

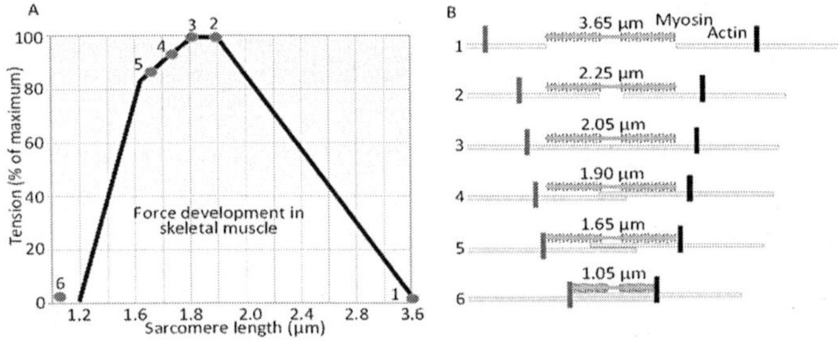

Figure 7.37 A) Diagram relating the force with the length of the sarcomere measured microscopically as the distance between the Z lines. B) Schematic of the degree of overlap of the actin and myosin filaments and the length of the sarcomere, highlighted by the grey dots in (A). In the diagram, the length of the myosin filament is 1.6 µm, and actin, 1.0 µm. The vertical bars indicate the position of the Z lines. (Modified from A. M. Gordon, 1966.)

The x-coordinate shows the distance between the Z-lines of a single sarcomere observed at the center of a single fiber. Here too, as in the muscle as a whole, there is a preferential length at which the fiber develops the maximum isometric force. The ideal distance between the Z-lines of the sarcomeres for maximum force to be exerted is between 2.0 and 2.2 µm. In the case of the single fiber, an excessive increase in length beyond 3.5 µm brings the force exerted by the contraction of the fiber to zero. The

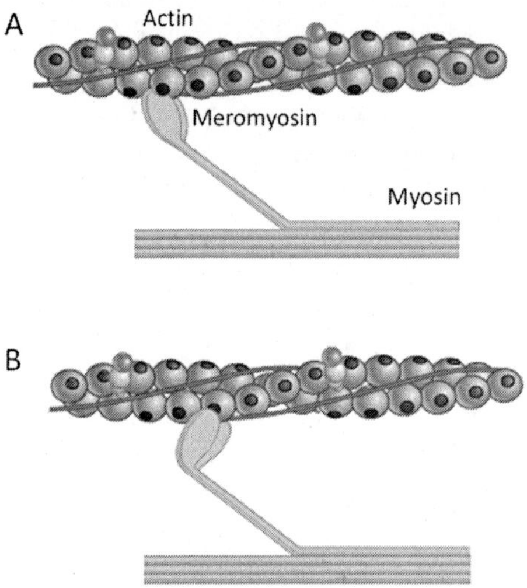

Figure 7.38 A) Diagram of the hypothetical mechanism of interaction between actin and myosin head and rotation of the latter. B) Diagram of the reciprocal sliding of actin with respect to myosin.

role of passive components is negligible in this preparation and elongation of the sarcomere above 3.5 μm results in irreversible damage to the cell.

By recording the tetanic force and observing the distance between the Z-lines of a single sarcomere, from which this force developed, it was possible to reconstruct the model represented in Figure 7.37B. The vertical bands delineate the sarcomere and represent the Z-lines, the thin filaments represent filamentous actin, while the thicker filaments, with outward protrusions, represent myosin filaments with heavy meromyosin heads making contact with the actin filament. The model demonstrates that the force recorded, shown in the force/length diagram, is directly proportional to the overlap between the two types of filaments. At 3.6 μm, the sum of the lengths of the two actin molecules, 1 μm each, and the myosin molecule, 1.6 μm, the force is zero (position 1 of Figure 7.37A). At lengths of 2.2 μm and 2.0 μm, there is maximum overlap between the actin and myosin filaments and consequently the maximum force developed by the fiber. At lengths shorter than 2.0 μm, the filaments interdigitate, effectively reducing the useful overlap. This experiment has shown that the force developed by the muscle motor depends on the degree of overlap between the two types of filaments. However, it is not yet possible to obtain any information about the dynamics of force generation from the interaction between the myosin heads and the actin filament. Morphological, pharmacological and molecular structure analyses provide information about the possibility that the myosin head can assume different positions, interact with the actin filament and that the interaction is subject to the presence of calcium ions and ATP. In particular, between the myosin heads and the filament there is a point called a *hinge* that is sensitive to papain. This enzyme is actually able to separate the myosin heads from the rest of the filament.

What the investigations just described are unable to establish is whether the myosin head is actually capable of performing a rotation as illustrated by Figure 7.38. In order to understand whether this rotation of the head occurs, a complex experimental procedure was used, which is partially technically explained in Data Sheet 7.2 and which will be described functionally here.

During a tetanic contraction in an isometric condition, the force rises in time with a continuous function, as would happen for the decrease in the distance of a single sarcomere's Z lines in the case of a concentric isotonic contraction. This continuous function is given by the repetitive and random interaction between the myosin heads and the actin filament. The number of interactions is so high that the contraction event, which is in fact digital, is perceived as an analogue event. In order to record a signal that is related to a single interaction, one has two choices. There is the possibility to isolate the single event or one must find a way to synchronize many identical events. At the moment, even the most advanced technology is unable to observe a dynamic biomechanical event at the molecular level. The attempt has been made, however, to synchronize as many interactions between actin and myosin as possible during muscle activity. The logic of the experimental design considers the plateau reached during the isometric tetanic contraction as a dynamic stationary state, in which all the bridges formed between actin and myosin are oriented in the same direction and exert static traction. The experiment, the original plot of which is shown in Figure 7.39, begins with the maintenance over time of the maximum value of force (initial horizontal stretch of P_0 in Figure 7.39A) under isometric conditions (horizontal stretch of l_0 in Figure 7.39B). The recording apparatus allows two important experimental conditions: a length feedback or a force feedback (tab 7.2). A sudden drop in tension (arrow in Figure 7.39A) is followed by a phase in which the force at the ends of the muscle is maintained at 95% of the maximum isometric force with force feedback (final horizontal stretch of P_0 in Figure 7.39A).

The recording has moved from a contraction isometric condition to an isotonic activity. Attention should be focused on the trace measuring the distance between the Z lines of a single sarcomere (Figure 7.39B). In line with the fall in force, there is phase 1, which consists of a shortening that responds to Hooke's law (Equation 7.1), i.e., it affects only the purely undamped component of the sarcomere, which shortens alongside the fall in force. This is followed by phase 2 of shortening, which is slower but occurs against a constant force, i.e., shortening by the contractile motor. Phase 3 represents an interval of time in which the distance between the Z-lines remains constant. Shortening then resumes in phase 4, which appears as a linear decrease.

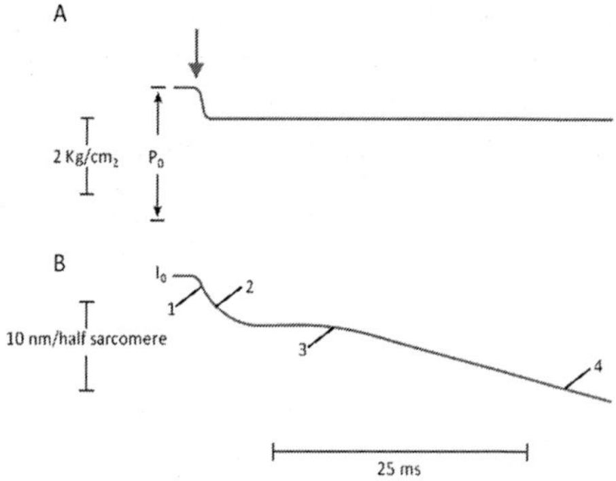

Figure 7.39 A) Trend of isometric force P0 measured in a single muscle fiber stimulated tetanically. B) Corresponding 4-step change in muscle fiber length observed when isometric force is rapidly reduced to an isotonic value corresponding to approximately 0.8-P0. (Modified from A.F. Huxley, 1974.)

The data from this experiment were interpreted by Huxley (1974) based on the concept that during muscle contraction, the bridges between actin and myosin attach and detach randomly. During the plateau phase of tetanic contraction, there is a momentary synchronization of all bridges that are attached at that moment. A sudden drop in tension, after an initial undamped shortening (phase 1), is followed by a phase of shortening due to the rotation of the heads, which, under these conditions, occurs synchronously and is therefore visible as the Z-lines of the single sarcomere moving nearer to each other (phase 2). This is followed by a phase in which the bridges detach in a still fairly synchronized manner (phase 3), which evolves into a return to the random interaction of the bridges, with shortening becoming constant and desynchronized. The measurement of the displacement of a Z-line during phase 2 of the first shortening attributed to head rotation is about 5 nm, fully compatible with histological measurements of possible myosin head oscillation.

From these experiments, a biomechanical model of the interaction between actin and myosin was derived, with a hypothesis for the mechanism of how this interaction could generate force at the ends of the muscle. Figure 7.40 shows a schematic of the site of interaction between myosin and actin. The myosin head is free to oscillate continuously along the major axis of the myosin filament, moved by thermal agitation but constrained in its movement by the two springs that support it. The actin filament has the site of interaction with myosin at a distance X from the equilibrium position of the myosin. At the moment when the muscle fiber is excited and intracellular calcium is released, the actin site changes conformation and the myosin head, oscillating randomly, moves from distance X to a position with high potential energy and a high probability of interacting effectively with the actin site. In this way, the potential energy stored in the spring is converted into kinetic energy and therefore into mechanical energy that moves the entire actin filament relative to the myosin filament (from A. Huxley, "Muscle Contraction", 1974). For more

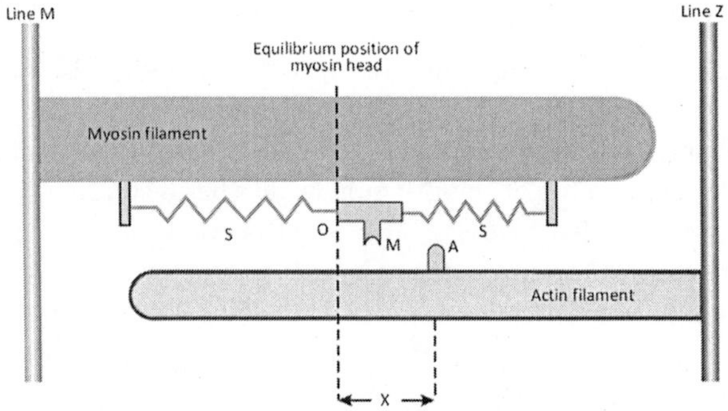

Figure 7.40 Biomechanical model of the interaction between actin and myosin. M = myosin head; A = actin interaction site with myosin; X = distance of the actin interaction site A from the equilibrium position of the myosin head O; S = springs binding the myosin head M. (Modified from A. F. Huxley, 1957.)

in-depth studies on the molecular mechanism of muscle contraction, please refer to the most recent scientific publications in the field (e.g., Reconditi M. et al. 2014, Brunello E. et al. 2009, Piazzesi G. et al. 1999 and the other studies of the Physiology Team of the Department of Physiological Sciences of the University of Florence, Italy).

7.3 The smooth muscle

Smooth muscle is made up of myocytes (muscle fibers) that are tens of micrometers long and approximately 5 µm in diameter. They are fusiform, mononucleated and have a thicker central region containing the nucleus and thinner ends. The myocyte has no transverse striations and are arranged in bundles that are out of phase with each other, so that the thicker central region of each is juxtaposed with the thinned ends of the adjacent cells (Figure 7.41). In some locations, such as the "tonaca propria" of the intestinal villi, muscle fibers may be isolated or in small groups surrounded by connective tissue. The sarcolemma of smooth muscle cells has a considerable number of invaginations called caveolae, probably the equivalent of the T-tubules of striated muscle. The caveolae are in close contact with tubular elements of the sarcoplasmic reticulum and may have a similar function to the sarcoplasmic reticulum of skeletal muscle. In the sarcoplasm, there is a small Golgi apparatus, a pair of centrioles, a small rough sarcoplasmic reticulum, free ribosomes (mainly concentrated in a conical region at each pole of the nucleus) and mitochondria scattered throughout the cytoplasm. At several points in the intercellular space, thin cellular branches are observed, through which adjacent cells connect by gap

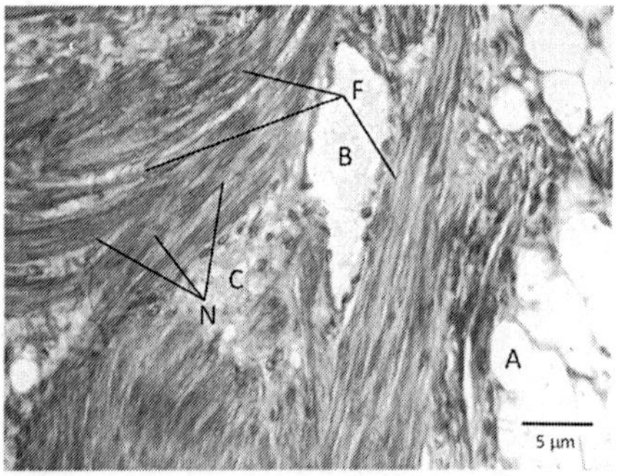

Figure 7.41 Light microscope image of a fragment of smooth muscle tissue. We can see the fibers (F) with the nuclei (N) in a central position, connective tissue (C), adipose tissue (A) and a blood vessel (B).

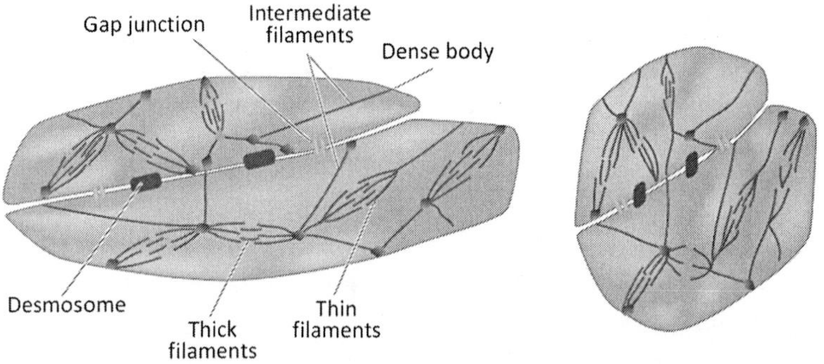

Figure 7.42 Diagram of the contraction of two adjacent smooth muscle fibers (A), after which the contracted fibers assume a spheroidal conformation (B).

junctions (tab. 6.1), creating, from an electrical point of view, a functional syncytium, as in cardiac muscle tissue (section 7.4).

The smooth muscle tissue forms the muscular "tonaca" of the wall of the digestive tract, the respiratory tract, the genital tract and the urinary tract and is also present in the walls of the arteries, veins and excretory ducts of the glands. In particular, in the digestive tract, the muscle bundles are distributed in two distinct layers, one longitudinal and one transverse, which, by contracting in a coordinated manner, determine the appearance of constrictions in the intestinal wall and promote the progression of the lumen contents. Unlike in skeletal muscle, where contractions are induced by action potentials (section 7.2.1), or in the heart, where contractions are controlled by the sinus node (Data Sheet 7.3), the contractions that form the peristaltic waves are stimulated by spontaneous oscillations of the membrane potential that induce currents of incoming calcium and also increase its intracellular concentration (Data Sheet 6.2).

The myocytes of smooth muscle tissue do not have the transverse striations of striated muscle, although they contain thin filaments composed of actin and fewer filaments composed of myosin. There are also characteristic dense bodies in the sarcoplasm that are crossed by myofilaments, which probably have a function similar to that of the Z-lines. Furthermore, the modalities of contraction are unique to this muscular tissue. In the case of muscle fibers and cardiomyocytes, the contractile proteins have an arrangement that imposes a preferential axis on the contractile process, which is parallel to the length of the same cell. In the case of smooth muscle myocytes, the contractile proteins are distributed in the cytoplasm in various directions (Figure 7.42A) and the contraction determines a change in shape along various cellular axes (Figure 7.42B).

7.3.1 Excitation-contraction coupling in smooth muscle

In the smooth muscle cell, similar to the skeletal muscle, an increase in the concentration of calcium is required for contraction. Such an increase is caused both by the calcium released by the sarcoplasmic reticulum, thanks to a channel coupled to a receptor for IP3 (inositol triphosphate), and by the entry of calcium from outside the cell by spontaneous depolarizations due to calcium currents that characterize this muscle tissue. Although thin filaments have tropomyosin, they lack the troponin complex, so they are always highly likely to interact with myosin, which is inactivated at rest. Calcium that is released into the cytoplasm by the electrical stimulus binds to intracellular calmodulin; the binding increases the likelihood of a conformational change. This makes the calcium-calmodulin complex capable of binding to the enzyme called myosin light chain kinase (MLCK), which activates (via phosphorylation) the myosin light chain and makes the myosin heads available for actin binding. In smooth muscle, calcium acts on myosin, not actin as in skeletal and cardiac muscle. At this point, transverse bridges are formed, causing the thin and thick filaments to slide apart, as in the case of the sarcomere (section 7.2). The contraction is stopped by the intervention of a phosphatase, which dephosphorylates the myosin that is detached from the actin. The phosphatase is an enzyme that is always present within the cell and competes with the kinase. However, its action becomes predominant over that of the kinase when the concentration of intracellular calcium is reduced by cessation of the electrical stimulus and recovery by the sarcoplasmic reticulum.

7.4 The heart muscle

The cardiac muscle or myocardium, which constitutes the major part of the heart, is formed by cardiomyocytes (Figure 7.43) of mesodermal origin. These single contractile cells are cylindrical in shape, 100 µm long, about 10 µm in diameter and are mononucleated, with a centrally arranged

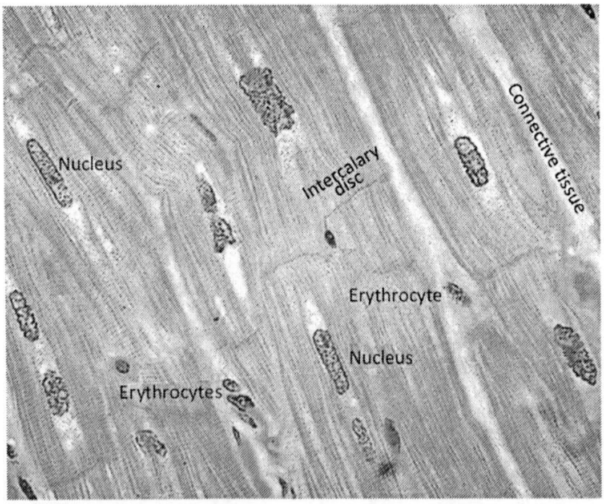

Figure 7.43 Light microscope image of a fragment of cardiac muscle tissue. The nucleus of the cardiomyocytes, the intercalary discs, the connective tissue and the vessels containing erythrocytes are highlighted.

nucleus. They are connected to each other, forming a three-dimensional network of cardiac myofibrils. The connections are formed by special connecting structures between adjoining cells called intercalary discs, which have areas similar in structure to the desmosomes and *zonula adherens* of the junctions between epithelial cells (Figure 6.1) and to which the contractile proteins and desmin filaments adhere. The desmosomes, *zonula adherens* and desmin probably have a mechanical function and ensure cohesion between the cellular elements of the myocardium.

At the level of the intercalary discs, and especially at the branches of different adjacent cells, there are numerous gap junctions (section 6.1) which form cytoplasmic bridges (section 6.2) and thus make the myocardium, from an electrical point of view, a functional syncytium. The cytoplasmic bridges formed by the gap junctions ensure electrical continuity between all of the cells, rapid diffusion of the impulse from one cardiomyocyte to another until it permeates the whole of the cardiac tissue and enable each region of the heart to contract synchronously.

Cardiomyocytes have more abundant sarcoplasm, more numerous mitochondria and the sarcoplasmic reticulum is less developed and less elaborate than skeletal muscle fibers. Subcellularly, cardiomyocytes have the T-tubule system and sarcoplasmic reticulum typical of skeletal muscle (Figure 7.11B), with the typical striations for each individual cell recognizable under the light microscope.

7.4.1 The cycle of stimulation and contraction in the heart

In an adult man, the heart makes an average of 60 beats every minute and sends 4 to 6 liters of blood into the bloodstream at rest. During exercise, the number of contractions per unit of time can increase up to around 3 fold. The human heart can make 3,600 beats every hour, 86,400 beats every day, 31,536,000 beats every year, and in 70 years of life, at least 2 billion beats.

Perfect integration between structure and function has meant that in the course of evolution, a vitally important organ has been selected. The heart has no chance of recovery in its work, as is the case for example, with the motor units of skeletal muscle (Data Sheet 7.1), and must therefore ensure continuous functioning. It is in fact thanks to the perfect correlation between the anatomical and histological organization of the heart, the electrical functions of generating and conducting stimuli and the mechanical function of generating blood flow that an organ such as the heart is able to work on a continuous cycle.

Two series of events are involved in cardiac contraction, which follow their own modes of action, but which are closely interconnected: the generation and propagation of the electrical impulse (Data Sheet 7.3) and the contraction of the various regions of cardiac muscle.

The first series of events includes: the generation of the I_f current, the regulation of the time to the peak of the action potential and its propagation to the atria by the sinoatrial node, the control of the speed of conduction of the impulse along the Hiss bundle (to create an appropriate delay that allows the completion of the contraction of the atria and the filling of the ventricles), and the regulation of the speed of conduction in the Purkinje fibers, so as to make the contraction of the ventricles start from the apex of the heart and spread, with appropriate timing, to the base (tab 7.3).

For the second series of events, refer to the diagram of the structure of the heart in Figure 7.44. The four cardiac cavities, right atrium, left atrium, right ventricle and left ventricle, pass cyclically from the relaxed state of diastole to the contracted state of systole (Figure 7.45). However, from the hydrodynamic point of view, the contraction of the atria, which are small

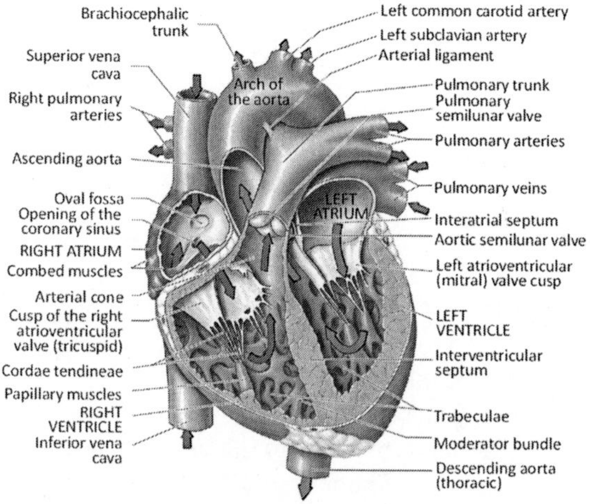

Figure 7.44 Schematic of the various structures of the mammalian heart, venous blood flows (arrows on the left side of the diagram) and arterial blood flows (arrows on the right side of the diagram).

in size and have little contractile tissue, gives a small contribution to the flow of blood, while the ventricles, especially the left, which are much larger than the atria and have thicker muscle tissue, are able to provide considerable flow (Figure 7.46).

The first event of the contraction cycle of the heart is the generation of an action potential by the sinoatrial node (Figure 7.47B). The action potential starts with a depolarization of the membrane potential, due to the intervention of the I_f current (paragraph 5.14.2; Chapter 6.5), which has the special property of being activated in hyperpolarization, up to the threshold

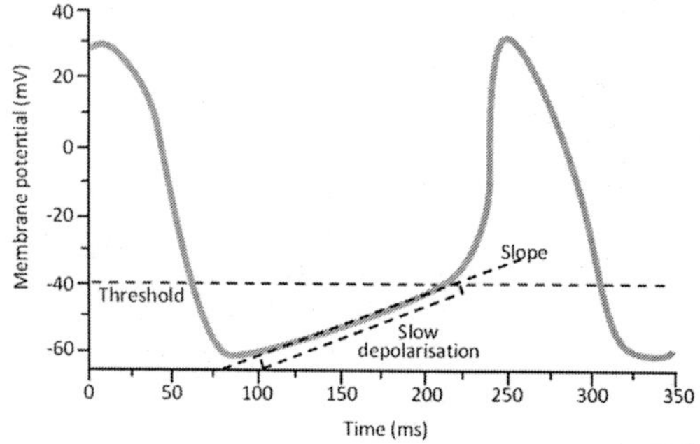

Figure 7.45 Action potential in the mammalian heart

potential of the action potential (about -40 mV in Figure 7.45). The slope of the depolarization due to the I_f current (dashed line in Figure 7.45) is modulated by acetylcholine, adrenaline and noradrenaline and determines the time to the duration of the action potential and therefore the heart rate (tab 7.3). In fact, a greater slope speeds the time to the peak and a smaller slope delays it, with an increase in the frequency in the first case and a decrease in the second.

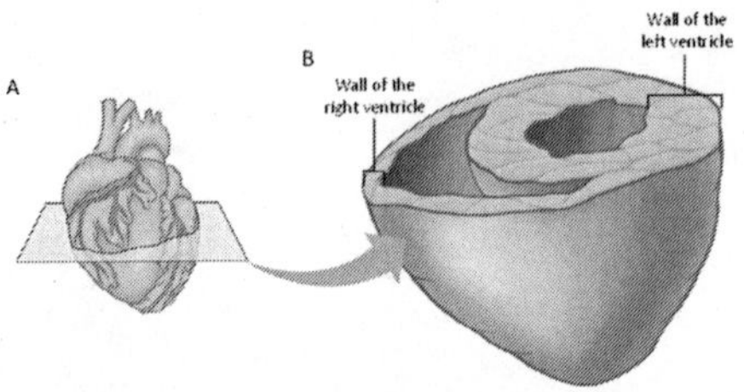

Figure 7.46 Cross-sectional drawing of the mammalian heart (B) according to the plane shown in A. The thickness of the muscular part of the left ventricle is much greater than that of the right ventricle.

In the meantime, the atria have passed from the phase of systole, i.e., contracted, to that of diastole, i.e., relaxed. This allows the right atrium to

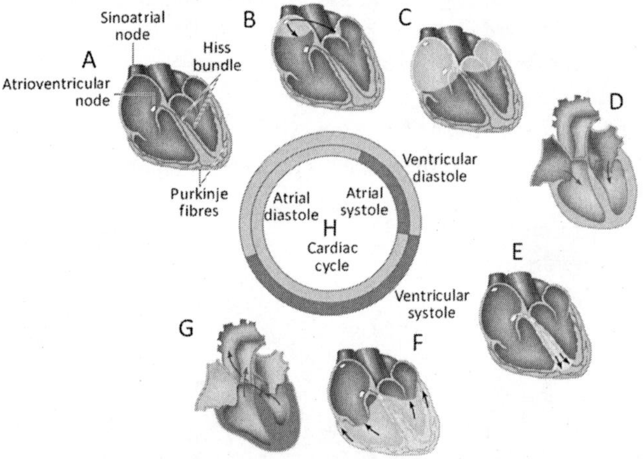

Figure 7.47 Schematic of the pacing and contraction cycle of the mammalian heart (A-G) and the alternation between systole and diastole of the four cardiac cavities (H). Arrows D and G represent arterial and venous blood flow. The clear cavities in B, C, D, E, F and G are excited by the action potential.

be filled with venous blood, coming from the superior and inferior hollow veins, and the left atrium with arterial blood coming from the pulmonary veins (Figure 7.47C). The ventricles, in diastole and with the valves between the atria and ventricles open, also fill up to about 75% with blood that passes directly through the atria due to the *vis a tergo* (venous return), originating from the difference between arterial and venous pressure. Complete filling of the ventricles will occur with the contraction of the atria.

During the filling of the atria, the action potential generated by the sinoatrial node directly permeates the right atrium and, conducted by Bachmann's bundle, the left atrium (Figure 7.47 B). This process exploits the particular structure of the muscular tissue, in which all of the cardiomyocytes of each myofibril are, from a functional point of view, an electrical syncytium, connected to each other by the numerous gap junctions (Figure S7.8 and Figure 6.3). At this point, we have the contraction of the atria themselves (Figure 7.47D) and the completion of the filling of the ventricles, which are already partially full.

The action potential generated by the sinoatrial node is also transmitted, through the internodal branches, to the atrioventricular node, from this, through Hiss's bundle, to the apex of the heart (Figure 7.46E) and subsequently, through the Purkinje fibers, to the base of the heart itself (Figure 7.46F).

The contraction of the ventricles begins at the apex, spreads radially and longitudinally to the base and closes the tricuspid and mitral valves. It effectively pushes the venous blood through the pulmonary valve, into the pulmonary arteries, then into the pulmonary circle, through the aortic valve into the aortic artery, then into the systemic circle (Figure 7.46G). The atria, in the meantime, have returned to diastole and are ready to initiate a new cycle of cardiac contraction (Figure 7.46A), a cycle that repeats in a human being, as mentioned, at least 2 billion times over the course of one’s existence.

Heart cells do not suffer from fatigue during normal activity because of the efficiency of the circulatory system formed by the coronary arteries. Moreover, recent studies have demonstrated the presence of cardiac stem cells, similar to those in skeletal muscle (section 7.1.1), that are capable of replacing aged or damaged myocytes (Beltrami AP et al. 2003; Bearzi C et al. 2005).

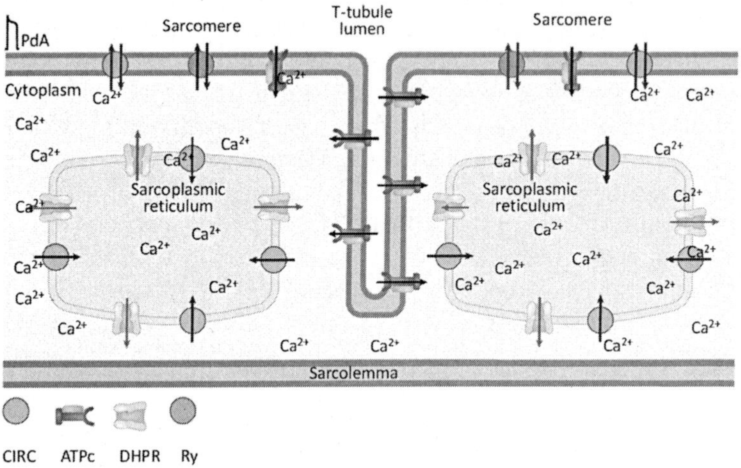

Figure 7.48 Schematic of the mechanism for the increase in intracellular calcium concentration in a cardiomyocyte. L-type calcium channels (DHPRs), ryanodine-sensitive calcium channels (Ry), Ca^{2+}-dependent ATPases (ATPc), calcium-sodium-dependent (ATPcn) and sodium-potassium-dependent (ATPnk) exchangers are shown.

7.4.2 Excitation-contraction coupling in the heart muscle

The excitation-contraction coupling mechanism in the heart shares many functional aspects of skeletal muscle. In cardiomyocytes, too, there is a T-tubule system for the propagation of excitation and a developed sarcoplasmic reticulum.

The action potential, generated by the sinoatrial node and propagated from cell to cell in a particular way to the atria and ventricles (Figure 7.3), reaches the cardiomyocytes (PdA of Figure 7.48) and depolarizes them. This causes a conformational change in the dihydropyridine-sensitive channels (DHPR), which in the cardiomyocytes are true L-type slow-activating calcium channels (Section 4.4.2). An incoming calcium current is activated (black arrows), which is what determines, in the atrial and ventricular action potential (AT and VEN traces), the plateau that prolongs depolarization and refractoriness of the cardiomyocytes (Data Sheet 7.3). The increase in intracellular calcium concentration activates, via the ryanodine-sensitive channels (Ry) or CICRC, a massive release of calcium from the reticulum to the sarcoplasm (arrows), changing the cytoplasmic concentration of the divalent ion from 10^{-7} - 10^{-8} M to 10^{-5}- 10^{-4} M. Intracellular calcium at a high concentration is highly likely to interact with troponin C of the troponin complex (section 7.1.2) and, as in skeletal muscle, initiate the cycle leading to sarcomere contraction.

When the action potential repolarizes (phase 3 of the AT and VEN traces in Figure S7.5B), the dihydropyridine-sensitive channels (DHPR in Figure 7.48) close, the flow of calcium ions ceases, and the ryanodine-sensitive channels (Ry) close due to the decrease in calcium concentration. At this point, the action of Ca^{2+}-dependent ATPases (ATPc), which bring calcium back into the sarcoplasmic reticulum, becomes predominant. In addition, calcium-sodium-dependent and sodium-potassium-dependent exchangers (ATPcn and ATPnk) remove calcium by exchanging it with sodium and remove sodium by exchanging it with potassium (green and blue arrows respectively), thus re-establishing resting ion concentrations.

Bibliography

Bearzi C, Cascapera S, Nascimbene A, Casarsa C, Rastaldo R, Hosoda T, De Angelis A, Rota M, Quaini F, Urbanek K, Leri A, Anversa P, Bolli R, Kajstura J. Late-Breaking Developments in Stem Cell Biology and Cardiac Growth Regulation. Circulation (2005) 111: 1720-1724.

Beltrami AP, Barlucchi L, Torella D, Baker M, Limana F, Chimenti S, Kasahara H, Rota M, Musso E, Urbanek K, Leri A, Kajstura J, Nadal-Ginard B, Anversa P. Adult cardiac stem cells are multipotent and support myocardial regeneration. Cell (2003) 114(6): 763-76.

Brunello E, Fusi L, Reconditi M, Linari M, Bianco P, Panine P, Narayanan T, Piazzesi G, Lombardi V, Irving M. Structural changes in myosin motors and filaments during relaxation of skeletal muscle. J Physiol. (2009) 587(Pt 18): 4509- 4521.

Cavagna GA, Dusman B, Margaria R. Positive work done by a previously stretched muscle. J. Appl. Physiol. (1968) 24: 21-32.

Ford LE, Huxley AF, Simmons RM. Tension responses to sudden lenght change in stimulated frog muscle fibres near slack lenght. J. Physiol. (1977) 269: 441-515.

Franzini-Armstrong C. The Membrane Systems of Muscle Cells. In Engel A.G. and Franzini-Armstrong C. Eds. Myology III Edition. McGrawHill NY (2004): 232-256.

Gordon AM, Huxley AF, Julian FJ. The variation in isometric tension with sarcomere lenght in vertebrate muscle fibres. J. Physiol. (1966) 184: 170-192.

Hanson J, Huxley HE. Structural basis of the cross-striations in muscle. Nature. (1953) 172(4377): 530- 532.

Herzog WJ. Neuroengineering Rehab (2017) 14: 98 DOI 10.1186/s12984-017-0310-6.

Huxley AF, Niedergerke R. Interference microscopy of living muscle fibres. Nature. (1954) 173(4412): 971-973.

Huxely AF. Muscle structure and theories of contraction. Prog. Biophys. biophys. Chem. (1957) 7: 255- 318.

Huxley AF. Muscle. Annu Rev Physiol. (1964) 26: 131-152.

Huxley AF. The activation of striated muscle and its mechanical response. Proc R Soc Lond B Biol Sci. (1971) 178 (50): 1-27.

Huxely AF. Muscle Contraction. Review. J. Physiol. (1974) 243: 1-43. Piazzesi G, Reconditi M, Dobbie I, Linari M, Boesecke P, Diat O, Irving M, Lombardi V. Changes in conformation of myosin heads during the development of isometric contraction and rapid shortening in single frog muscle fibres. J Physiol. (1999) 514(Pt 2): 305-312.

Reconditi M, Brunello E, Fusi L, Linari M, Martinez MF, Lombardi V, Irving M, Piazzesi G. Sarcomere- length dependence of myosin filament structure in skeletal muscle fibres of the frog. J. Physiol. (2014) 592(5): 1119-1137.

Data Sheet 7.1

The motor unit

Generally, the muscle is made up of a number of motor units. These are formed by a motor neuron, which emerges from the anterior

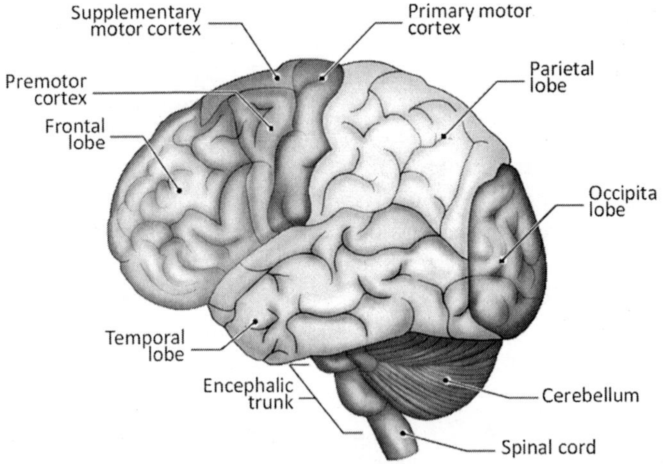

Figure S7.1 Reproduction of the motor cortex; left lateral view.

roots of the spinal cord and by a more or less branched synaptic termination, which forms the neuromuscular plaques. Smaller muscles, such as the extrinsic muscles of the eyes, have a relatively high number of motor units for fine control of movement, but a motor unit comprises a maximum of just 10-20 fibers. Larger muscles, for example those of the lower limbs, like the gastrocnemius, which must make large movements and develop a great deal of force, can have up to 2000 fibers per motor unit.

Muscle contraction is under strict control of the motor cortex, located in the posterior part of the frontal lobe (Figure S7.1). By processing signals from the sensory systems (Chapter 8), the brain programs voluntary movements, integrates them with memorized experiences and controls their correct execution. The motor cortex is composed of the primary motor cortex (Figure S7.1), which directly controls the execution of movements, and the secondary motor cortex. The latter in turn is subdivided into (1) the premotor cortex, which controls the movements of the proximal muscles and trunk, programs each movement and acts on the primary motor cortex or directly on the effector organs through the motor neurons; and (2) the supplementary motor area, which controls the movements of the distal limbs and coordinates complex movements, i.e., those that involve a sequence of movements.

Action potentials, which constitute the primary signal for muscle contraction (Chapter 5), are generated at an appropriate frequency by the motor cortex. They are propagated by a myelinated neuron (section 6.2.4.3), which runs in the corticospinal bundle of the spinal cord (Figure S7.2) to the grey matter, where it forms an axodendritic synapse with the cell body of the motor neuron. The motor neuron emerges from the anterior horns of the spinal cord and runs in the spinal nerve, together with the corresponding sensory neuron, to the muscle, where it forms the appropriate number of neuromuscular synapses. When the muscle has to contract, it is the motor cortex that decides how many and which motor

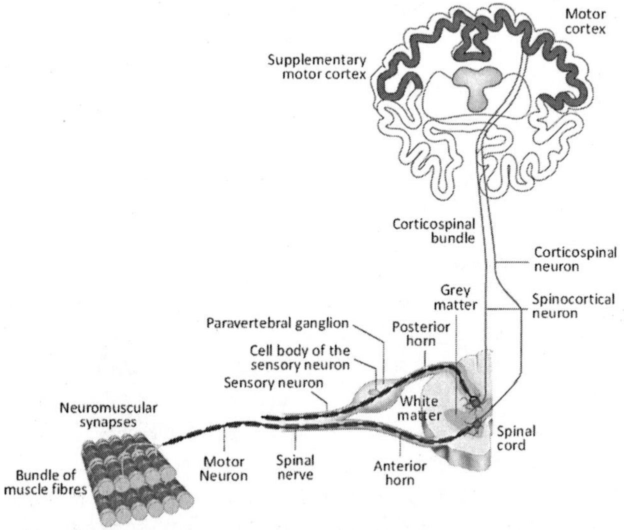

Figure S7.2 Schematic of a motor unit, formed by a motor neuron, several neuromuscular synapses and a bundle of muscle fibers, as well as its own corticospinal neuron and the axodendritic synapse.

units must be stimulated and with what frequency of action potentials. However, if the contraction of a motor unit lasts too long, its muscle fibers may suffer from fatigue. Fatigue is caused by a reduced amount of ATP that the cellular metabolism does not have time to replenish.

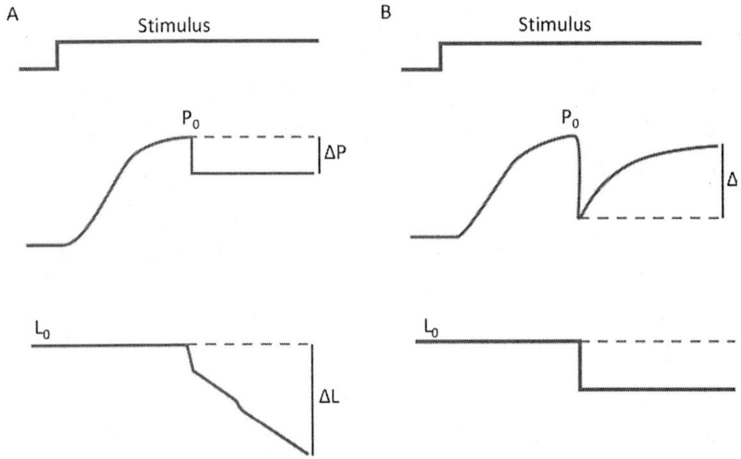

Figure S7.3 Biomechanical recordings from striated muscle fibers under force feedback (A) and length feedback (B). P0 = force developed by the muscle fiber; ΔP = change in force; L0 = length of the muscle fiber; ΔL = change in length.

To avoid the phenomenon of fatigue, the motor cortex intervenes, recruiting other motor units that were inactive until that moment, allowing the tired ones to rest.

Data Sheet 7.2

Force and length feedback recordings

Muscle contraction is developed in such a way that during activity, the muscle fibers change their length and consequently the force exerted at the ends of the muscle. Experimentally, on isolated muscle, it is possible to reproduce the two main conditions under which the muscle performs its function: isometric contraction and isotonic contraction. Recording data on force development and change in muscle length is not a problem. There are force transducers to 4 decimal places and very accurate methods of measuring length. The problem arises when one wishes to switch between length and force *feedback* during an experiment. In practice, it is necessary to set up an apparatus which allows one to change the load applied and observe how this is reflected in the length of the muscle, in particular in the distance between the Z-lines of the sarcomeres. Similarly, again starting from an isometric contraction, it must be possible to shorten the muscle abruptly and observe how the force behaves at the two ends of the muscle.

The examples in Figure S7.3 illustrate the two types of experiments in which recordings are made in a regime of force feedback and length feedback. In A, a load is applied to the ends of a muscle fiber during a maximal isometric contraction L_0. The force reaches the plateau of the tetanic contraction P_0, which causes the developed force of the quantity ΔP to vary up to 95% of its maximal force. After an initial period of isometric contraction, the trace of the distance between the Z-lines, following the drop in force, shows a shortening of ΔL, necessary to counterbalance the applied weight: the experiment has changed from an isometric contraction to an isotonic contraction. In the case shown in panel B, the start of the experiment is similar, but in this case, instead of the applied weight, the length of the muscle is suddenly changed by 5% (ΔL). The recording shows how, following the sudden shortening, the force goes to zero due to the complete unloading of the undamped structures. The force then recovers due to the action of the myosin heads, which shorten the muscle with attack-detachment kinetics. The force will reach a new plateau, with a smaller amplitude, because now the distance between the Z-lines of the sarcomeres will no longer be optimal.

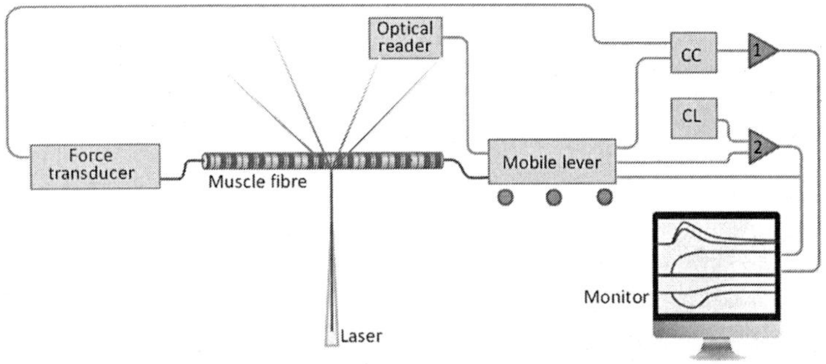

Figure S7.4 Simplified schematic of the force and length recording apparatus from a single muscle fiber of the anterior tibial muscle of *Rana temporaria* in a regime of force feedback and length feedback. CF = command center for force; CL = command center for length; 1 and 2 = differential amplifiers 1 and 2, respectively. (Modified from L.E. Ford, 1977.)

The experimental apparatus which allows this type of measurement is schematized in Figure S7.4. The single muscle fiber is supported in Ringer's solution immersion by two aluminum foil clips around the residual tendon, which is left from the isolation of the fiber. The ends of the fiber are secured to two glass hooks, which in turn are connected on one side to a force transducer and on the other to a micrometric lever controlled by a computer by means of a solenoid. The displacement trace is generated by an optical diode that receives the diffraction of light from a laser positioned in the center of the fiber. The aligned Z-lines are able to modify the path of the monochromatic light and change the diffraction angle as their position changes. This allows an accurate measurement of the displacement of a single Z-line and thus the shortening by half a sarcomere (Huxley 1974). The system is based on feedback circuits. The force transducer and the optical reader have continuous monitoring of the force and displacement of the Z-lines and are connected to the lever, which makes all the necessary adjustments. A control center for force and a control center for length complete the circuit. In the case of an isometric measurement, the distance between the Z-lines can be adjusted beforehand with the mobile lever. The stimulation of the fiber will thus initiate an isometric contraction. In the case of a force feedback, the computer commands the system, via the force control center, to bring the force value, for example, to 95% of the maximum isometric force. The lever will move accordingly, shortening the muscle until the desired force matches the force recorded by the differential amplifier 1, in order to maintain a constant load at the ends of the fiber. The displacement of the Z-lines is in this case recorded by the optical reader.

If, however, length feedback is operated, the force is only monitored continuously. In this case, the comparison is made between the length, set by the corresponding command, and the actual measurement of the displacement of the Z-lines recorded by the differential amplifier 2. If it is desired that the distance between the sarcomeres is 1.8 μm instead of 2 μm, the mobile lever will shorten the muscle until the optical reader records a 10% displacement of the Z-lines and will then keep their position constant while recording the force.

Data Sheet 7.3

Electrical conduction in the heart

Three aspects – the biophysical characteristics of the cardiac conduction system, in particular the different conduction velocities, the different delays in the onset of action potentials and the different durations of refractoriness in the various regions – mean that the contraction of the heart occurs with an extremely precise temporal sequence. The components involved in this function derive directly from the neural crest. During embryogenesis, they are inserted into the myocardium, interdigitating with the normal cardiac cells. The components are (Figure S7.5): the sinoatrial node or Keith-Flack node, the anterior, middle and posterior internodal branches, Bachmann's bundle, the atrioventricular node or Ashoff-Tawara node, the common branch of the Hiss bundle, the left and right branches of the Hiss bundle and the left and right branches of the Purkinje fibers.

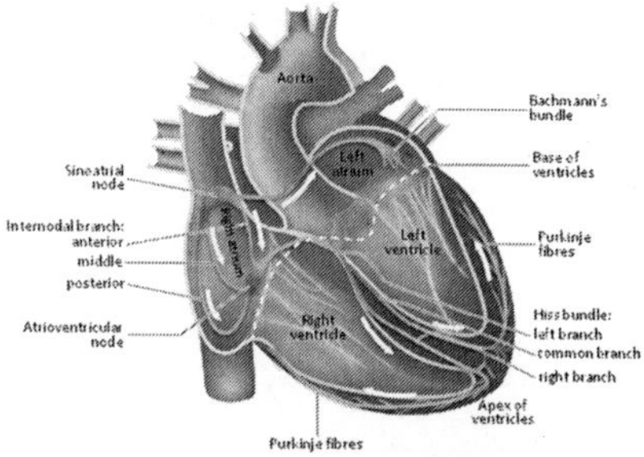

Figure S7.5 Schematic of the conduction system (grey lines) and the way action potentials are propagated (grey arrows) in mammalian hearts.

Sinoatrial node. It is flattened, ellipsoidal, approximately 3 mm wide, 15 mm long and 1 mm thick. It is located in the upper lateral part of the right atrium, just below the outlet of the superior vena cava, and is formed by cardiomyocytes (section 7.4), with a diameter varying between 3 and 5 μm. It is modified in such a way as to almost completely lose contractile capacity. It performs the essential function of a *pacemaker*, i.e., rhythmic stimulation of the heart without the intervention of external stimuli, owing to the particular property that the I_f current is activated by membrane hyperpolarization (section 5.14.2; Chapter 6.5). The I_f current of sodium ions (phase 4 of NS of Figure S7.6) depolarizes the membrane to the

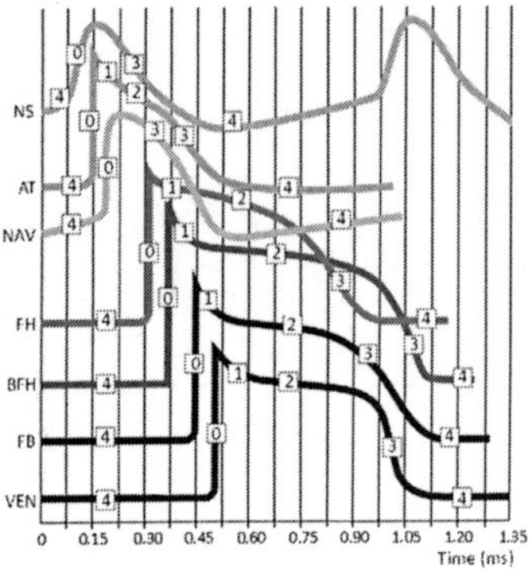

Figure S7.6 Alignment with respect to the time axis of action potentials recorded in the sinus node (SN), atria myocytes (AM), atrioventricular node (AN), common branch of the Hiss bundle (HF), left and right branches of the Hiss bundle (BHF), Purkinje fibres (PF) and ventricular myocytes (VEN). The numbers 0 to 4 identify the phases of the action potentials.

threshold value of the T-type calcium current, about -40 mV, which in turn brings the membrane potential to the threshold of the L-type calcium current, about -55 mV. Calcium current depolarizes the membrane to the peak of the action potential (phase 0 of NS) instantaneously in all active cells of the sinoatrial node. At this time, an outgoing current of potassium ions of the delayed rectifier type (section 5.8) becomes prevalent, which hyperpolarizes the membrane (phase 3 of NS) and activates the I_f current again. This generates a new action potential 1 second after the previous one (if the heart rate is 60 beats per minute), when the cycle of cardiac contraction is completed, and so on every second throughout the life of the organism (Figure S7.7). The duration of the

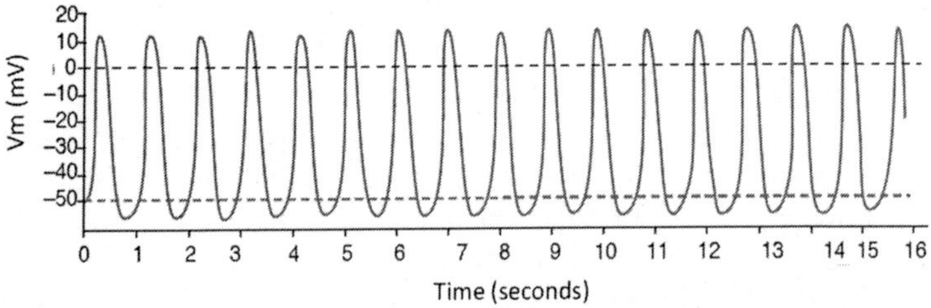

Figure S7.7 16-second microelectrode recording (Data Sheet 5.5) of the activity of a sinoatrial node cell producing action potentials at a rate of approximately 68 per minute.

action potential of the sinus node is produced by the slow depolarization generated by the I_f current (phase 4 of NS) and the subsequent slow depolarization to the peak due to the predominant involvement of calcium currents (phase 0 of NS). This leads to an action potential and

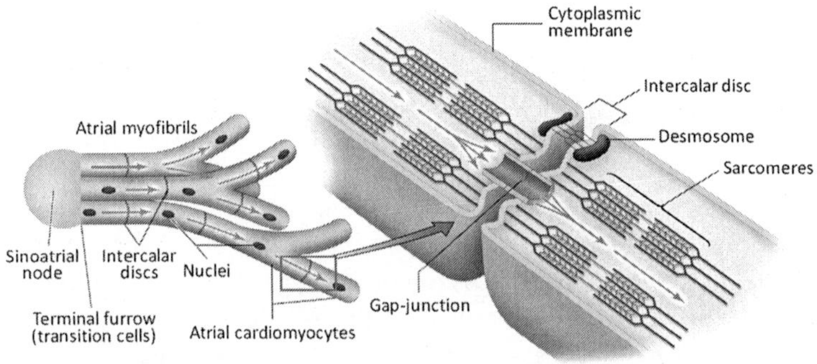

Figure S7.8 Schematic of the mode of diffusion of action potentials (thin arrows) from the sinoatrial node to the right atrium through the transition cells, located near the terminal groove, and the gap junctions that electrically connect the cardiomyocytes of each atrial myofibril.

refractoriness of longer duration than in other tissues. Atrial myofibrils. The cells of the sinoatrial node are directly connected through the transition cells located near the terminal groove to the atrial myofibrils (Figure S7.8 and paragraph 7.4). They are arranged in such a way as to form a sort of preferential conduction pathway: the stimuli from the sinoatrial node permeate, with zero delay (AT of Figure S7.6), the whole of the right atrium (Figure 7.46B), through Bachmann's bundle (Figure S7.5) and the left atrium (Figure 7.46C). The contraction of the atria then occurs (Figure 7.46D). Phase 4 of the atrial action potential, predominantly due to an inward rectifier type potassium current, is followed by phase 0, which leads to a peak in a few microseconds for the voltage-dependent sodium current. Repolarization is caused by an A-type potassium current (phase 1), followed by the plateau of the L-type calcium current (phase 2). This prolongs depolarization and refractoriness, preventing the return to the origin of the action potential, increases the intracellular calcium concentration and activates the contraction cycle (section 7.4.2). The cycle is finally completed by a delayed rectifier potassium current (phase 3) which returns the potential to the resting value.

Atrioventricular node. The stimulus produced by the sinoatrial node is carried to the atrioventricular node along the anterior, middle and posterior internodal branches (Figure S7.5). The branches are formed by modified atrial myocytes arranged in series, lacking contractile activity and organized to form a sort of preferential pathway, with a conduction velocity that creates a delay of approximately 91.91 ms. The atrioventricular node has an ovoid shape, about 3 mm wide, 5 mm long and 1 mm thick, and is located in a subendocardial position near the opening of the coronary venous sinus (Figure S7.5). It has transition cells both in the area of the internodal branches, to receive information from the sinoatrial node, and towards the Hiss bundle. An important function is to prevent the passage of stimuli from the ventricles to the atria, which is accomplished by the presence of a layer of insulating fibrous tissue. In addition, due to the presence of the I_f current, it can replace, for a short time and at a frequency of about 40 beats per minute, the pacemaker function of the sinoatrial node in the event of its malfunction. The action potential of the atrioventricular node is given, as in the sinoatrial node, by the I_f current, the T-type calcium current, the L-type calcium current and the delayed rectifier potassium current (NAV phases 4, 0 and 3 respectively of Figure S7.6).

Hiss bundle and Purkinje fibers. Both conduction systems are formed by modified cardiac myocytes that have almost completely lost their contractile capacity, although the action potential they generate is superimposed on that of atrial myocytes. In fact, in phase 2, the L-type calcium current is present, which gives a prolonged plateau (FH, BFH and FP of Figure S7.6). However, it does not serve to activate contraction, but to prolong the duration of the action potential and refractoriness. As in atrial myocytes, phase 4 is due to the inward rectifier potassium current, phase 0 to the voltage-dependent sodium current, phase 1 to the A-type potassium current and phase 3 to the delayed rectifier potassium current, which returns the potential to the resting value. The action potential of the atrioventricular node stimulates, through the transition cells, the common trunk of the Hiss bundle. The common trunk, about 1 cm long and about 2 mm wide, descends along the ventricular septum and divides into right and left bundles (Figure S7.5). The left bundles, in turn, divide into anterior fasciculus, thicker, and posterior fasciculus, thinner, with a delay of about 124.86 ms (FH in Figure S7.6). The specific biophysical properties and kinetics of the ion channels involved mean that the common trunk of the Hiss bundle stimulates, with a delay of 53.76 ms (BFH of Figure S7.6), the branches and the apex of the heart (Figure S7.5) and, with a delay of about 75.43 ms, the Purkinje fibers (P of Figure S7.6).

Ventricular myofibrils. The ventricular action potential (VEN of Figure S7.6) is given by the currents already described for the atria, Hiss bundle and Purkinje fibers, but with a plateau that, compared to the others already mentioned, remains constant for a long time. This increases the entry of calcium into the cell and the effectiveness of the contraction. The action potential coming from the Hiss bundle is conducted by the Purkinje fibers, progressively and at a suitable speed, from the apex to the base of the heart, so that in the area of the base, it arrives with a delay of 59.83 ms. This creates a contraction wave that starts at the apex and arrives at the base of the ventricles: the blood is pushed with maximum efficiency into the circulatory system.

CHAPTER 8

Transformers 2: Sensors

The contractile apparatus transforms chemical energy into mechanical work (Chapter 7), but transformers can also be defined as all the biological apparatuses able to transform different forms of energy into electrical messages to be interpreted by the nervous system. This is the case of the different sensory systems present in every organism that can receive stimuli from different origins. Mechanical, chemical and visual stimuli, among others, are environmental cues that any biological organism can interpret to different extents.

The process of energy transformation occurs every time a signal from the environment needs to be interpreted by the biological organism, usually via the nervous system. The interpretation and the organization of an adequate response also implies a final step in which electrical signals are converted in mechanical energy. Reception of the different stimuli is organized in special structures able to convert the intensity of an interaction into action potential frequency in the peripheral sensory neurons. The response always implies a conversion of action potential firing into muscle contraction, as we have seen in Chapter 7. From a functional point of view, the sensory endings have a cellular structure and organization capable of perceiving analogue information, i.e., stimuli of variable intensities, and converting them into digital signals. As already seen in Chapter 6, the language used by the nervous system consists of all-or-nothing events, action potentials, which are frequency-coded. Communication, both in and out of the biological organism, uses the same system. Analogue signals from the external environment are transformed into digital signals (neuronal firing, section 6.4), which in turn are transformed back into analogue signals to the outside of the organism in the case of a response to a stimulus.

Sensors have two modes of operation: an instinctive type and a voluntary type. The first is essentially an automatic and immediate response to an external stimulus, produced by a set of elements that form a circuit affecting only a limited part of the organism. An example of this behavior is that of the reflex arc in reaction to a painful stimulus, one of the primary defense mechanisms of organisms. The sensors transform the analogue pain sensation into a digital signal which, through a preferential pathway, reaches the central nervous system. The nervous system, in turn, generates an appropriate defense action that directly reaches the muscular apparatus. The motor system is immediately activated and an analogue response is generated. In this case, between the sensory stimulation(s) and the muscular response, there is an intermediate phase that passes through interconnections between digital elements in the central nervous system that integrate and process the information.

In general, transformers appear to be very complex structures, whose cells are extremely specialized in order to act as an interface between the external environment and the organisms' internal information and interpretation circuits. The differences between the various sensors lie in the different forms of energy they must acquire and transform. The diversity of forms of energy has led to the selection of structures that are also profoundly different according to their specific function.

In vertebrates, and particularly in mammals and humans, there are sensors for smell, taste, touch, sound, balance and light, with an organization that is always very similar in the various organisms, but often with different ranges of sensitivities.

8.1 Olfactory sensors

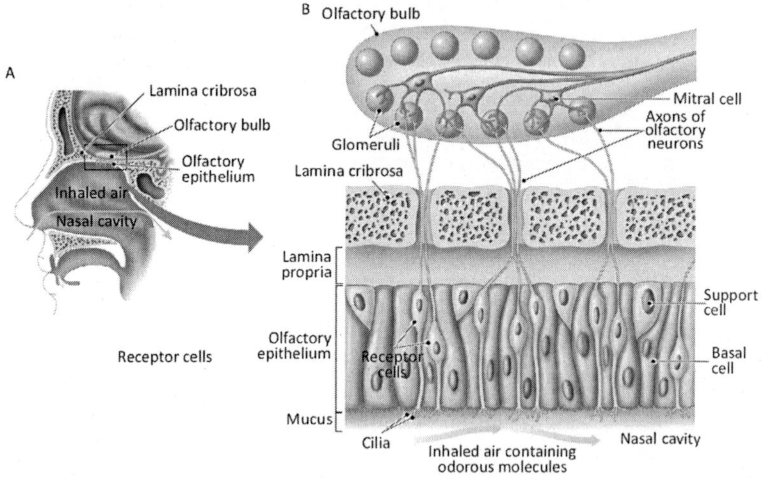

Figure 8.1 Anatomical layout of the olfactory system (A) and schematic of the olfactory bulb and epithelium (B).

The olfactory epithelium (Figure 8.1B) is generally located in the roof of the nasal cavity (Figure 8.1A) and consists of supporting cells and receptor neuronal cells, which are continuously replaced by differentiating progenitor basal cells. The axons of the receptors ascend towards the olfactory bulb and synapse with the dendrites of the mitral cells in a structure called the glomerulus; the axons of the mitral cells form the olfactory nerve, which runs in the first pair of cranial nerves up to the olfactory tubercle. Afferents depart from the olfactory tubercle towards the thalamus and from there towards the olfactory cortex. The receptor cells are equipped with cilia which increase the contact surface of the cell with odorous molecules and are exstended in the mucus that covers the part of the epithelium that faces the nasal cavity.

8.1.1 Transformation of the chemical stimulus into an electrical signal.

Airborne odor molecules diffuse through the mucous layer, either directly or via specific transporter molecules in the mucus, to receptor proteins that are only present in the cilia membrane. The binding of an odorous molecule to its receptor causes a G-protein (Figure 8.2) typical of olfactory neurons to be activated; the βγ subunit remains attached to the receptor while the α subunit binds to a GTP molecule and activates adenylate cyclase. To increase the likelihood of this sequence of chemical reactions, the receptor with the specific G-protein and adenylate cyclase are placed very close together in the membrane. Active adenylate cyclase converts ATP into cAMP, which acts as a second messenger and activates cAMP-dependent cation channels. Calcium and sodium ions enter through the membrane and the cell depolarizes, also due to the release of chloride through specific calcium-activated channels. Depolarization, if above the threshold, generates action potentials along the axons of the receptor cells and finally along the axons of the olfactory nerve, with a frequency proportional to the number of odor molecules that have interacted with the receptor molecules.

The higher the concentration of odor molecules, the greater the likelihood of receptor cells sending action potentials to the glomeruli; each receptor cell will be more depolarized and send action potentials to the cortex at a higher frequency.

When the binding of odor molecules to their receptors ceases, the activity of a phosphodiesterase that hydrolyses cAMP to 5'-AMP prevails in the cell: the decreased concentration of cAMP stops calcium and sodium from entering the cell due to the closure of the cation channels. In addition, the calcium that previously entered and bound to calmodulin (Ca^{2+}-CAM of Figure 8.2) reduces the affinity of Ca^{2+}-dependent chloride channels for calcium. Finally, calcium concentration is reduced by a membrane-dependent Na^+/Ca^{2+} ATPase that excretes calcium by exchanging it for sodium.

The increase in intracellular calcium can also be caused by another enzymatic pathway not shown in Figure 8.2. After binding of the odor molecule to the receptor and activation of the olfactory G-protein, the GTP-bound α-subunit activates phospholipase C, which cleaves phosphatidylinositol 4,5-biphosphate (PIP2) in the plasma membrane into diacylglycerol (DAG) and inositol triphosphate (IP3). DAG in turn activates protein kinase C, while IP3 opens the calcium channels on the endoplasmic reticulum membrane and releases the divalent ions from the stores.

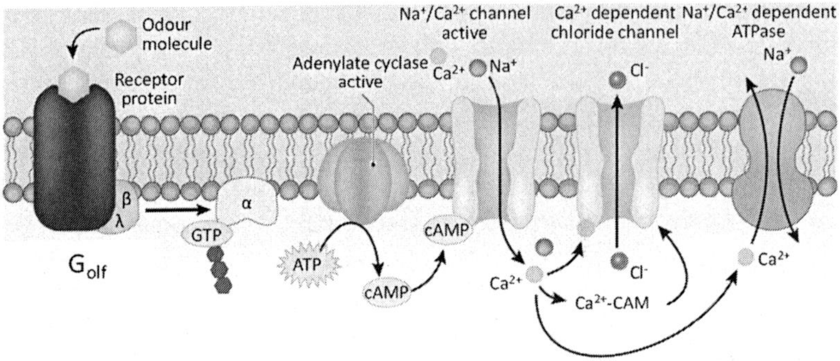

Figure 8.2 Schematic representation of the transformation of the chemical signal into a nervous signal in the olfactory system. G_{olf} = olfactory G-protein.

In the olfactory epithelium, there are all the same receptor cells, but with as many types of receptor proteins as there are classes of odorous substances to be identified. When an odorous molecule binds to its receptor molecule, the same enzymatic cascade is activated for all receptor cells (Figure 8.2). The recognition of the odorous substance is done at the level of the olfactory cortex, where the specificities of all the receptor cells are mapped during development, so that the activation of a certain cell, or certain group of cells, is interpreted as the presence of a particular odorous molecule. Another interesting aspect that characterizes the olfactory apparatus is that of adaptation, whereby the continuous and highly concentrated presence of a certain odorous molecule leads to a decrease in the frequency of the action potentials, which reduces the subject's perception of that odor.

8.2 Taste sensors

The function of the taste sensors is to perceive the flavors of ingested food, provided these are in solution with saliva, so that they can easily diffuse towards the receptor cells. This function is carried out by specialized structures located in the epithelium of the upper face of the tongue and in the epiglottis, called taste buds. The taste buds are of three basic types: the circumvallate papillae, arranged in a V shape in the posterior part of the tongue, the foliate papillae, located at the lateral margins of the tongue, and the fungiform papillae scattered in the anterior part of the tongue (Figure 8.3A).

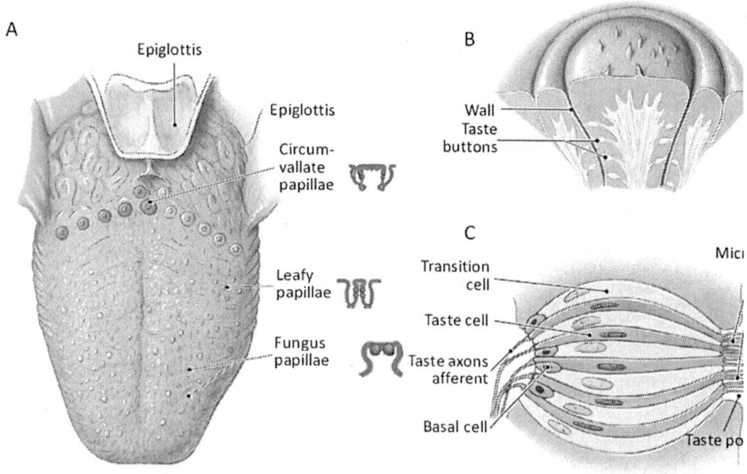

Figure 8.3 Schematic of the taste sensory system. A) Distribution of taste buds in the human tongue. B) A circumvallate papilla. C) A taste bud.

It is believed that there are six primary tastes: sour, sweet, salty, bitter, fatty and umami, the Japanese word for 'savory', which refers to the taste of glutamate, an amino acid found particularly in meat, cheese and protein-rich foods in general. Humans, however, can perceive and distinguish a much wider and more complex range of flavors, because flavors stimulate the various taste buds in different ways and the central nervous system has developed the ability to assess the combination of various flavors and their intensity.

The gustatory papillae, including the circumvallates, comprise about 10,000 taste buds of about 50 µm in diameter. Each has a gustatory pore with microvilli in the apical part of the true gustatory cells that form synapses directly with gustatory afferent axons. Also present in the gustatory button are basal cells, which originate from the epithelial cells surrounding the papilla and differentiate into new gustatory cells to replace those that are aging.

The gustatory afferent nerve arrives at the gustatory nucleus of the bulb with the 7th, 9th and 10th pairs of cranial nerves. Second-order neurons then arrive in the thalamus where they synapse with third-order neurons that terminate in the gustatory cortex.

8.2.1 The transformation of the chemical stimulus into an electrical signal

How the chemical signal is transformed into an electrical signal varies according to the perceived taste. For clarity, consider four of the six main tastes: sour, salty, sweet and bitter (Figure 8.4).

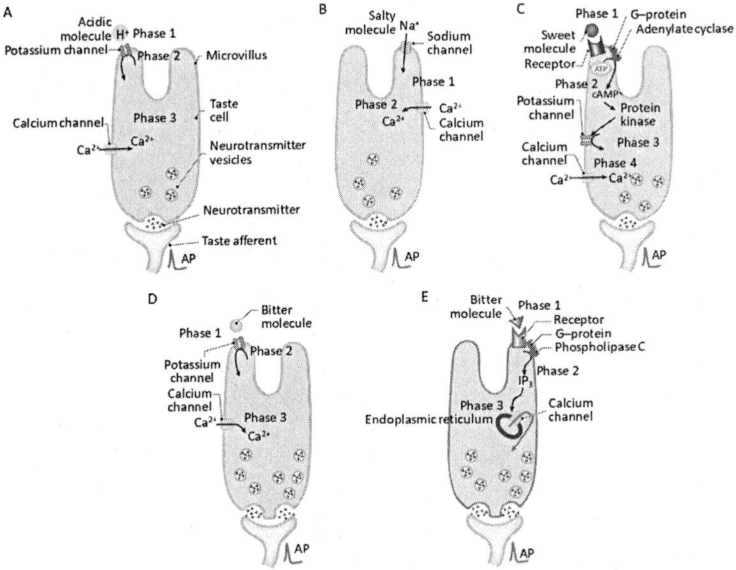

Figure 8.4 Schematic of the transformation by specific taste cells of the chemical signals acidic (A), salty (B), sweet (C) and bitter (D, E) into action potentials (AP) along the gustatory afferent axon.

Acidic taste is due to the presence of hydrogen ions in solution that bind to and block specific potassium channels (phase 1 and phase 2 of

Figure 8.4A). Therefore, the hyperpolarizing potassium current, which maintains the membrane potential at negative values, decreases until it stops. This results in depolarization of the receptor cell, activation of voltage-dependent calcium channels, entry of calcium and an increase in its intracellular concentration (phase 3 of Figure 8.4A).

Salty taste is due to the presence of sodium ions in solution. These enter the cell through specific channels present in the membrane of the taste cell (phase 1 of Figure 8.4B). There is then depolarization of the receptor cell, activation of the voltage-dependent calcium channels, entry of calcium and an increase in its intracellular concentration (phase 2 of Figure 8.4B).

Sweet taste is due to the presence in solution of organic molecules similar in structure to sucrose. The sweet molecule binds to specific membrane receptors (step 1 in Figure 8.4C) that activate a specific G-protein, gustoducin, which in turn activates adenylate cyclase, which converts ATP into cAMP (step 2 in Figure 8.4C). A protein kinase uses cAMP to phosphorylate the membrane potassium channels, which are blocked (phase 3 of Figure 8.4C), resulting in depolarization due to a decrease in the hyperpolarizing outgoing current of potassium and the entry of calcium (phase 4 of Figure 8.4C).

Bitter taste is due to many different nitrogen-containing molecules, some of which are also toxic, to the extent that animals generally tend not to ingest them. They interact with the taste cell by two distinct mechanisms. Like hydrogen ions (Figure 8.4A), a group of bitter molecules, including quinine, for example, binds to potassium channels (phase 1 of Figure 8.4D), blocking them (phase 2 of Figure 8.4D); depolarization and calcium entry then occur (phase 3 of Figure 8.4D). Other bitter molecules bind, in contrast, to specific membrane receptors (phase 1 of Figure 8.4E), and the receptor activates an enzymatic cascade with the activation of a specific G-protein, transducin. The cascade in turn activates phospholipase C, which catalyzes the formation of inositol triphosphate (IP3, phase 2 of Figure 8.4E), which induces the release of calcium from the endoplasmic reticulum (phase 3 of Figure 8.4E).

A particular behavior seems to be the umami taste, associated, as mentioned, with glutamate, which does not seem to provide a real taste, but rather accentuates others. It seems to bind to specific sodium channels in taste cells, inducing the ion's entry and depolarization of the cell, thus opening voltage-dependent calcium channels and increasing intracellular calcium.

Whatever the mechanism with which the gustatory molecule interacts with the taste cells, there is an increase in the concentration of intracellular calcium that activates the release of the excitatory neurotransmitter. This release, in turn, depolarizes the postsynaptic cell, which generates action potentials along the axon of the afferent neuron (Figure 8.4) with a frequency that is proportional to the magnitude of the chemical signal perceived by the taste cell.

8.3 Touch sensors

Touch sensors are mechanoreceptors located mainly in the skin, where they perceive non-painful mechanical deformations caused by local variations in pressure or tension. There are four main types (Figure 8.5): Pacinian corpuscles, Meissner's corpuscles, Merkel's discs and Ruffini's corpuscles.

Tactile sensations are carried from the skin to the central nervous system by four afferent nerve fibers, each of which corresponds to one of the four main types of receptors and conveys information from several tactile receptors of the same type.

8.3.1 Pacinian corpuscles

The Pacinian corpuscles (Figure 8.5A), named after their discoverer, the Italian anatomist Filippo Pacini (Pistoia 1812-Florence 1883), are formed by the unmyelinated initial part of an afferent axon covered by about seventy concentric layers of connective tissue, arranged like the layers of an onion, with solution between the layers of the same composition as that outside the cell.

They are located in the deepest portion of the skin, with large receptive fields and blurred edges. They are rapidly adapting receptors, larger than other receptors, and are the most sensitive of the tactile sensors, capable of detecting skin depressions in the order of a micrometer and high-frequency vibration applied tangentially to the skin with an amplitude of only 10 nm. The particular sensitivity of Pacinian corpuscles means that the afferent axon has action potentials with a frequency proportional to the frequency of the stimulus, while the intensity or amplitude of the stimulus is encoded by the number of activated receptors.

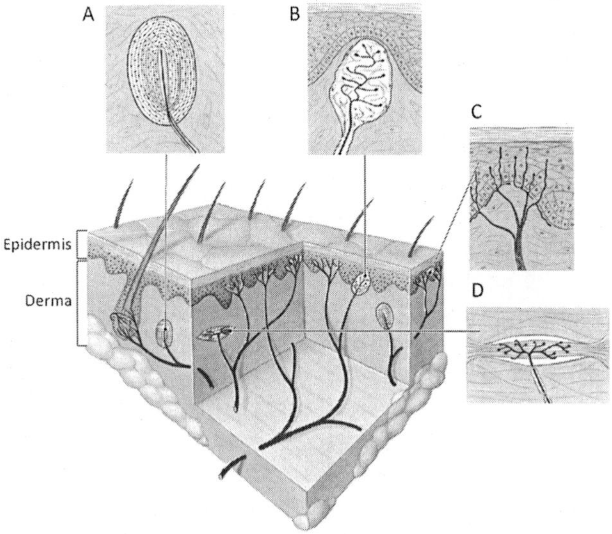

Figure 8.5 Schematic of the outermost part of the skin and Pacinian corpuscles (A), Meissner's corpuscles (B), Merkel's discs (C) and Ruffini's corpuscles (D).

8.3.2 Meissner's corpuscles

Meissner's corpuscles (Figure 8.5B), named after the German anatomist Georg Meissner (Hanover, 1829-1905), are composed of encapsulated oval-shaped nerve endings, located in the superficial layer of the dermis. They have small receptive fields, with a transverse diameter of less than 5 mm and sharp margins. They are particularly numerous in areas of thick, hairless skin. They have an almost cylindrical shape, with a length of about 80 μm and a diameter of about 30 μm, and are made up of a connective lining, inside of which are epithelial cells enveloped by the afferent terminal. They are rapidly adapting receptors, i.e., in the presence of continuous stimulation of the skin, the axon responds only at the beginning and end to changes in the pressure exerted. These receptors are sensitive to low-frequency vibrations generated by direct, active contact with objects, and are of great importance in controlling the force with which an object to be lifted is grasped. They are also involved in the reception of fine movement and the smallest perceptible irregularities on the surface of an object.

8.3.3 Merkel's discs

Merkel's discs (Figure 8.5 C), also called Merkel's corpuscles after the German anatomist and histologist Friedrich Sigmund Merkel (Nuremberg 1845-1919), are slow-adapting mechanoreceptors located superficially (close to the epidermis). They are formed by the expansions of nerve endings and appear to be globular epithelial cells arranged in clumps surrounding the afferent endings. They have small receptive fields and are selectively sensitive to local deformations of the skin caused by contact with tips, edges and curved surfaces and are responsible for fine, small tactile sensitivity. Unlike other types of sensors, made of modified nerve cells that form synapses with axons of afferent neurons, all touch sensors are made directly from axons of afferent nerve cells coated with myelin (Figure 6.11), which is generally lost near the actual receptor zone. The afferent nerve endings are embedded in a specific structure closely related to function, such as the connective lining and epithelial cells for Meissner's corpuscles, the numerous branched afferent nerve endings in Ruffini's corpuscles, the concentric layers of connective tissue for Pacinian corpuscles and the globular epithelial cells in Merkel's discs. The afferent fibers have a diameter of between 5 and 15 μm and a conduction velocity of 35 m/s to 75 m/s. They are the peripheral branch of T-shaped neurons, with a cell body in the dorsal root ganglia and project, via the dorsal roots of the spinal cord and specific spino-thalamic pathways, to the ventro-postero-lateral nucleus of the thalamus. In turn, the thalamus projects to the postcentral circumvolution of the ipsilateral parietal lobe, which contains the primary receiving areas for touch sensitivity, then to the insula of the cerebral cortex and to the frontal lobe.

8.3.4 Ruffini’s corpuscles

Ruffini’s corpuscles (Figure 8.5D), which are slow-adapting sensors, i.e., they continue to send impulses for as long as they are stimulated, are located in the deeper layers of the dermis and are named after the Italian physician and histologist Angelo Ruffini (Arquata del Tronto 1864-

Baragazza 1929), who discovered them. Spindle-shaped, they are formed by enlarged, highly-differentiated nerve endings with complex branching. They are surrounded by Schwann cells and composed of a variable number of cylinders, each of which consists of a branch of the afferent nerve ending.

Ruffini's corpuscles are very sensitive and are stimulated by stretching the skin. They are present at high density at the joints and near the nails and are characterized by large receptive fields – therefore being implicated in coarse tactile perception. They are thought to be useful in controlling the sliding of objects along the surface of the skin, thus allowing the modulation of the grip on an object.

They also act as thermoreceptors, specializing in thermal sensitivity to heat for temperatures between 20°C and 45°C, with a maximum sensitivity between 37°C and 40°C. This sensitivity decreases dramatically at 45°C, but increases again for temperatures between 46°C and 50°C. This particular property leads to the phenomenon of the so-called 'cold paradox', whereby for water temperatures around 45°C, one does not feel that the water is too hot.

8.3.5 The transformation of mechanical stimulus into electrical signal

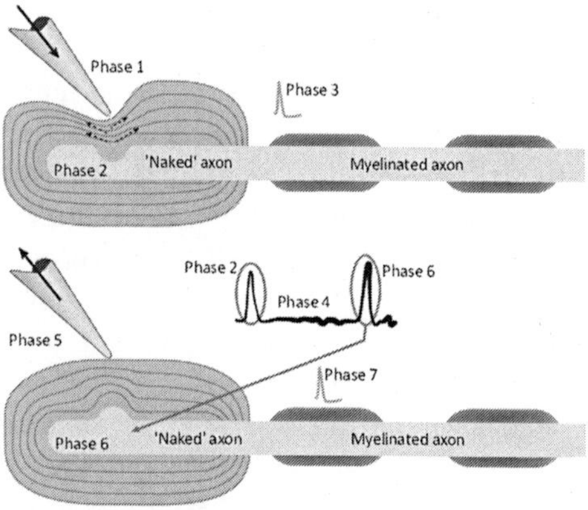

Figure 8.6 Schematic of the Pacinian corpuscle. Both mechanical stimulation (phase 1) and its cessation (phase 6) produce a transient change in membrane potential and an action potential along the axon (phase 3 and phase 7).

A fairly simple mechanism to transform a mechanical stimulus into an electrical signal is used by the Pacinian corpuscle (Figure 8.6). The pressure exerted on the skin, for example by an inhomogeneity on a surface or an object, is transmitted to the connective tissue lamellae (phase 1). As the lamellae are crushed, the

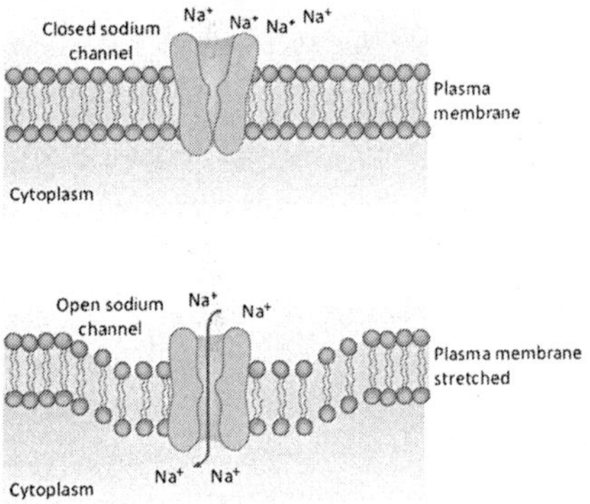

Figure 8.7 The deformation of the plasma membrane following stimulation of the Pacinian corpuscle causes the entry of sodium ions, which depolarize the cell.

hydrostatic pressure of the solution increases, which leads to deformation of the plasma membrane of the axon (Figure 8.7). A very important role is played by the cytoskeleton, which is anchored to the membrane and the ion channels, generally called stretch-activated ion channels and specifically, stretch-sensitive sodium channels. Deformation of the plasma membrane causes sodium channels to open directly through the cytoskeleton. These let in sodium, which depolarizes the membrane with an electrotonic current (phase 2) to the first node of Ranvier where the action potential is formed (phase 3). Depending on the intensity of the mechanical stimulus, bursts of action potentials will occur, with a frequency directly proportional to the electrotonic current induced by the activation of stretch-activated ion channels. The signal generated propagates along the afferent axon. For persistent signals, there is the phenomenon of adaptation (phase 4): the maintained mechanical stimulus causes depolarization and an increase in cytoplasmic calcium, but when the latter reaches a particular concentration, it activates calcium-dependent potassium channels that repolarize the cell, effectively terminating the tactile stimulus. When the object placed on the skin is moved away (phase 5), the lamellae return to their resting position due to their elasticity, the fluid undergoes a depression, and the axon membrane deforms towards the lamellae and causes the sodium channels to reopen. This leads to a new depolarization (phase 6) and the formation of a new action potential that is conducted towards the central nervous system (phase 7).

8.4 Sound sensors

In the vertebrate ear, and in particular in humans, we distinguish three functional zones, each with particular structures and functions: the outer ear, the middle ear and the inner ear (Figure 8.8).

The auricle and the external acoustic meatus, or acoustic canal, form the outer ear, which is filled with air. It has the function of collecting and conveying sound waves, which are caused by a series of compressions and decompressions of the air, to the inner parts of the ear.

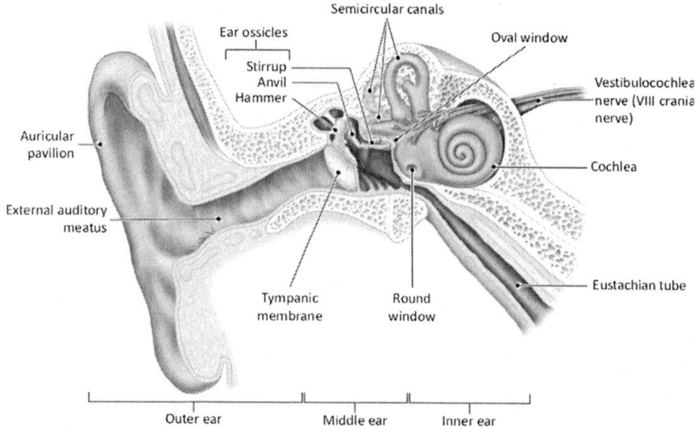

Figure 8.8 Schematic anatomy of the human ear.

The middle ear, filled with air, is formed by the tympanic membrane, which separates it from the outer ear. Its function is to oscillate in the direction of the cranium when the sound wave is in the compression phase and in the distal direction when the sound wave is in the decompression phase (Figure 8.9). The oscillations to the chain are transmitted to the combination of small bones (ossicles), i.e., the hammer, the anvil and the stirrup, which amplify them and transmit them to the inner ear. In the middle ear, there is also the mouth of the Eustachian tube, a canal that flows into the pharynx with the function of compensating for any excessive changes in pressure due to sound waves.

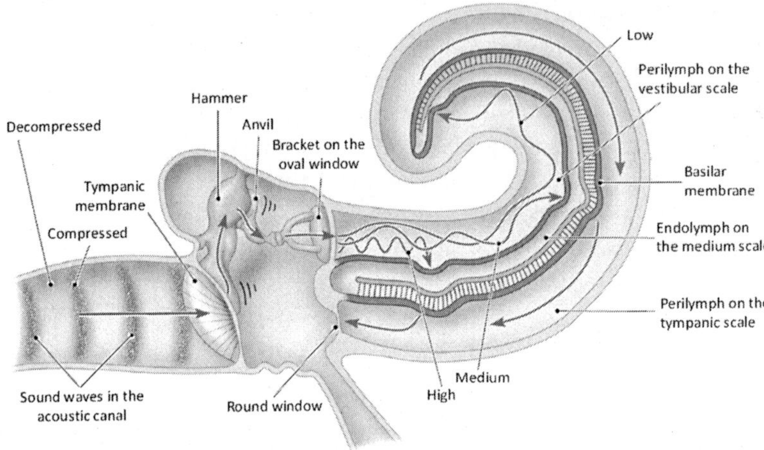

Figure 8.9 Schematic organization of the cochlea with arrows highlighting the path of the sound wave in the middle ear, the path of the oscillations produced in the scala vestibuli and the perilymph flowing up to the round window. High, medium and low refer to the frequencies of the perceived sounds.

The inner ear includes the oval window, which receives the oscillations from the chain of ossicles, the round window, which has the function of compensating for the pressure of the inner ear, and the cochlea, the seat of the transformation of the mechanical signal due to sound into an electrical signal. The inner ear also contains the semicircular canals of the vestibular apparatus, which are involved in the mechanisms for regulating balance (section 8.5).

The cochlea (Figure 8.8) is a spiral structure with a pseudo-cylindrical shape formed by two external canals in chemical continuity with each other, the scala vestibuli and the scala tympani. These structures contain the perilymph, which has a composition similar to that of extracellular fluid, i.e., 140 mM sodium ions, 120 mM chloride ions and a small amount of protein, potassium ions and glucose (Figure 8.9). The third channel forming the cochlea is the scala media containing the endolymph, a physiological solution consisting of potassium at a concentration ranging from 150 to 180 mM, 150 mM chloride, about 1 mM sodium, calcium ranging from 0.02 to 0.20 mM, 0.01 mM magnesium, less than 0.6 mM glucose and 0.6 g/L of protein.

The neurons of the cochlear nerve emerging from the organ of Corti (Figure 8.10) project into the cochlear nuclei of the brainstem, from which neurons depart towards the medial geniculate body of the thalamus. From there, other neurons project towards the auditory cortex of the temporal lobe of the brain.

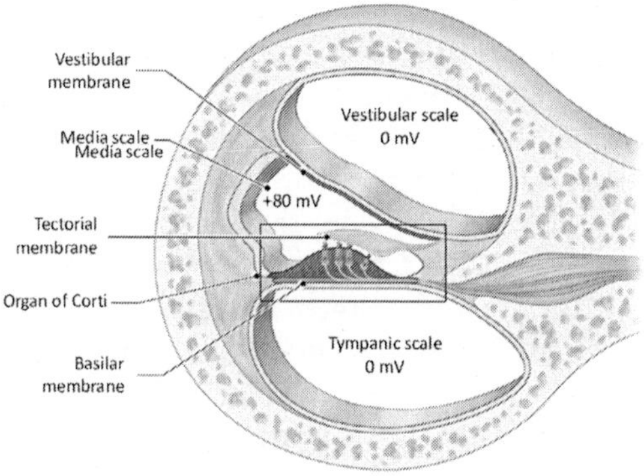

Figure 8.10 Schematic cross-section of the cochlea. The vestibular scale tympani and media scale with the organ of Corti are highlighted.

8.4.1 The transformation of the sound stimulus into an electrical signal

The scala media is the site of the transformation of sound into an electrical signal. It is formed by a vestibular membrane, which separates it from the scala vestibuli, a basilar membrane towards the scala tympani, and the cochlear nerve that emerges from the organ of Corti (Figure 8.10), which comprises the Corti hair cells, the tectorial membrane and supporting cells. All these structures are parallel to each other and run parallel to the scala vestibuli and scala tympani along the length of the cochlea (Figure 8.9).

The cells of Corti, which are the true sensors for sound, are cells that have two rows of cilia neatly arranged in a V (Figure 8.11) inserted firmly into the tectorial membrane, so that the major axis of the V coincides with its transversal axis (white arrow of Figure 8.11). The set of cilia of each cell is composed of a kinocilium, the longest and positioned at the apex of the

V, and numerous stereocilia, of decreasing length, arranged in two rows starting from the kinocilium.

The pressure oscillations transmitted from the ossicular chain to the oval window are in turn transmitted to the perilymph and produce a series of swellings and narrowing of the scala vestibuli. These, in turn, cause the lowering or raising of the scala media (grey double arrow in Figure 8.12) and, because of its particular disposition in relation to the rest of the structure, there is an oscillation of the tectorial membrane. The oscillation occurs in one direction or the other (black double arrow of Figure 8.12) depending on whether there is a swelling or a narrowing in the scala vestibuli. Following the direction of the oscillations, stereocilia bend towards the kinocilia or in the opposite direction (white double arrow of Figure 8.11). The cilia of Corti cells, which form synapses with the axons of afferent neurons, have specific, stretch-activated potassium channels, i.e., channels that are sensitive to membrane stretching (Figure 8.13). When the cell is at rest in the absence of sound (Figure 8.13A), there are a number of open potassium channels, and potassium enters the cell by a

Figure 8.11 Scanning electron microscope image of a fragment of the organ of Corti showing the Corti cells, the direction of the transverse axis of the tectorial membrane (white double arrow) and the V-arrangement of the cilia (in close proximity to the double arrow).

concentration gradient (section 8.4), depolarizing the membrane and activating calcium channels present in the cell body membrane. The entry of calcium releases neurotransmitter that depolarizes the axon of the afferent neuron, with production of action potentials with a basic frequency. In the presence of sound, the oscillation of the tectorial membrane (Figure 8.12) makes the cilia move towards the kinocilia or in the opposite direction. In the first case, compared to when there is no sound, the number of open potassium channels increases, the cell depolarizes and calcium permeates through calcium channels, increasing the intracellular concentration. The consequence is an increase in neurotransmitter release that induces a higher firing frequency in the afferent fiber (Figure 8.13B). In the second case, there is less depolarization due to a decrease in active potassium channels, lower intracellular calcium concentration and downregulation of synaptic release, reducing the action potential frequency (Figure 8.13C).

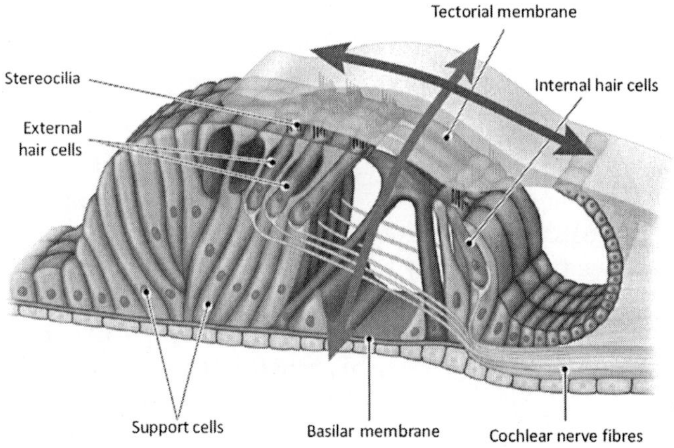

Figure 8.12 Schematic representation of the organ of Corti.

The frequencies that the human ear is able to perceive are between 20 and 20,000 Hz (Hertz = oscillations/second), with greater acoustic sensitivity between 1000 and 4000 Hz. The cochlea responds to the high frequencies in the area close to the oval window of the first branch of the scala vestibuli, which presents a more rigid and narrower basilar membrane (high in Figure

8.9). The low frequencies are perceived towards the terminal zone of the first branch of the scala vestibuli, with a less rigid and wider basilar membrane (low in Figure 8.9). The response to intermediate frequencies occurs in the intermediate stretch (medium in Figure 8.9).

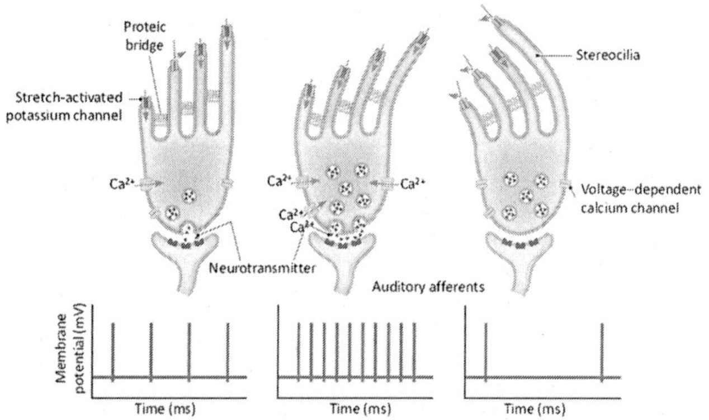

Figure 8.13 Corti cell. Schematic representation of the effect of mechanical stimulation. Explanations are in the text.

In the auditory cortex, mapping of the afferents of the organ of the Corti make it possible to recognize the characteristics of the perceived sound, while the intensity of the sound is encoded by the frequency of action potentials. Firing frequency is higher for more intense sounds, which have deformed the organ of the Corti more, have made the tectorial membrane oscillate to a greater extent, and have moved the cilia of the Corti cells more, with greater depolarization, greater release of neurotransmitter and high-frequency action potentials.

8.5 Balance sensors

The balance sensors of the vestibular apparatus have the function of sensing the acceleration of the body and the position of the head to maintain balance and equilibrium of the body.

The vestibular apparatus (Figure 8.14), located in a cavity of the temporal bone, is formed by three semicircular canals, in addition to the utricle and the sacculus. The endolymph, a physiological solution containing, among other things, potassium at a concentration that varies

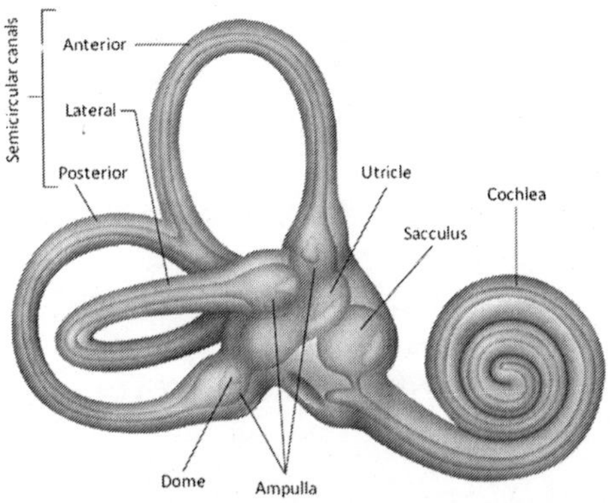

Figure 8.14 Schematic of the vestibular apparatus showing the reciprocal position of the three semicircular canals. The cochlea of the auditory apparatus has a close relationship with the vestibular apparatus.

from 150 to 180 mM. Canals are arranged in such a way that each is on a plane perpendicular to the other two, so the organism can perceive the rotational acceleration of the head in the three planes of space and, for each plane, in two directions. The utricle, on the other hand, detects the linear acceleration of the head forward and backward or left and right in the frontal plane. The sacculus detects linear acceleration upwards and downwards on the frontal plane. Figure 8.15 shows the arrangement of the rows of hair cells in the semicircular canals (A, B, C), in the sacculus (D) and in the utricle (E) and the reciprocal position in space of their ampullae, which allows total control of the position of the body.

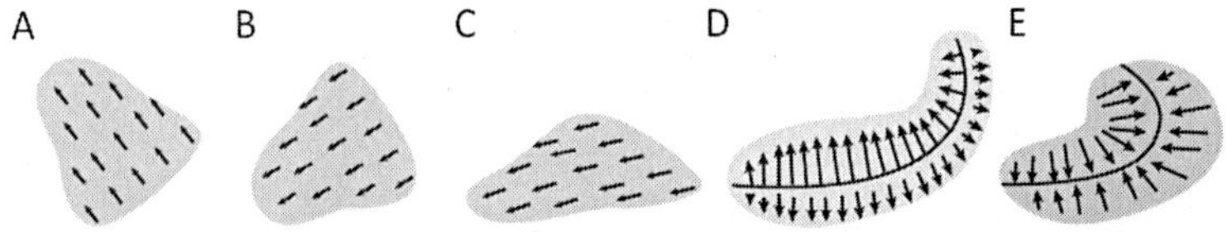

Figure 8.15 Mutual position in space of the ampullae of the three semicircular canals (A, B, C), the sacculus (D) and the utricle (E). The arrows indicate the orientation of the cell cilia.

At the base of each semicircular canal (Figure 8.16), there is the ampulla comprising the dome, a gelatinous structure separated from the endolymph by a membrane and, intercalated with supporting cells, the hair cells. Each hair cell has a number of stereocilia and a larger kinocilium, as in the organ of the Corti (paragraph 8.4.1).

The utricle and the sacculus are relatively large structures, located between the semicircular canals and the cochlea (Figure 8.14). They have well-ordered rows of cells inside, with stereocilia and a kinocilium that extend inside a gelatinous substance. On top, there are the otoliths, calcium carbonate crystals that increase the mass of the gelatinous substance. The hair cells of the utricle are arranged horizontally with respect to the main axis of the head, with the stereocilia oriented vertically, while the hair cells of the sacculus are arranged vertically with the stereocilia oriented horizontally. This arrangement causes the utricle to perceive linear acceleration forwards and backwards, while the sacculus perceives linear acceleration upwards and downwards. The vestibular afferents (Figure 8.16), in the form of the vestibular nerve, reach the vestibular nuclei of the brainstem, from which some proceed directly to the cerebellum to provide immediate information on balance and the body's center of gravity in order to coordinate motor activity. However, the bulk of the sensory information is carried from the vestibular nuclei towards the cortex by secondary sensory neurons.

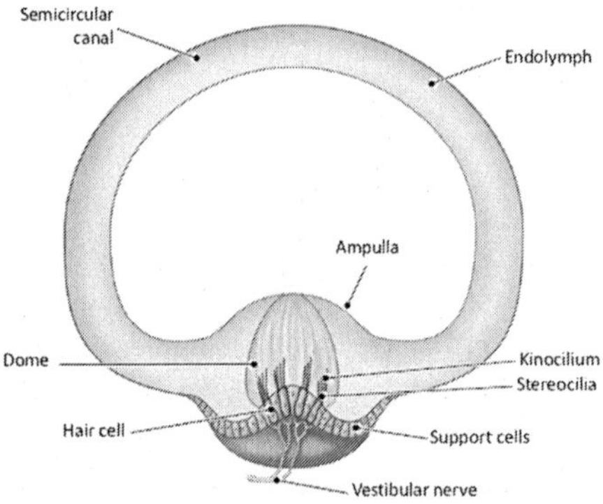

Figure 8.16 Schematic representation of a semicircular canal.

8.5.1 The transformation of mechanical stimulus into electrical signal

The hair cells of the vestibular apparatus have cilia with the same functional characteristics as the Corti cells (Figure 8.13), with potassium channels (known to be the Kv7.4 type) that pass from the closed state to the open state and vice versa depending on the degree of deformation of the cilia membranes, caused by the linear acceleration of the head. If the

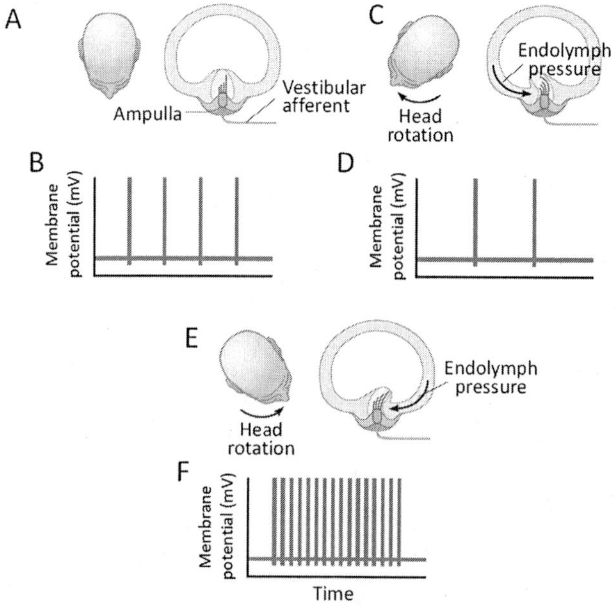

Figure 8.17 A) Resting position of cilia in the absence of head rotation. Movement patterns of cilia in the presence of head rotation (C, E) and electrical responses (B, D, F) in vestibular afferents.

head is stationary (Figure 8.17A), irrespective of its position in space, the cilia are in a resting position. There is a low probability that the potassium channels are open, and thus a slight basal depolarization and neurotransmitter release at a level such that the vestibular afferent sends action potentials with a low frequency (Figure 8.17B). On the other hand, when the head rotates to the right (Figure 8.17C), the endolymph moves in the direction of the black arrow. The cilia in the ampulla flex, because they are partially blocked by the gelatinous substance and the otoliths. Flexion is in the opposite direction of the stereocilia. The probability that the potassium channels are open decreases (Figure 8.13 C), depolarization and the release of neurotransmitter both decrease and the vestibular afferent sends action potentials with a very low frequency (Figure 8.17D). If the head rotates to the left (Figure 8.17E), the endolymph moves in the direction of the black arrow, the cilia in the ampulla flex in the direction of the stereocilia and the probability of the potassium channels being open increases (Figure 8.13B). Depolarization and neurotransmitter release increase and the vestibular afferent sends action potentials with a very high frequency (Figure 8.17F).

Obviously, if the rotation of the head stops abruptly, the hair cells undergo opposite flexions to those caused by the start of the rotation. For a few moments, there is the sensation that the head, instead of stopping, is moving in the opposite direction than previously.

8.6 Light sensors

In the eye, there are two fundamental zones made up of particular structures that carry out highly specialized functions (Figure 8.18). The anterior part of the eye consists of the cornea, which is transparent and allows light to penetrate, and the anterior cavity, which is full of aqueous humor, supplying nutrients to the cornea and the crystalline lens. In front of the crystalline lens is the iris, formed by two layers of pigmented muscle cells responsible for the color of the eyes. The iris is a dynamic diaphragm able to modulate the pupil diameter. The pupil, a hole located in the center of the iris, regulates the amount of light to be sent towards the retina. Finally, image focusing is determined by the ciliary muscles of the ciliary body.

The body of the eye consists of 3 main parts. The vitreous chamber is the prominent part of the eye that contains the vitreous humor, a gelatinous substance that helps to maintain the spherical structure of the eye. The sclera is an opaque connective tissue that has the function of support and containment; the anterior part is continuous with the cornea. At the back of the vitreous chamber is the retina. The light coming from the pupil is conveyed to the fovea, the central point of the retina and the area with the greatest visual acuity. The retina has a dark layer, the "pigmented epithelium", due to the presence of melanin, which absorbs light that has passed through the retina and any light reflected from the back of the eye. In the lower part of the retina is the optic disc, the point at which the optic nerve and blood vessels cross the walls of the eye. The optic disc is a blind spot because it has no photoreceptors. The last element of the retina is the choroid, a layer of highly vascularized tissue, which performs a trophic function for the photoreceptors.

A light ray is made up of photons, particles comparable in their physical properties to electrons, which oscillate in space and propagate in a straight line at a constant speed of $3\cdot10^5$ km/s. It is well known that photons oscillate at a frequency ranging from 1.33 MHz, corresponding to the 750 nm wavelength of red, to 2.86 MHz, corresponding to the 350 nm wavelength of violet.

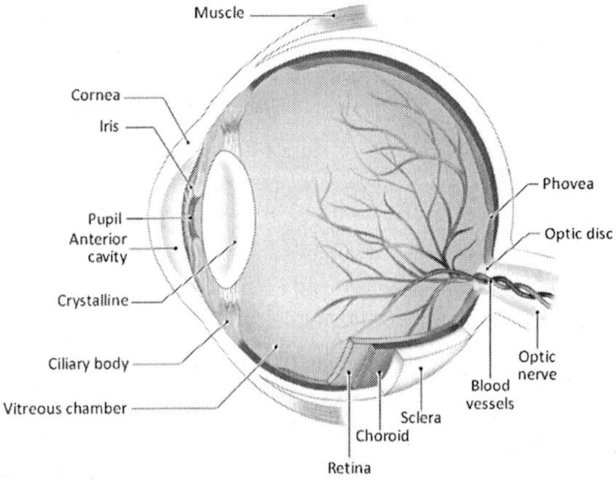

Figure 8.18 Schematic anatomy of the eye.

The light ray passes freely through the cornea, anterior cavity, pupil and iris without being altered by the very low refractive index of the aqueous humor and vitreous humor. The light is focused by the crystalline lens in the retina, especially in the fovea area, the richest in photoreceptors, and is absorbed by the choroid.

The optic nerves, formed by the axons of the ganglion cells (Figure 8.18), run from each eye in the second pair of cranial nerves to the base of the brain, in front of the brainstem, where they form the optic chiasma. They then continue as the optic tract to the lateral geniculate body of the thalamus, from which ascending neurons run to the primary visual cortex in the occipital lobe of the brain. A particular area located to the right of the image perceived by the eyes simultaneously affects the visual half-field towards the nose of the right eye and the visual half-field towards the temple of the left eye. Conversely, a different area located to the left of the image perceived by the eyes simultaneously affects the visual half-field towards the temple of the right eye and the visual half-field towards the nose of the left eye. For this reason, at the level of the optic chiasma, the axons coming from the visual half-field of the retina of the left eye, which is located towards the nasal area, cross and join the axons coming from the visual half-field of the retina of the right eye located towards the temporal area, and vice versa for the right eye. Therefore, the two images of the same area, which affect the visual half-fields in different positions in the two eyes, converge in the same area of the visual cortex, where they are processed to give the subject the sense of three-dimensionality.

8.6.1 The retina

The retina (Figure 8.19) is neural ectoderm, i.e., a peripheral projection of the central nervous system. In the cells of the retina, initial processing of the light signal takes place. It is a structure that could be seen as a very complex set of connections between cells, but is arranged in ordered layers. An inner layer, sitting on the choroid, is made up of photoreceptors of two types: the cones (involved in chromatic daytime vision when the light is most intense, as they are sensitive to light stimuli of more than 100 photons) and the rods (involved in monochromatic vision in dim light or twilight conditions with a sensitivity of one photon). The middle layer consists of bipolar cells and the outer layer consists of ganglionic nerve cells. In addition to these three layers are the horizontal cells, which project dendrites to different photoreceptors and modulate their

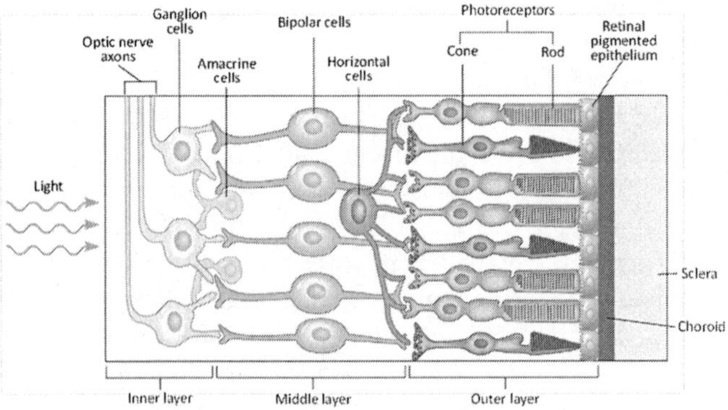

Figure 8.19 Schematic structure of the retina.

response, and the amacrine cells, which form synapses via dendrites between various ganglion cells and modulate their activity.

The modulation of the photoreceptor response, which constitutes peripheral processing of the central nervous signal responsible for vision, is possible because the electrical signals involving photoreceptors, bipolar cells, horizontal cells, amacrine cells and the cell body of the ganglion cells are of an electrotonic type. This means that modulation is possible (section 6.2.4.1), since the action potential is generated only at the cone of insertion of the axons of the ganglion cells (if the potential arriving there after processing by the other cells of the retina exceeds the threshold value).

A special feature of the retina is that light has to pass through all of the layers of cells to reach the photoreceptors (light in Figure 8.19).

The distribution of cones and rods in the retina is not homogeneous. At the level of the optic disc, the area from which the optic nerve and vessel system emerge from the eye (Figure 8.18), the photoreceptors are completely absent, while at the optic center of the retina, the macula lutea is formed. This structure includes the fovea, which is the area with the greatest visual acuity due to the presence only of the cones, and where the bipolar and ganglion cells are folded radially so as not to interfere with the light. Moving away from the fovea area, the number of rods progressively increases and the number of cones decreases, up to the peripheral part of the retina, where only rods are present.

Initial processing of the light signal takes place in the retina at four different levels. The ability to modify and modulate the signal coming from outside makes the retina the outermost projection of the central nervous system. The first level is the neurotransmitter glutamate released by the photoreceptor in the dark (section 8.6.2), which can hyperpolarize or depolarize a bipolar cell, depending on the type of postsynaptic receptor. In the case of hyperpolarization, it inhibits the excitability of the bipolar cell, which transmits the dark message to the central nervous system. In the case of depolarization, the bipolar cell is excited, which in turn excites the neighboring ganglion cell, which transmits the light message to the central nervous system and increases, in contrast, the dark sensation of the photoreceptor. The second and third levels concern horizontal cells and amacrine cells which, by forming synapses with several photoreceptors and ganglion cells, respectively, can modulate their activity. Finally, the fourth level concerns the phenomenon of synaptic convergence, whereby several photoreceptors form synapses with a single bipolar cell, resulting in increased sensitivity. Signals processed by several receptors simultaneously are conveyed to a bipolar cell and a ganglion cell. This gives greater sensitivity, but at the same time reduces visual acuity. The only light signal transmitted to the central nervous system is from a relatively large area of the retina. This processing mode is virtually absent at the fovea, where the cones are concentrated and where there is a need for visual acuity at the expense of sensitivity. This gives the rods the functional characteristic of having greater sensitivity and the cones, greater visual acuity.

8.6.2 Cones and rods

In the cones and rods, there is a clear distinction (Figure 8.20) between an external segment, characterized by the presence of numerous discs that occupy the entire cytoplasm, and an internal segment, with the nucleus, the synaptic terminal and the neurotransmitter vesicles, which form synapses with the bipolar cell. While the discs of the cones are formed by simple introflexions of the plasma membrane, such that between one disc and the other there is extracellular solution. In the case of the rods, the discs are closed vesicles, also derived from introflexions of the membrane, but completely internal to the cytoplasm. This structure confers the high sensitivity.

Figure 8.20 Diagram of the organization of a rod and cone of a mammalian retina.

The cones and rods contain about 108 molecules each of the protein rhodopsin, the photopigment capable of processing the light stimulus. Rhodopsin is inserted into the membrane of the stacked discs in the cytoplasm of the outer segment of the rod, while in the cones it is found on the invaginations of the plasma membrane (Figure 8.20).

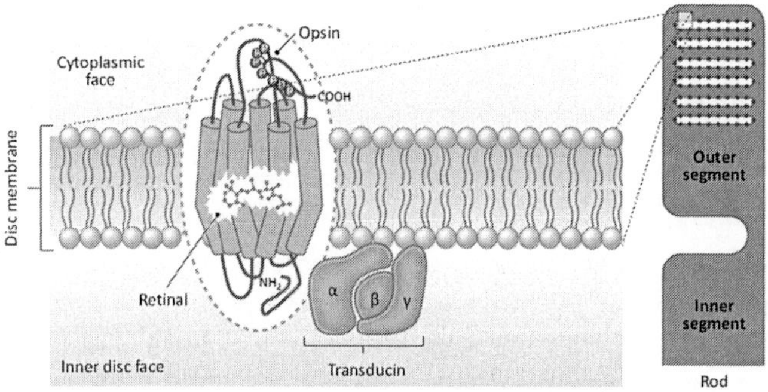

Figure 8.21 The photopigment rhodopsin, formed by opsin and retinal, is embedded in the disc membrane and is associated with a specific G-protein, transducin.

Rhodopsin consists of a compound called retinal, which is common to all photoreceptors, and a protein part called opsin, which is different depending on the type of receptor (Figure 8.21). The retinal can assume two conformations: in the dark, it is in the 11-cis configuration, while in light it changes to all-trans retinal (Figure 8.26).

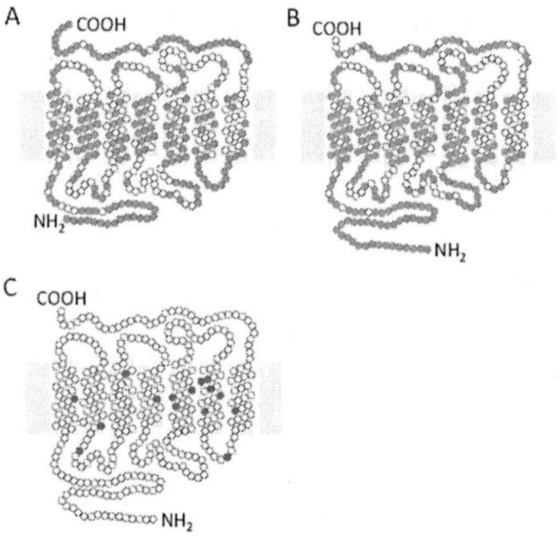

Figure 8.22 Schematic of the structure of the three types of retinal cones in the mammalian retina. The dots represent the amino acids that make up the peptide. The dark dots indicate the amino acids that vary from the rod retinal and give the different

In the human retina, there is only one type of rod with its own rhodopsin, characterized by a particular type of opsin for night and black-and-white twilight vision, with an absorption spectrum at 496 nm. In contrast, color vision is ensured by three different types of cones characterized by three different types of opsins (Figure 8.22), with a maximum absorption spectrum at 419 nm corresponding to the color blue, at 531 nm, corresponding to green, and at 559 nm, corresponding to red. Blue, green and red are therefore the three fundamental colors whose different combinations and intensities give the perception of all the colors of the rainbow, from red to violet.

8.6.3 The transformation of light stimulus into an electrical signal

Direct measurement of the resting potential of a rod (Figure 8.23A) shows that the cell in the dark is depolarized and is at a potential of -30 mV. If a flash of light is directed at the outer segment of the rod, it will cause transient hyperpolarization that will be more pronounced with a more intense light stimulus (Figure 8.23B). This behavior is due to an inhibition of the sodium permeabilities present on the membrane of the outer segment of the rods.

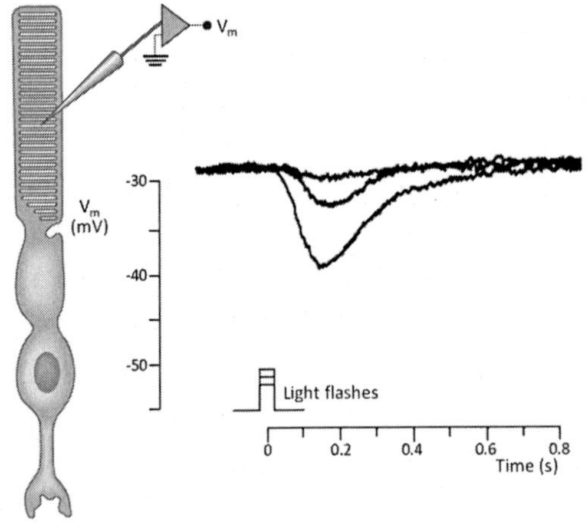

Figure 8.23 If, while measuring the membrane potential Vm of the outer segment of a rod with a microelectrode (A), there are flashes of light of varying amplitude, a transient hyperpolarization that is proportional to the intensity of the light is recorded (B).

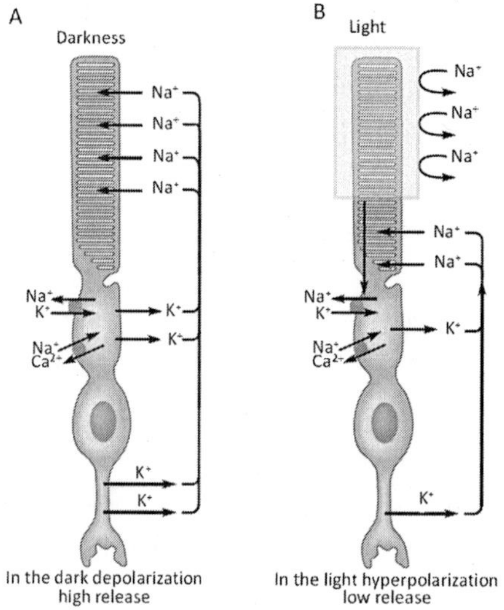

Figure 8.24 Schematic representation of the pattern of sodium and potassium currents and of sodium, potassium and calcium transports in both cones (not shown in the figure), and in the mammalian rod in the dark (A) and struck by a light beam (B).

In the dark, the cell potential is kept depolarized by the activation of sodium channels that are modulated by the action of adenylate cyclase, which continuously converts linear guanosynmonophosphate, 5'-GMP, into cyclic guanosynmonophosphate, cGMP. In practice, the current enters from the outer segment through the cGMP-dependent sodium channels and exits from the inner segment through the potassium channels, which are abundant in this part of the cell (Figure 8.24A). When light enters the outer segment of the rod, the sodium channels close and the cell is repolarized solely by the outgoing potassium current (Figure 8.24B).

The complete cascade of reactions, starting with the interaction of the photon with rhodopsin and ending with the inhibition of synaptic release, is illustrated in Figure 8.25A. When hit by light, retinal peptide changes its conformation to metarhodopsin II. Metarhodopsin II activates the G-protein-specific transducin, which is cleaved into the βγ and α subunits. While the βγ subunit remains attached to rhodopsin, the α subunit activates phosphodiesterase, which is inserted into the membranes of the cone and rod discs in the vicinity of each transducing molecule. The action of the activated phosphodiesterase drastically lowers the concentration of cGMP, transforming it into linear guanosynmonophosphate (5'-GMP). The

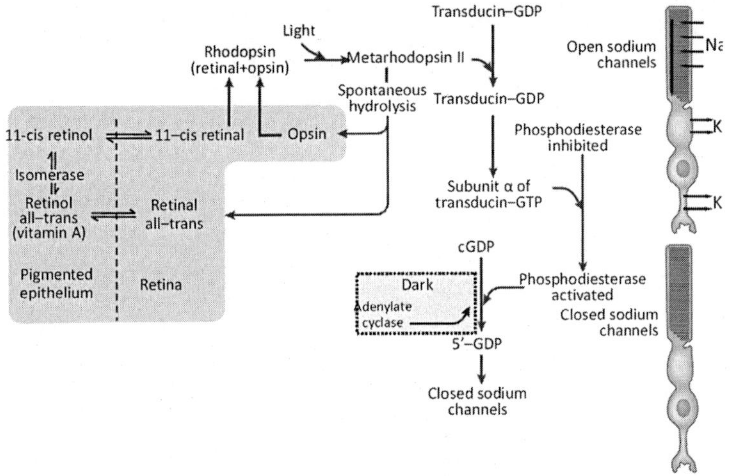

Figure 8.25 Complete sequence of reactions (A) which, in the mammalian photoreceptor (B), begins with the interaction of the photon with rhodopsin and ends with the closure of the sodium channels. Mechanism of rhodopsin replenishment (C) through recovery of opsin and retinol from spontaneous hydrolysis of metarhodopsin. In the dark (grey area), adenylate cyclase maintains a high concentration of cyclic guanosine monophosphate (cGMP) at the expense of linear guanosine monophosphate (5'-GMP).

cAMP-dependent sodium current at this point decreases, allowing the potassium current to hyperpolarize the cell. This phenomenon increases with increasing light intensity, i.e., the number of photons per unit time. Hyperpolarization of the cell dramatically decreases the calcium current near the presynaptic termination of the rod, which leads to a reduction in glutamate release. Decreased synaptic activity reduces stimulation of bipolar cells, with which the photoreceptor forms synapses, with an inhibitory action.

In addition, metarhodopsin II, which is unstable, dissociates by spontaneous hydrolysis into opsin and all-trans retinal (Figure 8.25C). In the pigmented epithelium, all-trans retinal is in equilibrium with all-trans retinol, which, in turn, is in equilibrium with all-trans retinol (vitamin A) from the blood capillaries of the choroid, and can be supplemented from the bloodstream. Through an isomerase, the all-trans retinol is in equilibrium with 11-cis retinol, which, in turn, is in equilibrium with 11-cis retinal in the retina (Figure 8.25).

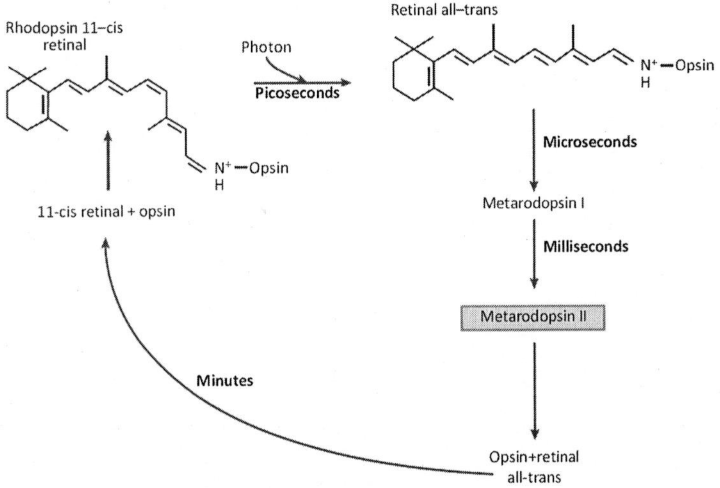

Figure 8.26 The timescales of a cycle of reactions found in a photoreceptor affected by light are shown in the Figure.

The durations of the reactions leading to sodium channel inhibition are shown in Figure 8.26. Once photons strike rhodopsin, there is a linearization of retinal, from 11-cis-retinal to all-trans retinal, with an immediate reaction (picoseconds). There are then two reactions of increasing duration leading to metarhodopsin I (microseconds) and metarhodopsin II (milliseconds), before having the final product, opsin plus all-trans retinal. However, the reaction takes several minutes to return to the initial conditions in the dark.